FPGA IMPLEMENTATIONS OF NEURAL NETWORKS

A C.I.P. Catalogue record for this book is available from the Library of Congress.

ISBN-13 978-1-4419-3942-5
ISBN-10 0-387-28487-7 (e-book)
ISBN-13 978-0-387-28487-3 (e-book)

Published by Springer,
P.O. Box 17, 3300 AA Dordrecht, The Netherlands.

www.springer.com

Printed on acid-free paper

Printed in the Netherlands.

FPGA Implementations
of Neural Networks

Edited by

AMOS R. OMONDI

Flinders University, Adelaide,
SA, Australia

and

JAGATH C. RAJAPAKSE

Nanyang Tecnological University,
Singapore

 Springer

Contents

Preface

During the 1980s and early 1990s there was significant work in the design and implementation of hardware neurocomputers. Nevertheless, most of these efforts may be judged to have been unsuccessful: at no time have have hardware neurocomputers been in wide use. This lack of success may be largely attributed to the fact that earlier work was almost entirely aimed at developing custom neurocomputers, based on ASIC technology, but for such niche areas this technology was never sufficiently developed or competitive enough to justify large-scale adoption. On the other hand, gate-arrays of the period mentioned were never large enough nor fast enough for serious artificial-neural-network (ANN) applications. But technology has now improved: the capacity and performance of current FPGAs are such that they present a much more realistic alternative. Consequently neurocomputers based on FPGAs are now a much more practical proposition than they have been in the past. This book summarizes some work towards this goal and consists of 12 papers that were selected, after review, from a number of submissions. The book is nominally divided into three parts: Chapters 1 through 4 deal with foundational issues; Chapters 5 through 11 deal with a variety of implementations; and Chapter 12 looks at the lessons learned from a large-scale project and also reconsiders design issues in light of current and future technology.

Chapter 1 reviews the basics of artificial-neural-network theory, discusses various aspects of the hardware implementation of neural networks (in both ASIC and FPGA technologies, with a focus on special features of artificial neural networks), and concludes with a brief note on performance-evaluation. Special points are the exploitation of the parallelism inherent in neural networks and the appropriate implementation of arithmetic functions, especially the sigmoid function. With respect to the sigmoid function, the chapter includes a significant contribution.

Certain sequences of arithmetic operations form the core of neural-network computations, and the second chapter deals with a foundational issue: how to determine the numerical precision format that allows an optimum tradeoff between precision and implementation (cost and performance). Standard single or double precision floating-point representations minimize quantization errors while requiring significant hardware resources. Less precise fixed-point representation may require less hardware resources but add quantization errors that may prevent learning from taking place, especially in regression problems. Chapter 2 examines this issue and reports on a recent experiment where we implemented a multi-layer perceptron on an FPGA using both fixed and floating point precision.

A basic problem in all forms of parallel computing is how best to map applications onto hardware. In the case of FPGAs the difficulty is aggravated by the relatively rigid interconnection structures of the basic computing cells. Chapters 3 and 4 consider this problem: an appropriate theoretical and practical framework to reconcile simple hardware topologies with complex neural architectures is discussed. The basic concept is that of *Field Programmable Neural Arrays* (FPNA) that lead to powerful neural architectures that are easy to map onto FPGAs, by means of a simplified topology and an original data exchange scheme. Chapter 3 gives the basic definition and results of the theoretical framework. And Chapter 4 shows how FPNAs lead to powerful neural architectures that are easy to map onto digital hardware. applications and implementations are described, focusing on a class

Chapter 5 presents a systolic architecture for the complete back propagation algorithm. This is the first such implementation of the back propagation algorithm which completely parallelizes the entire computation of learning phase. The array has been implemented on an Annapolis FPGA based coprocessor and it achieves very favorable performance with range of 5 GOPS. The proposed new design targets Virtex boards. A description is given of the process of automatically deriving these high performance architectures using the systolic array design tool MMALPHA, facilitates system-specification This makes it easy to specify the system in a very high level language (ALPHA) and also allows perform design exploration to obtain architectures whose performance is comparable to that obtained using hand optimized VHDL code.

Associative networks have a number of properties, including a rapid, compute efficient best-match and intrinsic fault tolerance, that make them ideal for many applications. However, large networks can be slow to emulate because of their storage and bandwidth requirements. Chapter 6 presents a simple but effective model of association and then discusses a performance analysis of the implementation this model on a single high-end PC workstation, a PC cluster, and FPGA hardware.

Chapter 7 describes the implementation of an artificial neural network in a reconfigurable parallel computer architecture using FPGA's, named Reconfigurable Orthogonal Memory Multiprocessor (REOMP), which uses p^2 memory modules connected to p reconfigurable processors, in row access mode, and column access mode. REOMP is considered as an alternative model of the neural network neocognitron. The chapter consists of a description of the RE-

OMP architecture, a the case study of alternative neocognitron mapping, and a performance performance analysis with systems systems consisting of 1 to 64 processors.

Chapter 8 presents an efficient architecture of Kohonen Self-Organizing Feature Map (SOFM) based on a new Frequency Adaptive Learning (FAL) algorithm which efficiently replaces the neighborhood adaptation function of the conventional SOFM. The proposed SOFM architecture is prototyped on Xilinx Virtex FPGA using the prototyping environment provided by XESS. A robust functional verification environment is developed for rapid prototype development. Various experimental results are given for the quantization of a 512 X 512 pixel color image.

Chapter 9 consists of another discussion of an implementation of SOFMs in reconfigurable hardware. Based on the universal rapid prototyping system, RAPTOR2000, a hardware accelerator for self-organizing feature maps has been developed. Using Xilinx Virtex-E FPGAs, RAPTOR2000 is capable of emulating hardware implementations with a complexity of more than 15 million system gates. RAPTOR2000 is linked to its host – a standard personal computer or workstation – via the PCI bus. A speed-up of up to 190 is achieved with five FPGA modules on the RAPTOR2000 system compared to a software implementation on a state of the art personal computer for typical applications of SOFMs.

Chapter 10 presents several hardware implementations of a standard Multi-Layer Perceptron (MLP) and a modified version called eXtended Multi-Layer Perceptron (XMLP). This extended version is an MLP-like feed-forward network with two-dimensional layers and configurable connection pathways. The discussion includes a description of hardware implementations have been developed and tested on an FPGA prototyping board and includes systems specifications using two different abstraction levels: register transfer level (VHDL) and a higher algorithmic-like level (Handel-C) as well as the exploitation of varying degrees of parallelism. The main test bed application is speech recognition.

Chapter 11 describes the implementation of a systolic array for a non-linear predictor for image and video compression. The implementation is based on a multilayer perceptron with a hardware-friendly learning algorithm. It is shown that even with relatively modest FPGA devices, the architecture attains the speeds necessary for real-time training in video applications and enabling more typical applications to be added to the image compression processing

The final chapter consists of a retrospective look at the REMAP project, which was the construction of design, implementation, and use of large-scale parallel architectures for neural-network applications. The chapter gives an overview of the computational requirements found in algorithms in general and motivates the use of regular processor arrays for the efficient execution of such

algorithms. The architecture, following the SIMD principle (Single Instruction stream, Multiple Data streams), is described, as well as the mapping of
some important and representative ANN algorithms. Implemented in FPGA,
the system served as an architecture laboratory. Variations of the architecture
are discussed, as well as scalability of fully synchronous SIMD architectures.
The design principles of a VLSI-implemented successor of REMAP-β are described, and the paper concludes with a discussion of how the more powerful
FPGA circuits of today could be used in a similar architecture.

AMOS R. OMONDI AND JAGATH C. RAJAPAKSE

Chapter 1

FPGA NEUROCOMPUTERS

Amos R. Omondi

School of Informatics and Engineering, Flinders University, Bedford Park, SA 5042, Australia

Amos.Omondi@flinders.edu.au

Jagath C. Rajapakse

School Computer Engineering, Nanyang Technogoical University, Singapore 639798

asrajapakse@ntu.edu.sg

Mariusz Bajger

School of Informatics and Engineering, Flinders University, Bedford Park, SA 5042, Australia

Mariusz.Bajger@flinders.edu.au

Abstract

This introductory chapter reviews the basics of artificial-neural-network theory, discusses various aspects of the hardware implementation of neural networks (in both ASIC and FPGA technologies, with a focus on special features of artificial neural networks), and concludes with a brief note on performance-evaluation. Special points are the exploitation of the parallelism inherent in neural networks and the appropriate implementation of arithmetic functions, especially the sigmoid function. With respect to the sigmoid function, the chapter includes a significant contribution.

Keywords: FPGAs, neurocomputers, neural-network arithmetic, sigmoid, performance-evaluation.

1.1   Introduction

In the 1980s and early 1990s, a great deal of research effort (both industrial and academic) was expended on the design and implementation of hardware neurocomputers [5, 6, 7, 8]. But, on the whole, most efforts may be judged

1

A. R. Omondi and J. C. Rajapakse (eds.), FPGA Implementations of Neural Networks, 1–36.
© 2006 *Springer. Printed in the Netherlands.*

to have been unsuccessful: at no time have have hardware neurocomputers been in wide use; indeed, the entire field was largely moribund by the end the 1990s. This lack of success may be largely attributed to the fact that earlier work was almost entirely based on ASIC technology but was never sufficiently developed or competetive enough to justify large-scale adoption; gate-arrays of the period mentioned were never large enough nor fast enough for serious neural-network applications.[1] Nevertheless, the current literature shows that ASIC neurocomputers appear to be making some sort of a comeback [1, 2, 3]; we shall argue below that these efforts are destined to fail for exactly the same reasons that earlier ones did. On the other hand, the capacity and performance of current FPGAs are such that they present a much more realistic alternative. We shall in what follows give more detailed arguments to support these claims.

The chapter is organized as follows. Section 2 is a review of the fundamentals of neural networks; still, it is expected that most readers of the book will already be familiar with these. Section 3 briefly contrasts ASIC-neurocomputers with FPGA-neurocomputers, with the aim of presenting a clear case for the former; a more significant aspects of this argument will be found in [18]. One of the most repeated arguments for implementing neural networks in hardware is the parallelism that the underlying models possess. Section 4 is a short section that reviews this. In Section 5 we briefly describe the realization of a state-of-the art FPGA device. The objective there is to be able to put into a concrete context certain following discussions and to be able to give grounded discussions of what can or cannot be achieved with current FPGAs. Section 6 deals with certain aspects of computer arithmetic that are relevant to neural-network implementations. Much of this is straightforward, and our main aim is to highlight certain subtle aspects. Section 7 nominally deals with activation functions, but is actually mostly devoted to the sigmoid function. There are two main reasons for this choice: first, the chapter contains a significant contribution to the implementation of elementary or near-elementary activation functions, the nature of which contribution is not limited to the sigmoid function; second, the sigmoid function is the most important activation function for neural networks. In Section 8, we very briefly address an important issue — performance evaluation. Our goal here is simple and can be stated quite succintly: as far as performance-evaluation goes, neurocomputer architecture continues to languish in the "Dark Ages", and this needs to change. A final section summarises the main points made in chapter and also serves as a brief introduction to subsequent chapters in the book.

---

[1]Unless otherwise indicated, we shall use *neural network* to mean *artificial neural network*.

1.2   Review of neural-network basics

The human brain, which consists of approximately 100 billion neurons that are connected by about 100 trillion connections, forms the most complex object known in the universe. Brain functions such as sensory information processing and cognition are the results of emergent computations carried out by this massive neural network. Artificial neural networks are computational models that are inspired by the principles of computations performed by the biological neural networks of the brain. Neural networks possess many attractive characteristics that may ultimately surpass some of the limitations in classical computational systems. The processing in the brain is mainly parallel and distributed: the information are stored in connections, mostly in myeline layers of axons of neurons, and, hence, distributed over the network and processed in a large number of neurons in parallel. The brain is adaptive from its birth to its complete death and learns from exemplars as they arise in the external world. Neural networks have the ability to learn the rules describing training data and, from previously learnt information, respond to novel patterns. Neural networks are fault-tolerant, in the sense that the loss of a few neurons or connections does not significantly affect their behavior, as the information processing involves a large number of neurons and connections. Artificial neural networks have found applications in many domains — for example, signal processing, image analysis, medical diagnosis systems, and financial forecasting.

The roles of neural networks in the afore-mentioned applications fall broadly into two classes: pattern recognition and functional approximation. The fundamental objective of pattern recognition is to provide a meaningful categorization of input patterns. In functional approximation, given a set of patterns, the network finds a smooth function that approximates the actual mapping between the input and output.

A vast majority of neural networks are still implemented on software on sequential machines. Although this is not necessarily always a severe limitation, there is much to be gained from directly implementing neual networks in hardware, especially if such implementation exploits the parellelism inherent in the neural networks but without undue costs. In what follows, we shall describe a few neural network models — multi-layer perceptrons, Kohonen's self-organizing feature map, and associative memory networks — whose implementations on FPGA are discussed in the other chapters of the book.

1.2.1   Artificial neuron

An artificial neuron forms the basic unit of artficial neural networks. The basic elements of an artificial neurons are (1) a set of input nodes, indexed by, say, 1, 2, ... I, that receives the corresponding input signal or pattern vector, say $\mathbf{x} = (x_1, x_2, \ldots x_I)^{\mathrm{T}}$; (2) a set of synaptic connections whose strengths are represented by a set of weights, here denoted by $\mathbf{w} = (w_1, w_2, \ldots w_I)^{\mathrm{T}}$; and (3) an activation function Φ that relates the total synaptic input to the output (activation) of the neuron. The main components of an artificial neuron is illustrated in Figure 1.

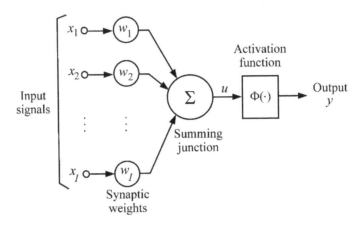

Figure 1: The basic components of an artificial neuron

The total synaptic input, u, to the neuron is given by the inner product of the input and weight vectors:

$$u = \sum_{i=1}^{I} w_i x_i \qquad (1.1)$$

where we assume that the threshold of the activation is incorporated in the weight vector. The output activation, y, is given by

$$y = \Phi(u) \qquad (1.2)$$

where Φ denotes the activation function of the neuron. Consequently, the computation of the inner-products is one of the most important arithmetic operations to be carried out for a hardware implementation of a neural network. This means not just the individual multiplications and additions, but also the alternation of successive multiplications and additions — in other words, a sequence of *multiply-add* (also commonly known as *multiply-accumulate* or *MAC*) operations. We shall see that current FPGA devices are particularly well-suited to such computations.

The total synaptic input is transformed to the output via the non-linear activation function. Commonly employed activation functions for neurons are

- the *threshold* activation function (unit step function or hard limiter):

$$\Phi(u) = \begin{cases} 1.0, & \text{when } u > 0, \\ 0.0, & \text{otherwise.} \end{cases}$$

- the *ramp* activation function:[2]

$$\Phi(u) = \max\{0.0, \min\{1.0, \ u + 0.5\}\}$$

- the *sigmodal* activation function, where the *unipolar sigmoid* function is

$$\Phi(u) = \frac{a}{1 + \exp(-bu)}$$

and the *bipolar sigmoid* is

$$\Phi(u) = a \left(\frac{1 - \exp(-bu)}{1 + \exp(-bu)} \right)$$

where a and b represent, repectively, real constants the *gain* or amplitude and the *slope* of the transfer function.

The second most important arithmetic operation required for neural networks is the computation of such activation functions. We shall see below that the structure of FPGAs limits the ways in which these operations can be carried out at reasonable cost, but current FPGAs are also equipped to enable high-speed implementations of these functions if the right choices are made.

A neuron with a threshold activation function is usually referred to as the *discrete perceptron*, and with a continuous activation function, usually a sig-moidal function, such a neuron is referred to as *continuous perceptron*. The sigmoidal is the most pervasive and biologically plausible activation function.

Neural networks attain their operating characteristics through learning or training. During training, the *weights* (or *strengths*) of connections are gradu-ally adjusted in either *supervised* or *unsupervised* manner. In supervised learn-ing, for each training input pattern, the network is presented with the *desired output* (or a teacher), whereas in unsupervised learning, for each training input pattern, the network adjusts the weights without knowing the correct target. The network *self-organizes* to classify similar input patterns into clusters in unsupervised learning. The learning of a continuous perceptron is by adjust-ment (using a gradient-descent procedure) of the weight vector, through the minimization of some error function, usually the square-error between the de-sired output and the output of the neuron. The resultant learning is known as

---

[2]In general, the slope of the ramp may be other than unity.

as *delta* learning: the new weight-vector, $\mathbf{w}^{\text{new}}$, after presentation of an input $\mathbf{x}$ and a desired output d is given by

$$\mathbf{w}^{\text{new}} = \mathbf{w}^{\text{old}} + \alpha\delta\mathbf{x}$$

where $\mathbf{w}^{\text{old}}$ refers to the weight vector before the presentation of the input and the error term, δ, is $(d - y)\Phi'(u)$, where y is as defined in Equation 1.2 and Φ' is the first derivative of Φ. The constant α, where $0 < \alpha \leq 1$, denotes the *learning factor*. Given a set of training data, $\Gamma = \{(\mathbf{x}_i, d_i); i = 1, \ldots n\}$, the complete procedure of training a continuous perceptron is as follows:

begin: /* *training a continuous perceptron* */
Initialize weights $\mathbf{w}^{\text{new}}$
Repeat
For each pattern $(\mathbf{x}_i, d_i)$ do
    $\mathbf{w}^{\text{old}} = \mathbf{w}^{\text{new}}$
    $\mathbf{w}^{\text{new}} = \mathbf{w}^{\text{old}} + \alpha\delta\mathbf{x}_i$
until convergence
end

The weights of the perceptron are initialized to random values, and the convergence of the above algorithm is assumed to have been achieved when no more significant changes occur in the weight vector.

1.2.2   Multi-layer perceptron

The *multi-layer perceptron* (MLP) is a feedforward neural network consisting of an *input layer* of nodes, followed by two or more layers of perceptrons, the last of which is the *output layer*. The layers between the input layer and output layer are referred to as *hidden layers*. MLPs have been applied successfully to many complex real-world problems consisting of non-linear decision boundaries. Three-layer MLPs have been sufficient for most of these applications. In what follows, we will briefly describe the architecture and learning of an L-layer MLP.

Let *0-layer* and *L-layer* represent the input and output layers, respectively; and let w_{kj}^{l+1} denote the synaptic weight connected to the k-th neuron of the $l + 1$ layer from the j-th neuron of the l-th layer. If the number of perceptrons in the l-th layer is N_l, then we shall let $\mathbf{W}_l \triangleq \{w_{kj}^l\}_{N_l \times N_{l-1}}$ denote the matrix of weights connecting to l-th layer. The vector of synaptic inputs to the l-th layer, $\mathbf{u}_l = (u_1^l, u_2^l, \ldots u_{N_l}^l)^{\text{T}}$ is given by

$$\mathbf{u}_l = \mathbf{W}_l\mathbf{y}_{l-1},$$

where $\mathbf{y}_{l-1} = (y_1^{l-1}, y_2^{l-1}, \ldots y_{N_{l-1}}^{l-1})^{\text{T}}$ denotes the vector of outputs at the $l-1$ layer. The generalized delta learning-rule for the layer l is, for perceptrons, given by

$$\mathbf{W}_l^{\text{new}} = \mathbf{W}_l^{\text{old}} + \alpha \boldsymbol{\delta}_l \mathbf{y}_{l-1}^{\text{T}},$$

where the vector of error terms, $\boldsymbol{\delta}_l^{\text{T}} = (\delta_1^l, \delta_2^l, \ldots, \delta_{N_l}^l)$ at the l th layer is given by

$$\delta_j^l = \begin{cases} 2\Phi_j^{l'}(u_j^l)(d_j - o_j), & \text{when } l = L, \\ \Phi_j^{l'}(u_j^l) \sum_{k=1}^{N_{l+1}} \delta_k^{l+1} w_{kj}^{l+1}, & \text{otherwise,} \end{cases}$$

where o_j and d_j denote the network and desired outputs of the j-th output neuron, respectively; and Φ_j^l and u_j^l denote the activation function and total synaptic input to the j-th neuron at the l-th layer, respectively. During training, the activities propagate forward for an input pattern; the error terms of a particular layer are computed by using the error terms in the next layer and, hence, move in the backward direction. So, the training of MLP is referred as error *back-propagation* algorithm. For the rest of this chapter, we shall generaly focus on MLP networks with backpropagation, this being, arguably, the most-implemented type of artificial neural networks.

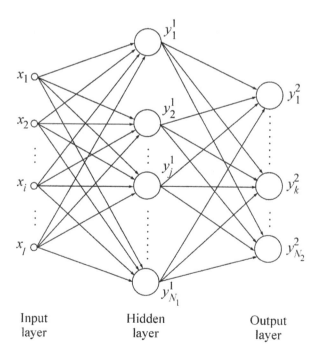

Figure 2: Architecture of a 3-layer MLP network

1.2.3   Self-organizing feature maps

Neurons in the cortex of the human brain are organized into layers of neurons. These neurons not only have bottom-up and top-down connections, but also have lateral connections. A neuron in a layer excites its closest neighbors via lateral connections but inhibits the distant neighbors. Lateral interactions allow neighbors to partially learn the information learned by a *winner* (formally defined below), which gives neighbors responding to similar patterns after learning with the winner. This results in topological ordering of formed clusters. The *self-organizing feature map* (SOFM) is a two-layer self-organizing network which is capable of learning input patterns in a topologically ordered manner at the output layer. The most significant concept in a learning SOFM is that of learning within a neighbourhood around a winning neuron. Therefore not only the weights of the winner but also those of the neighbors of the winner change.

The *winning neuron*, m, for an input pattern $\mathbf{x}$ is chosen according to the total synaptic input:

$$m = \arg \max_{j} \mathbf{w}_j^{\mathrm{T}} \mathbf{x},$$

where $\mathbf{w}_j$ denotes the weight-vector corresponding to the j-th output neuron. $\mathbf{w}_m^{\mathrm{T}} \mathbf{x}$ determines the neuron with the shortest Euclidean distance between its weight vector and the input vector when the input patterns are normalized to unity before training.

Let $\mathcal{N}_m(t)$ denote a set of indices corresponding to the neighbourhood size of the current winner m at the training time or iteration t. The radius of $\mathcal{N}_m$ is decreased as the training progresses; that is, $\mathcal{N}_m(t_1) > \mathcal{N}_m(t_2) > \mathcal{N}_m(t_3) \ldots$, where $t_1 < t_2 < t_3 \ldots$. The radius $\mathcal{N}_m(t = 0)$ can be very large at the beginning of learning because it is needed for initial global ordering of weights, but near the end of training, the neighbourhood may involve no neighbouring neurons other than the winning one. The weights associated with the winner and its neighbouring neurons are updated by

$$\Delta \mathbf{w}_j = \alpha(j, t) \left(\mathbf{x} - \mathbf{w}_j \right) \text{ for all } j \in \mathcal{N}_m(t),$$

where the positive learning factor depends on both the training time and the size of the neighbourhood. For example, a commonly used neighbourhood function is the Gaussian function

$$\alpha(N_m(t), t) = \alpha(t) \ \exp\left(-\frac{\|\mathbf{r}_j - \mathbf{r}_m\|^2}{2\sigma^2(t)} \right),$$

where $\mathbf{r}_m$ and $\mathbf{r}_j$ denote the positions of the winning neuron m and of the winning neighbourhood neurons j, respectively. $\alpha(t)$ is usually reduced at a rate that is inversely proportional to t. The type of training described above is known as *Kohonen's algorithm* (for SOFMs). The weights generated by the above algorithms are arranged spatially in an ordering that is related to the features of the trained patterns. Therefore, the algorithm produces topology-preserving maps. After learning, each input causes a localized response with positions on the output layer that reflects dominant features of the input.

1.2.4   Associative-memory networks

Associative memory networks are two-layer networks in which weights are determined in order to store a set of pattern associations, say, $\{(s_1, t_1), (s_2, t_2), \ldots (s_k, t_k), \ldots (s_n, t_n)\}$, where input pattern s_k is associated with output pattern t_k. These networks not only learn to produce associative patterns, but also are able to recall the desired response patterns when a given pattern is *similar* to the stored pattern. Therefore they are referred to as *content-addressible* memory. For each association vector (s_k, t_k), if $s_k = t_k$, the network is referred to as *auto-associative*; otherwise it is *hetero-associative*. The networks often provide input-output descriptions of the associative memory through a linear transformation (then known as linear associative memory). The neurons in these networks have linear activation functions. If the linearity constant is unity, then the output layer activation is given by

$$y = Wx,$$

where W denotes the weight matrix connecting the input and output layers. These networks learn using the Hebb rule; the weight matrix to learn all the associations is given by the batch learning rule:

$$W = \sum_{k=1}^{n} t_k s_k^T.$$

If the stored patterns are orthogonal to each other, then it can be shown that the network recalls the stored associations perfectly. Otherwise, the recalled patterns are distorted by cross-talk between patterns.

1.3   ASIC vs. FPGA neurocomputers

By far, the most often-stated reason for the development of custom (i.e. ASIC) neurocomputers is that conventional (i.e. sequential) general-purpose processors do not fully exploit the parallelism inherent in neural-network models and that highly parallel architectures are required for that. That is true as far as it goes, which is not very far, since it is mistaken on two counts [18]: The first is that it confuses the final goal, which is high performance — not merely parallelism — with artifacts of the basic model. The strong focus on parallelism can be justified only when high performance is attained at a reasonable cost. The second is that such claims ignore the fact that conventional microprocessors, as well as other types of processors with a substantial user-base, improve at a much faster rate than (low-use) special-purpose ones, which implies that the performance (relative to cost or otherwise) of ASIC neurocomputers will always lag behind that of mass-produced devices – even on special applications. As an example of this misdirection of effort, consider the latest in ASIC neurocomputers, as exemplified by, say, [3]. It is claimed that "with relatively few neurons, this ANN-dedicated hardware chip [Neuricam Totem] outperformed the other two implementations [a Pentium-based PC and a Texas Instruments DSP]". The actual results as presented and analysed are typical of the poor benchmarking that afflicts the neural-network area. We shall have more to say below on that point, but even if one accepts the claims as given, some remarks can be made immediately. The strongest performance-claim made in [3], for example, is that the Totem neurochip outperformed, by a factor of about 3, a PC (with a 400-MHz Pentium II processor, 128 Mbytes of main memory, and the neural netwoks implemented in Matlab). Two points are pertinent here:

- In late-2001/early 2002, the latest Pentiums had clock rates that were more than 3 times that of Pentium II above and with much more memory (cache, main, etc.) as well.

- The PC implementation was done on top of a substantial software (base), instead of a direct low-level implementation, thus raising issues of "best-effort" with respect to the competitor machines.

A comparison of the NeuriCam Totems and Intel Pentiums, in the years 2002 and 2004 will show the large basic differences have only got larger, primarily because, with the much large user-base, the Intel (x86) processors continue to improve rapidly, whereas little is ever heard of about the neurocomputers as PCs go from one generation to another.

So, where then do FGPAs fit in? It is evident that in general FPGAs cannot match ASIC processors in performance, and in this regard FPGAs have always lagged behind conventional microprocessors. Nevertheless, if one considers FPGA structures as an alternative to software on, say, a general-purpose processor, then it is possible that FPGAs may be able to deliver better cost:performance ratios on given applications.[3] Moreover, the capacity for reconfiguration means that may be extended to a range of applications, e.g. several different types of neural networks. Thus the main advantage of the FPGA is that it may offer a better cost:performance ratio than either custom

---

[3]Note that the issue is *cost:performance* and not just *performance*

ASIC neurocomputers or state-of-the art general-purpose processors and with more flexibility than the former. A comparison of the NeuriCam Totem, Intel Pentiums, and M FPGAs will also show that improvements that show the advantages of of the FPGAs, as a consequence of relatively rapid changes in density and speed.

It is important to note here two critical points in relation to custom (ASIC) neurocomputers versus the FPGA structures that may be used to implement a variety of artificial neural networks. The first is that if one aims to realize a custom neurocomputer that has a signficiant amount of flexibility, then one ends up with a structure that resembles an FPGA — that is, a small number of different types functional units that can be configured in different ways, according to the neural network to be implemented — but which nonetheless does not have the same flexibility. (A particular aspect to note here is that the large variety of neural networks — usually geared towards different applications — gives rise a requirement for flexibility, in the form of either programmability or reconfigurability.) The second point is that raw hardware-performance alone does not constitute the entirety of a typical computing structure: software is also required; but the development of software for custom neurocomputers will, because of the limited user-base, always lag behind that of the more widely used FPGAs. A final drawback of the custom-neurocomputer approach is that most designs and implementations tend to concentrate on just the high parallelism of the neural networks and generally ignore the implications of Amdahl's Law, which states that ultimately the speed-up will be limited by any serial or lowly-parallel processing involved. (One rare exception is [8].)[4] Thus non-neural and other serial parts of processing tend to be given short shrift. Further, even where parallelism can be exploited, most neurocomputer-design seem to to take little account of the fact that the degrees of useful parallelism will vary according to particular applications. (If parallelism is the main issue, then all this would suggest that the ideal building block for an appropriate parallel-processor machine is one that is less susceptible to these factors, and this argues for a relatively large-grain high-performance processor, used in smaller numbers, that can nevertheless exploit some of the parallelism inherent in neural networks [18].)

All of the above can be summed up quite succintly: despite all the claims that have been made and are still being made, to date there has not been a custom neurocomputer that, on artificial neural-network problems (or, for that matter, on any other type of problem), has outperformed the best conventional computer of its time. Moreover, there is little chance of that happening. The

---

[4]Although not quite successful as a neurocomputer, this machine managed to survive longer than most neurocomputers — because the flexibility inherent in its design meant that it could also be useful for non-neural applications.

promise of FPGAs is that they offer, in essence, the ability to realize "semi-custom" machines for neural networks; and, with continuing developments in technology, they thus offer the best hope for changing the situation, as far as possibly outperforming (relative to cost) conventional processors.

1.4   Parallelism in neural networks

Neural networks exhibit several types of parallelism, and a careful examination of these is required in order to both determine the most suitable hardware structures as well as the best mappings from the neural-network structures onto given hardware structures. For example, parallelism can be of the SIMD type or of the MIMD type, bit-parallel or word-parallel, and so forth [5]. In general, the only categorical statement that can be made is that, except for networks of a trivial size, fully parallel implementation in hardware is not feasible — virtual parallelism is necessary, and this, in turn, implies some sequential processing. In the context of FPGa, it might appear that reconfiguration is a silver bullet, but this is not so: the benefits of dynamic reconfigurability must be evaluated relative to the costs (especially in time) of reconfiguration. Nevertheless, there is litle doubt that FPGAs are more promising that ASIC neurocomputers. The specific types of parallelism are as follows.

- *Training parallelism*: Different training sessions can be run in parallel, e.g. on SIMD or MIMD processors. The level of parallelism at this level is usually medium (i.e. in the hundreds), and hence can be nearly fully mapped onto current large FPGAs.

- *Layer parallelism*: In a multilayer network, different layers can be processed in parallel. Parallelism at this level is typically low (in the tens), and therefore of limited value, but it can still be exploited through pipelining.

- *Node parallelism*: This level, which coresponds to individual neurons, is perhaps the most important level of parallelism, in that if fully exploited, then parallelism at all of the above higher levels is also fully exploited. But that may not be possible, since the number of neurons can be as high as in the millions. Nevertheless, node parallelism matches FPGAs very well, since a typical FPGA basically consists of a large number of "cells" that can operate in parallel and, as we shall see below, onto which neurons can readily be mapped.

- *Weight parallelism*: In the computation of an output

$$y = \Phi\left(\sum_{i=1}^{n} w_i x_i\right),$$

where x_i is an input and w_i is a weight, the products x_iw_i can all be computed in parallel, and the sum of these products can also be computed with high parallelism (e.g. by using an adder-tree of logarithmic depth).

■ *Bit-level parallelism*: At the implementation level, a wide variety of parallelism is available, depending on the design of individual functional units. For example, bit-serial, serial-parallel, word-parallel, etc.

From the above, three things are evident in the context of an implementation. First, the parallelism available at the different levels varies enormously. Second, different types of parallelism may be traded off against others, depending on the desired cost:performance ratio (where for an FPGA cost may be measured in, say, the number of CLBs etc.); for example, the slow speed of a single functional unit may be balanced by having many such units operating concurrently. And third, not all types of parallelism are suitable for FPGA implementation: for example, the required routing-interconnections may be problematic, or the exploitation of bit-level parallelism may be constrained by the design of the device, or bit-level parallelism may simply not be appropriate, and so forth. In the Xilinx Virtex-4, for example, we shall see that it is possible to carry out many neural-network computations without using much of what is usually taken as FPGA fabric.[5]

1.5 Xilinx Virtex-4 FPGA

In this section, we shall briefly give the details an current FPGA device, the Xilinx Virtex-4, that is typical of state-of-the-art FPGA devices. We shall below use this device in several running examples, as these are easiest understood in the context of a concrete device. The Virtex-4 is actually a family of devices with many common features but varying in speed, logic-capacity, etc.. The Virtex-E consists of an array of up to 192-by-116 *tiles* (in generic FPGA terms, *configurable logic blocks* or *CLBs*), up to 1392 Kb of *Distributed-RAM*, upto 9936 Kb of *Block-RAM* (arranged in 18-Kb blocks), up to 2 PowerPC 405 processors, up to 512 Xtreme DSP slices for arithmetic, input/ouput blocks, and so forth.[6]

A tile is made of two *DSP48 slices* that together consist of eight function-generators (configured as 4-bit lookup tables capable of realizing any four-input boolean function), eight flip-flops, two fast carry-chains, 64 bits of Distributed-RAM, and 64-bits of shift register. There are two types of slices:

---

[5] The definition here of *FPGA fabric* is, of course, subjective, and this reflects a need to deal with changes in FPGA realization. But the fundamental point remains valid: bit-level parallelism is not ideal for the given computations and the device in question.

[6] Not all the stated maxima occur in any one device of the family.

SLICEM, which consists of logic, distributed RAM, and shift registers, and SLICEL, which consists of logic only. Figure 3 shows the basic elements of a tile.

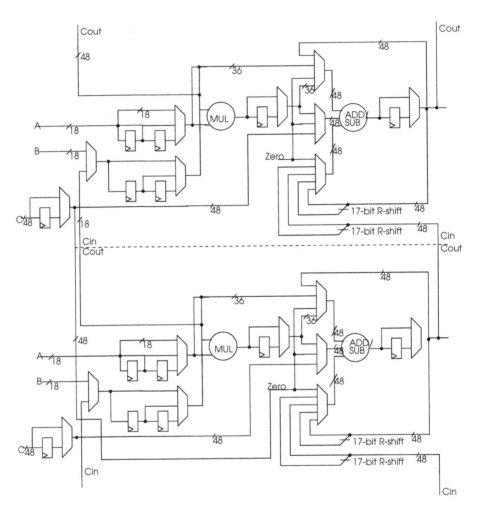

Figure 3: DSP48 tile of Xilinx Virtex-4

Blocks of the Block-RAM are true dual-ported and recofigurable to various widths and depths (from $16K \times 1$ to 512×36); this memory lies outside the slices. Distributed RAM are located inside the slices and are nominally single-port but can be configured for dual-port operation. The PowerPC processor core is of 32-bit Harvard architecture, implemented as a 5-stage pipeline. The significance of this last unit is in relation to the comment above on the serial parts of even highly parallel applications — one cannot live by parallelism alone. The maximum clock rate for all of the units above is 500 MHz.

Arithmetic functions in the Virtex-4 fall into one of two main categories: arithmetic within a tile and arithmetic within a collection of slices. All the slices together make up what is called the *XtremeDSP* [22]. DSP48 slices are optimized for multipliy, add, and mutiply-add operations. There are 512 DSP48 slices in the largest Virtex-4 device. Each slice has the organization shown in Figure 3 and consists primarily of an 18-bit×18-bit multiplier, a 48-bit adder/subtractor, multiplexers, registers, and so forth. Given the importance of inner-product computations, it is the XtremeDSP that is here most crucial for neural-network applications. With 512 DSP48 slices operating at a peak rate of 500 MHz, a maximum performance of 256 Giga-MACs (multiply-accumlate operations) per second is possible. Observe that this is well beyond anything that has so far been offered by way of a custom neurocomputer.

1.6   Arithmetic

There are several aspects of computer arithmetic that need to be considered in the design of neurocomputers; these include data representation, inner-product computation, implementation of activation functions, storage and update of weights, and the nature of learning algorithms. Input/output, although not an arithmetic problem, is also important to ensure that arithmetic units can be supplied with inputs (and results sent out) at appropriate rates. Of these, the most important are the inner-product and the activation functions. Indeed, the latter is sufficiently significant and of such complexity that we shall devote to it an entirely separate section. In what follows, we shall discuss the others, with a special emphasis on inner-products. Activation functions, which here is restricted to the sigmoid (although the relevant techniques are not) are sufficiently complex that we have relegated them to seperate section: given the ease with which multiplication and addition can be implemented, unless sufficient care is taken, it is the activation function that will be the limiting factor in performance.

Data representation: There is not much to be said here, especially since existing devices restrict the choice; nevertheless, such restrictions are not absolute, and there is, in any case, room to reflect on alternatives to what may be on offer. The standard representations are generally based on two's complement. We do, however, wish to highlight the role that residue number systems (RNS) can play.

It is well-known that RNS, because of its carry-free properties, is particularly good for multiplication and addition [23]; and we have noted that inner-product is particularly important here. So there is a natural fit, it seems. Now, to date RNS have not been particularly successful, primarily because of the difficulties in converting between RNS representations and conventional ones. What must be borne in mind, however, is the old adage that computing is about insight, not numbers; what that means in this context is that the issue of conversion need come up only if it is absolutely necessary. Consider, for example, a neural network that is used for classification. The final result for each input is binary: either a classification is correct or it is not. So, the representation used in the computations is a side-issue: conversion need not be carried out as long as an appropriate output can be obtained. (The same remark, of course, applies to many other problems and not just neural networks.) As for the constraints of off-the-shelf FPGA devices, two things may be observed: first, FPGA cells typically perform operations on small slices (say, 4-bit or 8-bit) that are perfectly adequate for RNS digit-slice arithmetic; and, second, below the level of digit-slices, RNS arithmetic will in any case be realized in a conventional notation.

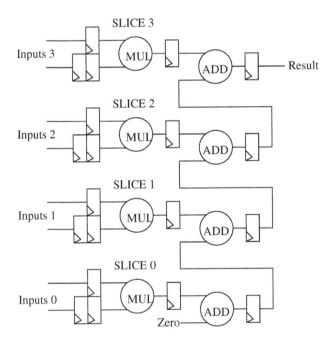

Figure 4: XtremeDSP chain-configuration for an inner-product

The other issue that is significant for representation is the precision used. There have now been sufficient studies (e.g. [17]) that have established 16 bits for weights and 8 bits for activation-function inputs as good enough. With this knowledge, the critical aspect then is when, due to considerations of performance or cost, lower precision must be used. Then a careful process of numerical analysis is needed.

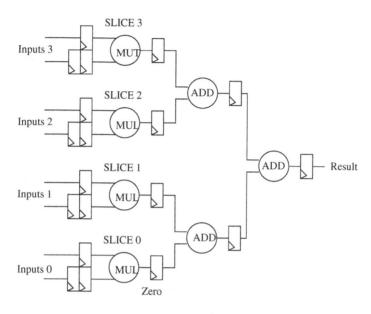

Figure 5: XtremeDSP tree-configuration for an inner-product

Sum-of-products computations: There are several ways to implement this, depending on the number of datasets. If there is just one dataset, then the operation is $\sum_{i=1}^{N} w_i X_i$, where w_i is a weight and X_i is an input. (In general, this is the matrix-vector computation expressed by Equation 1.1.) In such a case, with a device such as the Xilinx Virtex-4, there are several possible implementations, of which we now give a few sketches. If N is small enough, then two direct implementations consist of either a chain (Figure 4) or a tree (Figure 5) of DSP48 slices. Evidently, the trade-off is one of latency versus effecient use of device logic: with a tree the use of tile logic is quite uneven and less efficient than with a chain. If N is large, then an obvious way to proceed is to use a combination of these two approaches. That is, partition the computation into several pieces, use a chain for each such piece, and then combine in a tree the results of these chains, or the other way around. But there are other possible approaches: for example, instead of using chains, one DSP48 slice could be used (with a feedback loop) to comute the result of each nominal chain, with all such results then combined in a chain or a tree. Of course, the latency will now be much higher.

With multiple datasets, any of the above approaches can be used, although some are better than others — for example, tree structures are more amenable to pipelining. But there is now an additional issue: how to get data in and out at the appropriate rates. If the network is sufficiently large, then most of the inputs to the arithmetic units will be stored outside the device, and the number of device pins available for input/output becomes a minor issue. In this case, the organization of input/output is critical. So, in general, one needs to consider both large datasets as well as multiple data sets. The following discussions cover both aspects.

Storage and update of weights, input/output: For our purposes, Distributed-RAM is too small to hold most of the data that is to be processed, and therefore, in general Block-RAM will be used. Both weights and input values are stored in a single block and simualtously read out (as the RAM is dual-ported). Of course, for very small networks, it may be practical to use the Distributed-RAM, especially to store the weights; but we will in general assume networks of arbitrary size. (A more practical use for Distributed-RAM is the storage of constants used to implement activation functions.) Note that the disparity (discussed below) between the rate of inner-product computations and activation-function computations means that there is more Distributed-RAM available for this purpose than appears at first glance. For large networks, even the Block-RAM may not be sufficient, and data has to be periodically loaded into and retrieved from the FPGA device. Given pin-limitations, careful consideration must be given to how this is done.

Let us suppose that we have multiple datasets and that each of these is very large. Then, the matrix-vector product of Equation 1.1, that is,

$$\mathbf{u} = \mathbf{W}\mathbf{y}$$

becomes a matrix-matrix product,

$$\mathbf{U} = \mathbf{W}\mathbf{Y},$$

where each column of Y is associated with one input dataset. The most common method used for matrix-matrix multiplication is the *inner-product method*; that is, each element of the output matrix is directly generated as an inner-product of two vectors of the input matrices. Once the basic method has been selected, the data must be processed — in particular, for large datasets, this includes bringing data into, and retrieving data from, the FPGA — exactly as indicated above. This is, of course, true for other methods as well.

Whether or not the inner-product method, which is a highly sequential method, is satisfactory depends a great deal on the basic processor microarchitecture, and there are at least two alternatives that should always be considered: the *outer-product* and the *middle-product* methods.[7] Consider a typical "naive" sequential implementation of matrix multiplication. The inner-product method would be encoded as three nested loops, the innermost of which computes the inner-product of a vector of one of the input matrices and a vector of the other input matrix:

for i:=1 **to** n **do**
   for j:=1 **to** n **do**
      for k:=1 **to** n **do**
         U[i,j]:=U[i,j]+W[i,k]*Y[k,j];

(where we assume that the elements of $U[i, j]$ have all been initialized to zero.) Let us call this the *ijk-method*[8], based on the ordering of the index-changes. The only parallelism here is in the multiplication of individual matrix element and, to a much lesser extent (assuming the tree-method is used instead of the chain method) in the tree-summation of these products). That is, for $n \times n$ matrices, the required n^2 inner-products are computed one at a time. The middle-product method is obtained by interchanging two of the loops so as to yield the *jki-method*. Now more parallelism is exposed, since n inner-products can be computed concurrently; this is the *middle-product* method. And the outer-product method is the *kij-method*. Here all parallelism is now exposed: all n^2 inner products can be computed concurrently. Nevertheless, it should be noted that no one method may be categorically said to be better than another — it all depends on the architecture, etc.

To put some meat to the bones above, let us consider a concrete example — the case of 2×2 matrices. Further, let us assume that the multiply-accumulate (MAC) operations are carried out within the device but that all data has to be brought into the device. Then the process with each of the three methods is shown in Table 1. (The horizontal lines delineate groups of actions that may take place concurrently; that is within a column, actions separated by a line must be performed sequentially.)

A somewhat rough way to compare the three methods is measure the ratio, $M : I$, of the number of MACs carried out per data value brought into the array. This measure clearly ranks the three methods in the order one would expect; also note that by this measure the kij-method is completely efficient ($M : I = 1$): every data value brought in is involved in a MAC. Nevertheless, it is not entirely satisfactory: for example, it shows that the kij-method to be better than the jki-method by factor, which is smaller that what our intuition

---

[7]The reader who is familiar with compiler technology will readily recognise these as *vecorization (parallelization)* by *loop-interchange*.

[8]We have chosen this terminology to make it convenient to also include methods that have not yet been "named".

would lead us to expect. But if we now take another measure, the ratio of $M : I$ to the number, S, of MAC-steps (that must be carried out sequentially), then the diference is apparent.

Lastly, we come to the main reason for our classifcation (by index-ordering) of the various methods. First, it is evident that any ordering will work just as well, as far as the production of correct results goes. Second, if the data values are all of the same precision, then it is sufficient to consider just the three methods above. Nevertheless, in this case dataflow is also important, and it easy to establish, for example, that where the jki-method requires (at each input step) one weight and two inout values, there is an ordering of indices that requires two weights and one input value. Thus if weights are higher precision, the latter method may be better.

Table 1: Matrix multiplication by three standard methods

Inner-Product	Middle-Product	Outer-Product
Input: $W_{1,1}, W_{1,2}, Y_{1,1}, Y_{2,1}$	Input: $W_{1,1}, Y_{2,1}, Y_{1,1}$	Input: $W_{1,1}, W_{2,1}, Y_{1,1}, Y_{1,2}$
MAC: $t_1 = t_1 + W_{1,1} * Y_{1,1}$	MAC: $t_1 = t_1 + W_{1,1} * Y_{1,1}$	MAC: $t_1 = t_1 + W_{1,1} * Y_{1,1}$
MAC: $t_1 = t_1 + W_{1,2} * Y_{2,1}$	MAC: $t_2 = t_2 + W_{1,1} * Y_{1,2}$	MAC: $t_2 = t_2 + W_{1,1} * Y_{1,2}$
Input: $W_{1,1}, W_{1,2}, Y_{1,2}, Y_{2,2}$	Input: $W_{1,2}, Y_{2,1}, Y_{2,2}$	MAC: $t_3 = t_3 + W_{2,1} * Y_{1,1}$
		MAC: $t_4 = t_4 + W_{2,1} * Y_{1,2}$
MAC: $t_2 = t_2 + W_{1,1} * Y_{1,2}$	MAC: $t_1 = t_1 + W_{1,2} * Y_{2,1}$	Input: $W_{1,2}, W_{2,2}, Y_{2,1}, Y_{2,2}$
MAC: $t_2 = t_2 + W_{1,2} * Y_{2,2}$	MAC: $t_2 = t_2 + W_{1,2} * Y_{2,2}$	
Input: $W_{2,1}, W_{2,2}, Y_{1,1}, Y_{2,1}$	Input: $W_{2,1}, Y_{1,1}, Y_{1,2}$	MAC: $t_1 = t_1 + W_{1,2} * Y_{2,1}$
		MAC: $t_2 = t_2 + W_{1,2} * Y_{2,2}$
MAC: $t_3 = t_3 + W_{2,1} * Y_{1,1}$	MAC: $t_3 = t_3 + W_{2,1} * Y_{1,1}$	MAC: $t_3 = t_3 + W_{1,1} * Y_{1,1}$
MAC: $t_3 = t_3 + W_{2,2} * Y_{2,1}$	MAC: $t_4 = t_4 + W_{2,1} * Y_{1,2}$	MAC: $t_4 = t_4 + W_{2,2} * Y_{2,2}$
Input: $W_{2,1}, W_{2,2}, Y_{1,2}, Y_{2,2}$	Input: $W_{2,2}, Y_{2,1}, Y_{2,2}$	
MAC: $t_4 = t_4 + W_{2,1} * Y_{1,2}$	MAC: $t_3 = t_3 + W_{2,2} * Y_{2,1}$	
MAC: $t_4 = t_4 + W_{2,2} * Y_{2,2}$	MAC: $t_4 = t_4 + W_{2,2} * Y_{2,2}$	
$M : I = 0.5$	$M : I = 0.667$	$M : I = 1.0$
$(M : I)/S = 0.125$ (S=8)	$(M : I)/S = 0.167$ (S=4)	$(M : I)/S = 0.5$ (S=2)

Learning and other algorithms: The typical learning algorithm is usually chosen on how quickly it leads to convergence (on, in most cases, a software platform). For hardware, this is not necessarily the best criteria: algorithms need to be selected on the basis on how easily they can be implemented in hardware and what the costs and performance of such implementations are. Similar considerations should apply to other algorithms as well.

1.7 Activation-function implementation: unipolar sigmoid

For neural networks, the implementation of these functions is one of the two most important arithmetic design-issues. Many techniques exist for evaluating such elementary or nearly-elementary functions: polynomial approximations, CORDIC algorithms, rational approximations, table-driven methods, and so forth [4, 11]. For hardware implementation, accuracy, performance and cost are all important. The latter two mean that many of the better techniques that have been developed in numerical analysis (and which are easily implemented in software) are not suitable for hardware implementation. CORDIC is perhaps the most studied technique for hardware implementation, but it is (relatively) rarely implemented: its advantage is that the same hardware can be used for several functions, but the resulting performance is usually rather poor. High-order polynomial approximations can give low-error implementations, but are generally not suitable for hardware implementation, because of the number of arithmetic operations (multiplications and additions) that must be performed for each value; either much hardware must be used, or performance be compromised. And a similar remark applies to pure table-driven methods, unless the tables are quite small: large tables will be both slow and costly. The practical implication of these constraints is as indicated above: the best techniques from standard numerical analysis are of dubious worth.

Given trends in technology, it is apparent that at present the best technique for hardware function-evaluation is a combination of low-order polynomials and small look-up tables. This is the case for both ASIC and FPGA technologies, and especially for the latter, in which current devices are equipped with substantial amounts of memory, spread through the device, as well as many arithmetic units (notably mulipliers and adders).[9] The combination of low-order polynomials (primarily linear ones) is not new — the main challenges has always been one of how to choose the best interpolation points and how to ensure that look-up tables remain small. Low-order interpolation therefore has three main advantages. The first is that exactly the same hardware structures can be used to realize different functions, since only polynomial coefficients (i.e. the contents of look-up tables) need be changed; such efficient reuse is not possible with the other techniques mentioned above. The second is that it is well-matched to current FPGA devices, which come with built-in multipliers, adders, and memory.

The next subsection outlines the basic of our approach to linear interpolation; the one after that discusses implementation issues; and the final subsection goes into the details of the underlying theory.

---

[9]This is validated by a recent study of FPGA implementations of various techniques [16].

1.7.1   Basic concepts

On the basis of the considerations above, we have chosen piecewise linear interpolation for the approximation of the sigmoid function.

For most functions, interpolation with uniformly-sized intervals (i.e. uniformly-spaced abscissae) is not ideal; in the case of the sigmoid, it is evident that, ideally, more intervals should be used as the magnitude of the argument increases. Nevertheless, for hardware implementation, the need to quickly map arguments onto the appropriate intervals dictates the use of such intervals. With this choice and linear interpolation, the critical issue then becomes that of what function-value to associate with each interval. The most common choice is to arbitrarily select the value at the midpoint of the interval — that is, if $x \in [L, U]$, then $f(x) = f(L/2 + U/2)$ — or to choose a value that minimizes absolute errors. [10] Neither is particularly good. As we shall show, even with a fixed number of intervals, the best function-value for an interval is generally not the midpoint. And, depending on the "curvature" of the function at hand, relative error may be more critical than absolute error. For example, for the sigmoid function, $f(x) = 1/(1 + e^{-x})$, we have a function that is symmetric (about the y-axis), but the relative error grows more rapidly on one side of the axis than the other, and on both sides the growth depends on the interval. Thus, the effect of a given value of absolute error is not constant or even linear.

The general approach we take is as follows. Let $I = [L, U]$ be a real interval with $L < U$, and let $f : I \to \mathbf{R}$ be a function to be approximated (where $\mathbf{R}$ denotes the set of real numbers). Suppose that $\widehat{f} : I \to \mathbf{R}$ is a linear function — that is, $\widehat{f}(x) = c_1 + c_2 x$, for some constants c_1 and c_2 — that approximates f. Our objective is to investigate the relative-error function

$$\varepsilon(x) = \frac{f(x) - \widehat{f}(x)}{f(x)}, \quad x \in I, \tag{Err}$$

and to find c_1 and c_2 such that $\varepsilon(x)$ is small. One way to obtain reasonably good values for c_1, c_2 is to impose the

$$f(L) = \widehat{f}(L), \quad f(U) = \widehat{f}(U) \tag{C}$$

and then compute values for c_1 and c_2. But a much better result can be obtained using the "improved" condition

$$|\varepsilon(L)| = |\varepsilon(U)| = |\varepsilon(x_{stat})|, \tag{IC}$$

where x_{stat} (stationary point) is the value of x for which $\varepsilon(x)$ has a local extremum. An example of the use of this technique to approximate reciprocals

---

[10]Following [12], we use *absolute error* to refer to the difference between the exact value and its approximation; that is, it is not the absolute value of that difference.

can be found in [4, 10] for the approximation of divisor reciprocals and square-root reciprocals. It is worth noting, however, that in [10], $\varepsilon(x)$ is taken to be the absolute-error function. This choice simplifies the application of (IC), but, given the "curvature" of these functions, it is not as good as the relative-error function above. We will show, in Section 7.3, that (IC) can be used successfully for sigmoid function, despite the fact that finding the exact value for x_{stat} may not be possible. We show that, compared with the results from using the condition (C), the improved condition (IC) yields a massive 50% reduction in the magnitude of the relative error. We shall also give the analytical formulae for the constants c_1 and c_2. The general technique is easily extended to other functions and with equal or less ease [13], but we shall here consider only the sigmoid function, which is probably the most important one for neural networks.

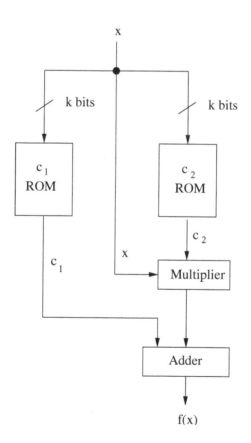

Figure 6: Hardware organization for piecewise linear interpolation

1.7.2   Implementation

It is well-known that use of many interpolation points generally results in better approximations. That is, subdividing a given interval into several subintervals and keeping to a minimum the error on each of the subintervals improves accuracy of approximation for the given function as a whole. Since for computer-hardware implementations it is convenient that the number of data points be a power of two, we will assume that the interval I is divided into 2^k intervals: $\left[L, L + \frac{\Delta}{2^k}\right), \left[L + \frac{\Delta}{2^k}, L + \frac{2\Delta}{2^k}\right), \ldots, \left[L + \frac{2^{k-1}\Delta}{2^k}, U\right]$, where $\Delta = U - L$. Then, given an argument, x, the interval into which it falls can readily be located by using, as an address, the k most significant bits of the binary representation of x. The basic hardware implementation therefore has the high-level organization shown in Figure 6. The two memories hold the constants c_1 and c_2 for each interval.

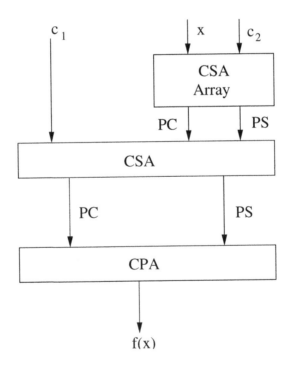

Figure 7: High-performance hardware organization for function evaluation

Figure 6 is only here to be indicative of a "naive" implementation, although it is quite realistic for some current FPGAs. For a high-speed implementation, the actual structure may differ in several ways. Consider for example the multiplier-adder pair. Taken individually, the adder must be a carry-propagate adder (CPA); and the multiplier, if it is of high performance will consist of an array of carry-save adders (CSAs) with a final CPA to assimilate the partial-sum/partial-carry (PC/PS) output of the CSAs. But the multiplier-CPA may be replaced with two CSAs, to yield much higher performance. Therefore, in a high speed implementation the actual structure would have the form shown in Figure 7.

Nevertheless, for FPGAs, the built-in structure will impose some constraints, and the actual implementation will generally be device-dependent. For example, for a device such as the Xilinx Virtex-4, the design of Figure 6 may be implemented more or less exactly as given: the DSP48 slice provides the multiply-add function, and the constants, c_1 and c_2, are stored in Block-RAM. They could also be stored in Distributed-RAM, as it is unlikely that there will be many of them. Several slices would be required to store the constants at the required precision, but this is not necessarily problematic: observe that each instance of activation-function computation corresponds to several MACs (bit slices).

All of the above is fairly straightforward, but there is one point that needs a particular mention: Equations 1.1 and 1.2 taken together imply that there is an inevitable disparity between the rate of inner-product (MACs) computations and activation-function computations. In custom design, this would not cause particular concern: both the design and the placement of the relevant hardware units can be chosen so as to optimize cost, performance, etc. But with FPGAs, this luxury does not exist: the mapping of a network to a device, the routing requirements to get an inner-product value to the correct place for the activation-function computation, the need to balance the disparate rates ... all these mean that the best implementation will be anything but straightforward.

1.7.3   Theory

We shall illustrate our results with detailed numerical data obtained for a fixed number of intervals. All numerical computations, were carried out in the computer algebra system MAPLE [24] for the interval[11] $I = [0.5, 1]$ and $k = 4$; that is, I was divided into the 16 intervals:

$$\left[\frac{1}{2}, \frac{17}{32}, \frac{9}{16}, \frac{19}{32}, \frac{5}{8}, \ldots, 1 \right].$$

We have used MAPLE to perform many of complex symbolic computations. Floating-point calculations in MAPLE are carried out in finite precision, with

---

[11]Note that evaluation on any other interval can be transformed into evaluation on the interval $[0.5, 1]$.

intermediate results rounded to a precision that is specified by MAPLE constant $Digits$. This constant controls the number of digits that MAPLE uses for calculations. Thus, generally, the higher the $Digits$ value is, the higher accuracy of the obtainable results, with roundoff errors as small as possible. (This however cannot be fully controlled in case of complex algebraic expressions). We set $Digits$ value to 20 for numerical computations. Numerical results will be presented using standard (decimal) scientific notation.

Applying condition (C), in Section 7.1, to the sigmoid function, we get

$$\begin{cases} c_1 + c_2 L &= \frac{1}{1+e^{-U}} \\ c_1 + c_2 U &= \frac{1}{1+e^{-L}}. \end{cases}$$

For simplicity, we will use θ to denote the expression

$$Ue^L - Le^U - Le^{(L+U)} + Ue^{(L+U)}.$$

Then the solution of the above system may be expressed as

$$c_1 = \frac{\theta}{\theta + U - L + Ue^U - Le^L}$$

$$c_2 = \frac{e^U - e^L}{\theta + U - L + Ue^U - Le^L}.$$

and the approximation function $\widehat{f}(x) = c_1 + c_2 x$ takes the form

$$\widehat{f}(x) = \frac{\theta + x\left(e^U - e^L\right)}{\theta + U - L + Ue^U + Le^L}, \quad x \in I.$$

The relative error is now

$$\varepsilon(x) = \frac{-Le^L - L + Ue^U + U - e^{-x}\theta - xe^U}{\theta + U - L + Ue^U + Le^L}$$
$$+ \frac{-xe^{U-x} + xe^L + xe^{(L-x)}}{\theta + U - L + Ue^U + Le^L}, \quad x \in I.$$

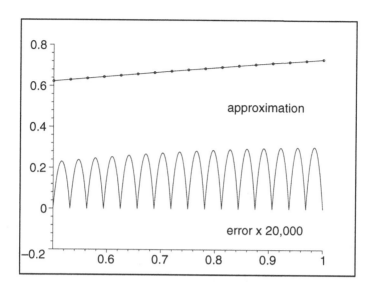

Figure 8: Error in piecewise-linear approximation of the sigmoid

Figure 8 shows the results for the 16-interval case. As the graphs show, the amplitude of the error attains a maximum in each of the sixteen intervals. To ensure that it is in fact so on any interval we investigate the derivatives of the error function.

The first derivative of the error function is

$$\varepsilon'(x) = \frac{\theta + e^L - e^U + x\left(e^{(U-x)} - e^{(L-x)}\right)}{\theta + (U - L) + Ue^U - Le^L}$$

$$+ \frac{e^{(L-x)} - e^{(U-x)} + \theta e^{-x}}{\theta + (U - L) + Ue^U - Le^L}, \quad x \in I.$$

A closer look at the formula for the derivative, followed by simple algebraic computations, reveals that the equation $\varepsilon(x) = 0$ is reducible to the equation

$$Ae^x = B + Cx, \quad \text{for some constants} \quad A, B, C.$$

The solution of this equation is the famous Lambert W function, which has been extensively studied in the literature; and many algorithms are known for the computation of its values.[12] Since the Lambert W function cannot be analytically expressed in terms of elementary functions, we leave the solution of

---

[12]The reader interested in a recent study of the Lambert W function is refereed to [9].

our equation in the form

$$x_{stat} = e^L \, LambertW \left(-e^{\left(U - \frac{\theta + \left(e^U - e^L \right)(U-1)}{e^U - e^L} \right)} \right)$$

$$+ e^U \, LambertW \left(-e^{\left(U - \frac{\theta + \left(e^U - e^L \right)(U-1)}{e^U - e^L} \right)} \right).$$

where *LambertW* is the MAPLE notation for the Lambert W function. There is no straightforward way to extend our results to an arbitrary interval I. So, for the rest of this section we will focus on the 16-interval case, where, with the help of MAPLE, we may accurately ensure validity of our findings. It should nevertheless be noted that since this choice of intervals was quite arbitrary (within the domain of the investigated function), the generality of our results are in no way invalidated. Figure 9 shows plots of the first derivative of the relative-error function on sixteen intervals, confirming that there exists a local maximum on each interval for this function.

From Figure 9, one can infer that on each interval the stationary point occurs somewhere near the mid-point of the interval. This is indeed the case, and the standard Newton-Raphson method requires only a few iterations to yield a reasonably accurate approximation to this stationary value. (To have a full control over the procedure we decided not to use the MAPLE's built-in approximation method for Lambert W function values.) For the 16-interval case, setting the tolerance to 10^{-17} and starting at the mid-point of each interval, the required level of accuracy is attained after only three iterations. For the stationary points thus found, the magnitude of the maximum error is

$$\varepsilon_{max} \quad = \quad 1.5139953883 \times 10^{-5} \tag{1.3}$$

which corresponds to 0.3 on the "magnified" graph of Figure 9.

We next apply the improved condition (IC) to this approximation. By (Err), we have

$$\varepsilon(x) \quad = \quad 1 - c_1 - c_1 e^{-x} - c_2 x - c_2 x e^{-x} \tag{1.4}$$

hence

$$\varepsilon(L) \quad = \quad 1 - c_1 - c_1 e^{-L} - c_2 L - c_2 L e^{-L} \tag{1.5}$$

$$\varepsilon(U) \quad = \quad 1 - c_1 - c_1 e^{-U} - c_2 U - c_2 U e^{-U}. \tag{1.6}$$

From Equations (1.5) and (1.6) we get an equation that we can solve for c_2:

$$c_2 = \frac{c_1\left(e^{-L} - e^{-U}\right)}{U + Ue^{-U} - L - Le^{-L}} \tag{1.7}$$

Substituting for c_2 in Equation (1.4) yields the final formula for the relative error

$$\varepsilon(x) = 1 - c_1 - c_1 e^{-x} + \frac{c_1\left(e^{-U} - e^{-L}\right)x}{U + Ue^{-U} - L - Le^{-L}}$$

$$+ \frac{c_1\left(e^{-U} - e^{-L}\right)xe^{-x}}{U + Ue^{-U} - L - Le^{-L}}, \quad x \in I.$$

To study the error magnitude we investigate its first derivative for $x \in I$:

$$\varepsilon'(x) = c_1 \frac{e^{-x}U + e^{(-x-U)}U - e^{-x}L - e^{(-x-L)}L + e^{-U}}{U + Ue^{-U} - L - Le^{-L}}$$

$$+ c_1 \frac{-e^{-L} + e^{(-x-U)} + e^{(-x-L)}}{U + Ue^{-U} - L - Le^{-L}}$$

$$+ c_1 \frac{-xe^{(-x-U)} + xe^{(-x-L)}}{U + Ue^{-U} - L - Le^{-L}}.$$

We may assume without loss of generality that c_1 is positive: the graph of the sigmoid function indicates that this is a valid assumption. For simplicity, let us assume that $c_1 = 1$ and see how the first derivative behaves on the sixteen intervals. Figure 10 shows the graphs of the first derivative of the new error function. From these plots we see that within each interval the derivative changes sign at a unique stationary point. Finding an exact analytical formula for the values of these points is not possible, because, as above, the equation $\varepsilon(x) = 0$, reduces to a Lambert-type of equation. So, once again, we apply the Newton-Raphson method to get some reasonably accurate estimate values. Starting the iteration from the mid-point we obtain good approximation after just a few iterations.

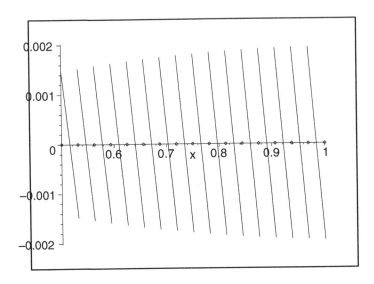

Figure 9: Plots of first derivative of sigmoid error-function

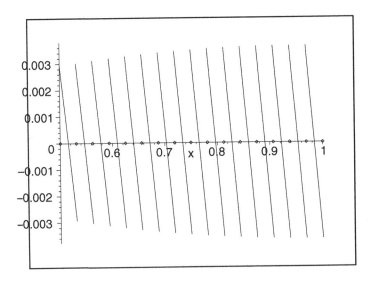

Figure 10: Plots of first derivative of improved sigmoid error-function

It is critical to note that although the Newton-Raphson procedure is easily (and frequently) implemented in hardware [4], in this case a software implementation is sufficient. The procedure is required only to obtain c_1 and c_2, and once these have been obtained off-line and stored in memory, the procedure is not relevant.

Let by x_a denote the approximate value at which the error has a local extremum. Then, by the the final formula for the relative error, we have

$$\varepsilon(x_a) = 1 - c_1 - c_1 e^{-x_a} + \frac{c_1\left(e^{-U} - e^{-L}\right) x_a}{U + Ue^{-U} - L - Le^{-L}}$$
$$+ \frac{c_1\left(e^{-U} - e^{-L}\right) xe^{-x_a}}{U + Ue^{-U} - L - Le^{-L}}$$

Since, by condition (IC), we must have $\varepsilon(x_a) = -\varepsilon(L)$, we end up with one equation with one variable c_1. Solving this equation gives us the required second parameter for our approximation function $\widehat{f}(x) = c_1 + c_2 x$. We omit tedious (but elementary) algebraic calculations, presenting only the final formula

$$c_1 = -2\left(U + Ue^{-U} - L - Le^{-L}\right)$$
$$/\Bigg(-2U - 2Ue^{-U} + 2L + 2Le^{-L} - e^{-x_a}U(1 + U)$$
$$+ e^{-x_a}L(1 + L) + x_a e^{-U} - x_a e^{-L} + x_a e^{-x_a - U}$$
$$+ x_a e^{-x_a - L} + x_a e^{-x_a - L} - e^{-L}U$$
$$- e^{-L-U}U + Le^{-U} + Le^{-L-U}\Bigg)$$

which, by Equation (1.7) yields the final formula for c_2. Finally, we substitute c_1, c_2 values into the relative-error formula, expressed by Equation (1.4). (Note that, x_a must be replaced by the corresponding approximate value from the Newton-Raphson procedure.) We do not present the final analytical formula for the error as it is quite complex and of little interest by itself. Figure 11 shows the results finally obtained. The magnitude of the maximum relative error is now

$$\varepsilon(max) = 7.5700342463 \times 10^{-6}$$

which, compared with (1.3) is a reduction of 50.00038%. This concludes the exposition.

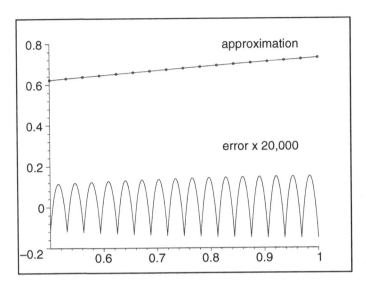

Figure 11: Plots of improved sigmoid error-function

1.8   Performance evaluation

Having outlined above the promise of realizing neural networks in hardware, we now come to an especially delicate point — that of actually showing that "semi-custom" (i.e. FPGA) or custom (i.e. ASIC) neurocomputers can actually deliver what is promised. In this respect, the neural-network community has not especially distinguished itself, which, in turn explains the dearth of many practical neurocomputers, despite the many years of research, the development of countless prototypes, and so forth. (This point can be readily appreciated by comparing the status of performance-evaluation for nuerocomputers with that for conventional computers.) We will not here aim to solve the underlying problems or even suggest specific concrete solutions — either being an enormous task that is outside the scope of this work — but it is our objective to sharply highlight them and to indicate general directions for their solution.

At the very least there are two issues that must be considered for a proper evaluation of performance: the *metrics* used and what *benchmarks* are used to obtain the measurements. (Both need to be such that they can, in the main, be easily agreed on and understood by most users.) The neural-network area is sorely lacking in both. The most commonly used metrics are *connections-per-second* (CPS), which is essentially the rate at which neuron multiply-add operations are carried out, and *connection-updates-per-second* (CUPS), which is essentialy the rate at which updates of weights are carried out. *Speedup* is also sometimes used but is of even less value than CPS and CUPS; and even worse is to rely on just time-complexity, e.g. [1]. The are three main problems with these metrics:

- MCPS and MCUPS are similar to, and suffer from the same drawbacks as, MIPS (Millions of Instructions Per Second) and MFLOPS (Millions of Floating-point Operations Per Second) that were long used for general-purpose computers and now largely discredited as useful measures.

- MCPS and MCUPS cannot always be meaningfully used with all types of networks; for example, they are of little worth with radial-basis-function networks.

- A large value of MCPS or MCUPS does not necessarily mean better performance (i.e. in the sense of a network being faster) when applied to different algorithms. In other words, interpreting the numbers is not straightforward.

As is the case for general-purpose computers, there is little doubt that ultimately the best measure of performance is actual execution-time [19].

As an example of the untidy state of affairs that curently exists, consider, for example [3], which has already been partially discussed above. In that work, execution-time is one of the main metrics used, which is acceptable as far as it goes. But things quickly become problematic. First, although some neural-network applications have been chosen for use as benchmarks, no basis is given for why and how they been chosen. Second, it is not quite the case that exactly the same programs are being used for benchmarks: the authors compare a Reactive Tabu Search learning algorithm (on their machine) against Back-Propagation learning (on two other machines). Third, given the variety and non-uniform use of various metrics, what the target is is far from clear: Is it, say, the the proportion of patterns that are correcly classified? If so, then that should be fixed and measurements than made of the other metrics (error, execution-time, etc.) and the results then used to compare the different machines. The same remark applies to, say, fixing error limits and then measuring execution time, or number of iterations/epcohs, or patterns-classified, and so forth. As it is, the authors allow all parameters to simultaneously vary in a manner that practically renders meaningless the use of execution-time as a reasonable metric. Fourth, details of algorithm-implementations on the two other machines are not given, which begs the question of whether they were the best possible. Contrast this with the standards of SPEC [20], in which "owners" (usually the manufacturers) of given machines run (under highly regulated conditions) specific programs, under tight constraints, and then publicly report the results, thus ensuring best-effort implementations.

In summary, the area of performance-evaluation — in particular the choice performance metrics and selection of benchmarks — is one that needs to be addressed urgently for neurocomputers. Neurocomputers (whether in ASIC of FPGA) will not achieve widespread use unless potential users can be convinced of their worth; given their history so far, it should now be clear that merely extolling their virtues is insufficient.

1.9   Conclusions

This chapter covers a broad range of topics, many of which are discussed in more detail in following chapters. In the case of arithmetic, we have high-lighted inner-product and activation-function computations. We have advanced the use of piecewise linear interpolation, but the story need not end there: although interpolations of a degree higher than three do do not appear to be appropriate for high-speed hardware implementations, there may be some profit in the search for second-order ones that are well-suited to FPGAs. Chapter 2 discussed further aspects of arithmetic.

We have discussed the main types of parallelism that are to be found in neural networks, but little of that discussion has addressed the matter of mappings. With very large networks, the mapping from network to FPGA is a major factor in performance. Chapters 3 and 4 are devoted to a suitable theoretical framework for the derivation of such mappings. Chapter 5 also deals with mappings but is limited to back-propagation in the context of an actual device; Chapter 6 is similar in that it is limited to associative memories.

Chapters 7 through 11 cover the FPGA-implementation of neural networks for several specific applications. The last chapter is a retrospective: it discusses various lessons learned from the realization of a custom neurocomputer and projects these to current FPGAs.

References

[1] U. Ruckert. 2002. ULSI architectures for artificial neural networks. *IEEE Micro* (May–June): 10–19.

[2] J. Buddefeld and K. E. Grosspietsch. 2002. Intelligent-memory architectures for artificial neural networks. *IEEE Micro* (May–June): 32–40.

[3] G. Danese, F. Leoporati, and S. Ramat. 2002. A parallel neural processor for real-time applications. *IEEE Micro* (May–June): 20–31.

[4] A. R. Omondi. *Computer Arithmetic Systems: Algorithms, Architecture, and Implementations*. Prentice-Hall, UK, 1994.

[5] T. Nordstrom and B. Svensson. 1991. Using and designing massively parallel computers for artificial neural networks. *Journal of Parallel and Distributed Computing*, **14**:260–285.

[6] Y. Hirai. 1993. Hardware implementations of neural networks in Japan. *Neurocomputing*, **5**:3–16.

[7] N. Sundarajan and P. Satchandran. 1998. *Parallel Architectures for Artificial Neural Networks*. IEE Press, California.

[8] D. Hammerstom. 1991. A highly parallel digital architecture for neural network simulation. In: J.D. Delgado-Frias and W.R. Moore, Eds., *VLSI for Artificial Intelligence and Neural Networks*, Plenum Press.

[9] R. M. Corless, G. H. Gonnet, D. E. G. Hare, and D. J. Jeffrey, and D. E. Knuth. 1996. On the Lambert W Function. *Advances in Computational Mathematics*, **12**:329–359.

[10] M. Ito, N. Takagi, and S. Yajima. 1997. Efficient initial approximation for multiplicative division and square-root by a multiplication with operand modification. *IEEE Transactions on Computers*, **46**(4):95–498.

[11] J. M. Muller. 1997. *Elementary Functions: Algorithms and Implementation*. Birkhauser, Boston, USA.

[12] S. M. Pizer and V. L. Wallace. 1983. *To Compute Numerically*. Little, Brown, and Co., Boston, USA.

[13] M. Bajger and A. R. Omondi. 2005. Low-cost, high-speed implementations of square-root, exponential and sigmoidal function-evaluations. Submitted for publication.

[14] S. Vassiliadis, M. Zhang, and J. G. Delgado-Frias. 2000. Elementary function generators for neural network emulators. *IEEE Transactions on Neural Networks*, **11**(6):1438–1449.

[15] K. Basterretxea, J. M. Tarela, and I. del Campo. 2004. Approximation of sigmoid function and the derivative for hardware implementation of artificial neurons. *IEEE Proceedings — Circuits, Devices, and Systems*, **151**(1):18–24.

[16] O. Mencer and W. Luk. 2004. Parameterized high throughput function evaluation for FPGAs. *Journal of VLSI Signal Processing*, **36**:17–25.

[17] J. L. Holt and J. N. Hwang. 1993. Finite-precision error analysis of neural network hardware implementations. *IEEE IEEE Transactions on Computers*, **42**(3):280–290.

[18] A. R. Omondi. 2000. Neurocomputers: a dead end? *International Journal of Neural Systems*, **10**(6):475–481.

[19] J. L. Hennessy and D. A. Patterson. 2002. *Computer Architecture: A Quantitative Approach*. Morgan Kaufmann.

[20] SPEC. Standard Performance Evaluation Corporation. (www.spec.org)

[21] Xilinx. 2004. Virtex-4 User Guide.

[22] Xilinx. 2004. XtremeDSP Design Considerations: User Guide.

[23] A. P. Preethy, D. Radhakrishnan, A. R. Omondi. Mar 2001. A high-performance residue-number-system multiply-accumulate unit. In: *11th ACM Great Lakes Symposium on VLSI* (Purdue, Indiana, USA), pp 145–149.

[24] Waterloo Maple Inc. *Maple 8 Programming Guide*, 2002.

Chapter 2

ON THE ARITHMETIC PRECISION FOR IMPLEMENTING BACK-PROPAGATION NETWORKS ON FPGA: A CASE STUDY

Medhat Moussa and Shawki Areibi and Kristian Nichols

University of Guelph
School of Engineering
Guelph, Ontario, N1G 2W1, Canada
mmoussa@uoguelph.ca   sareibi@uoguelph.ca   knichols@uoguelph.ca

Abstract   Artificial Neural Networks (ANNs) are inherently parallel architectures which represent a natural fit for custom implementation on FPGAs. One important implementation issue is to determine the numerical precision format that allows an optimum tradeoff between precision and implementation areas. Standard single or double precision floating-point representations minimize quantization errors while requiring significant hardware resources. Less precise fixed-point representation may require less hardware resources but add quantization errors that may prevent learning from taking place, especially in regression problems. This chapter examines this issue and reports on a recent experiment where we implemented a Multi-layer perceptron (MLP) on an FPGA using both fixed and floating point precision. Results show that the fixed-point MLP implementation was over 12x greater in speed, over 13x smaller in area, and achieves far greater processing density compared to the floating-point FPGA-based MLP.

Keywords: Reconfigurable Computing, Back-propagation Algorithm, FPGAs, Artificial Neural Networks

2.1   Introduction

Artificial neural networks (ANNs) have been used in many applications in science and engineering. The most common architecture consists of multi-layer perceptrons trained using the error back-propagation algorithm (MLP-BP) [37]. One of the main problems in training a BP Network is the lack of a clear methodology to determine the network topology before training starts. Experimenting with various topologies is difficult due to the long time required

A. R. Omondi and J. C. Rajapakse (eds.), FPGA Implementations of Neural Networks, 37–61.
© 2006 *Springer. Printed in the Netherlands.*

for each training session especially with large networks. Network topology is an important factor in the network's ability to generalize after training is completed. A larger than needed network may over-fit the training data and result in poor generalization on testing data, while a smaller than needed network may not have the computational capacity to approximate the target function. Furthermore, in applications where online training is required, training time is often a critical parameter. Thus it is quite desirable to speed up training. This allows for reasonable experimentation with various network topologies and ability to use BP networks in online applications.

Since Neural Networks in general are inherently parallel architectures [55], there have been several earlier attempts to build custom ASIC based boards that include multiple parallel processing units such as the NI1000. However, these boards suffered from several limitations such as the ability to run only specific algorithms and limitations on the size of a network. Recently, much work has focused on implementing artificial neural networks on reconfigurable computing platforms. Reconfigurable computing is a means of increasing the processing density (i.e greater performance per unit of silicon area) above and beyond that provided by general-purpose computing platforms. Field Programmable Gate Arrays (FPGAs) are a medium that can be used for reconfigurable computing and offer flexibility in design like software but with performance speeds closer to Application Specific Integrated Circuits (ASICs).

However, there are certain design tradeoffs which must be dealt with in order to implement Neural Networks on FPGAs. One major tradeoff is *area vs. precision*. The problem is how to balance between the need for numeric precision, which is important for network accuracy and speed of convergence, and the cost of more logic areas (i.e. FPGA resources) associated with increased precision. Standard precisions floating-point would be the ideal numeric representation to use because it offers the greatest amount of precision (i.e. minimal quantization error) and matches the representation used in simulating Neural Networks on general purpose microprocessors. However, due to the limited resources available on an FPGA, standard floating-point may not be as feasible compared to more area-efficient numeric representations, such as 16 or 32 bit fixed-point.

This chapter explores this design trade-off by testing an implementation of an MLP-BP network on an FPGA using both floating-point and fixed-point representations. The network is trained to learn the XOR problem. The study's goal is to provide experimental data regarding what resources are required for both formats using current FPGA design tools and technologies. This chapter is organized as follows: In Section 2.2, background material on the area vs precision range trade-off is presented as well as an overview of the back-propagation algorithm and FPGA architectures. Section 2.3 provides details about the architecture design used to implement a BP network on FPGA. In section 2.4 the XOR problem is presented. Finally validation of the proposed implementations, and benchmarked results of floating-point and fixed-point arithmetic functions implemented on a FPGA are discussed in Section 2.5.

2.2   Background

One way to help achieve the density advantage of reconfigurable computing over general-purpose computing is to make the most efficient use of the hardware area available. In terms of an optimal *range-precision vs area trade-off*, this can be achieved by determining the *minimum allowable precision* and *minimum allowable range*, where their criterion is to minimize hardware area usage without sacrificing quality of performance. These two concepts combined can also be referred to as the *minimum allowable range-precision*.

2.2.1   Range-Precision vs. Area Trade-off

A reduction in precision usually introduces many errors into the system. Determining the minimum allowable precision is actually a question of determining the maximum amount of uncertainty (i.e. quantization error due to limited precision) an application can withstand before performance begins to degrade. It is often dependent upon the algorithm used and the application at hand.

For MLP using the BP algorithm, Holt and Baker [41] showed using simulations and theoretical analysis that *16-bit fixed-point* (1 bit sign, 3 bit left and 12 bit right of the radix point) was the minimum allowable range-precision for the back-propagation algorithm assuming that both input and output were normalized between [0,1] and a sigmoid transfer function was used.

Ligon III *et al.* [45] have also shown the density advantage of fixed-point over floating-point for older generation Xilinx 4020E FPGAs, by showing that the space/time requirements for 32-bit fixed-point adders and multipliers were less than that of their 32-bit floating-point equivalents.

Other efforts focused on developing a complete reconfigurable architecture for implementing MLP. Eldredge [3] successfully implemented the back-propagation algorithm using a custom platform he built out of Xilinx XC3090 FPGAs, called the Run-Time Reconfiguration Artificial Neural Network (RRANN). He showed that the RRANN architecture could learn how to approximate centroids of fuzzy sets. Heavily influenced by the Eldredge's RRANN architecture, Beuchat *et al.* [13] developed a FPGA platform, called RENCO–a REconfigurable Network COmputer. As it's name implies, RENCO contains four Altera FLEX 10K130 FPGAs that can be reconfigured and monitored over any LAN (i.e. Internet or other) via an on-board 10Base-T interface. RENCO's intended application was hand-written character recognition. Ferrucci and Martin [14, 15] built a custom platform, called Adaptive Connectionist Model Emulator (ACME) which consists of multiple Xilinx XC4010 FPGAs. They validated ACME by successfully carrying out a 3-input, 3-hidden unit, 1-output network used to learn the 2-input XOR problem. Skrbek's FPGA platform [26], called the ECX card, could also implement Radial Basis Function (RBF) neural networks, and was validated using pattern recognition applications such as parity problem, digit recognition, inside-outside test, and sonar signal recognition.

Since the size of an FPGA-based MLP-BP is proportional to the multiplier used, it is clear that given an FPGA's finite resources, a 32-bit signed (2's complement) *fixed-point* representation will allow larger [54] ANNs to be implemented than could be accommodated when using a 32-bit IEEE (a 32-bit floating point multiplier can be implemented on a Xilinx Virtex-II or Spartan-3 FPGA using four of the dedicated multiplier blocks and CLB resources) *floating-point*. However, while 32 fixed-point representation allows high processor density implementation, the quantization error of 32 floating-point representation is negligible. Validating an architecure on an FPGA using 32-bit floating point arithmetic might be easier than fixed point arithmetic since a software version of the architecture can be run on a Personal Computer with 32-bit floating point arithmetic. As such its use is justifiable if the relative loss in processing density is negligible in comparison.

FPGA architectures and related development tools have become increasingly sophisticated in more recent years, including improvements in the space/time optimization of arithmetic circuit designs. As such, the objective of this study is to determine the feasibility of floating-point arithmetic in implementing MLP-BP using today's FPGA design tools and technologies. Both floating-point and fixed-point precision are considered for implementation and are classified as *amplitude-based* digital numeric representations. Other numeric representations, such as digital *frequency-based* [42] and analog were not considered because they promote the use of low precision, which is often found to be inadequate for **minimum allowable range-precision**.

2.2.2   Overview of Back-propagation Algorithm

It is helpful before proceeding to discuss architecture design to give a brief review of MLP and the error Back-propagation algorithm. The general structure of a Multi-layer perceptron (MLP) neural network is shown in Figure 2.1, where layers are numbered 0 to M, and neurons are numbered 1 to N.

A MLP using the back-propagation algorithm has five steps of execution:

(1) Initialization

The following parameters must be initialized before training starts: (i) $w_{kj}^{(s)}(n)$ is defined as the *synaptic weight* that corresponds to the connection from neuron unit j in the $(s-1)^{th}$ layer, to k in the sth layer. This weight

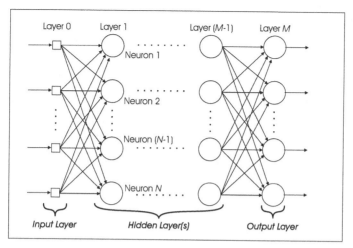

Figure 2.1. Generic structure of a feedforward ANN

is updated during the nth iteration, where $n = 0$ for initialization. (ii) η is defined as the *learning rate* and is a constant scaling factor used to control the step size in error correction during each iteration of the back-propagation algorithm. (iii) $\theta_k^{(s)}$ is defined as the *bias* of a neuron, which is similar to synaptic weight in that it corresponds to a connection to neuron unit k in the sth layer. Statistically, biases can be thought of as noise, which better randomizes initial conditions, and increases the chances of convergence.

(2) Presentation of Training Examples

Available training data are presented to the network either individually or as a group (a.k.a. epoch).

(3) Forward Computation

During the forward computation, data from neurons of a lower layer (i.e. $(s-1)^{th}$ layer), are propagated forward to neurons in the upper layer (i.e. s^{th} layer) via a feed-forward connection network. The computation performed by each neuron (in the hidden layer) is as follows:

$$H_k^{(s)} = \sum_{j=1}^{N_{s-1}} w_{kj}^{(s)} o_j^{(s-1)} + \theta_k^{(s)} \tag{2.1}$$

where $j < k$ and $s = 1, \ldots, M$
$H_k^{(s)} = $ *weighted sum* of the kth neuron in the sth layer
$w_{kj}^{(s)} = $ synaptic weight sd defined above
$o_j^{(s-1)} = $ *neuron output* of the jth neuron in the $(s-1)^{th}$ layer

$\theta_k^{(s)}$ = bias of the kth neuron in the sth layer.
On the other hand for the output layer neurons the computation is as follows:

$$o_k^{(s)} = f(H_k^{(s)}) \qquad (2.2)$$

where $k = 1, \ldots, N$ and $s = 1, \ldots, M$
$o_k^{(s)}$ = neuron output of the kth neuron in the sth layer
$f(H_k^{(s)})$ = *activation function* computed on the weighted sum $H_k^{(s)}$
Note that a unipolar sigmoid function is often used as the nonlinear activation function, such as the following *logsig* function:

$$f(x)_{logsig} = \frac{1}{1 + \exp(-x)} \qquad (2.3)$$

(4) Backward Computation

In this step, the weights and biases are updated. The learning algorithm's goal is to minimize the error between the expected (or teacher) value and the actual output value that was determined in the Forward Computation. The following steps are performed:

1 Starting with the output layer, and moving back towards the input layer, calculate the local gradients, as follows:

$$\varepsilon_k^{(s)} = \begin{cases} t_k - o_k^{(s)} & s = M \\ \sum_{j=1}^{N_{s+1}} w_{kj}^{s+1} \delta_j^{(s+1)} & s = 1, \ldots, M-1 \end{cases} \qquad (2.4)$$

where
$\varepsilon_k^{(s)}$ = *error term* for the kth neuron in the sth layer; the difference between the teaching signal t_k and the neuron output $o_k^{(s)}$
$\delta_j^{(s+1)}$ = *local gradient* for the jth neuron in the $(s+1)^{th}$ layer.

$$\delta_k^{(s)} = \varepsilon_k^{(s)} f'(H_k^{(s)}) \quad s = 1, \ldots, M \qquad (2.5)$$

where $f'(H_k^{(s)})$ is the derivative of the activation function.

2 Calculate the weight (and bias) changes for all the weights as follows:

$$\Delta w_{kj}^{(s)} = \eta \delta_k^{(s)} o_j^{(s-1)} \quad k = 1, \ldots, N_s \\ j = 1, \ldots, N_{s-1} \qquad (2.6)$$

where $\Delta w_{kj}^{(s)}$ is the change in synaptic weight (or bias) corresponding to the gradient of error for connection from neuron unit j in the $(s-1)^{th}$ layer, to neuron k in the sth layer.

3 Update all the weights (and biases) as follows:

$$w_{kj}^s(n+1) = \Delta w_{kj}^{(s)}(n) + w_{kj}^{(s)}(n) \qquad (2.7)$$

where $k = 1, \ldots, N_s$ and $j = 1, \ldots, N_{s-1}$

$w_{kj}^s(n+1)$ = updated synaptic weight (or bias) to be used in the $(n+1)^{th}$ iteration of the Forward Computation

$\Delta w_{kj}^{(s)}(n)$ = change in synaptic weight (or bias) calculated in the nth iteration of the Backward Computation, where n = the current iteration

$w_{kj}^{(s)}(n)$ = synaptic weight (or bias) to be used in the nth iteration of the Forward and Backward Computations, where n = the current iteration.

(5) Iteration

Reiterate the Forward and Backward Computations for each training example in the epoch. The trainer can continue to train the MLP using one or more epochs until some stopping criteria is met. Once training is complete, the MLP only needs to carry out the Forward Computation when used in applications.

2.2.3   Field Programmable Gate Arrays

FPGAs are a form of programmable logic, which offer flexibility in design like software, but with performance speeds closer to Application Specific Integrated Circuits (ASICs). With the ability to be reconfigured an endless number of times after having been manufactured, FPGAs have traditionally been used as a prototyping tool for hardware designers. However, as growing die capacities of FPGAs have increased over the years, so has their use in reconfigurable computing applications too.

The fundamental architecture of Xilinx FPGAs consists of a two-dimensional array of programmable logic blocks, referred to as *Configurable Logic Blocks (CLBs)*. Figure 2.2 shows the architecture of a CLB from the Xilinx Virtex-E family of FPGAs, which contains four *logic cells (LCs)* and is organized in two similar *slices*. Each LC includes a 4-input look-up table (LUT), dedicated fast carry-lookahead logic for arithmetic functions, and a storage element (i.e. a flip-flop). A CLB from the Xilinx Virtex-II family of FPGAs, on the other hand, contains eight 4-input LUTs, and is over twice the amount of logic as a Virtex-E CLB. As we will see, the discrepancies in CLB architecture from one family to another is an important factor to take into consideration when comparing the spatial requirements (in terms of CLBs) for circuit designs which have been implemented on different Xilinx FPGAs.

2.3   Architecture design and implementation

There has been a rich history of attempts at implementing ASIC-based approaches for neural networks - traditionally referred to as *neuroprocessors* [29]

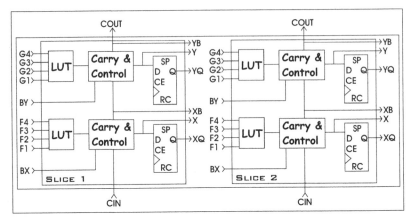

Figure 2.2.   Virtex-E Configurable Logic Block

or *neurochips*. FPGA-based implementations, on the other hand, are still a fairly new approach which has only been in effect since the early 1990s. The type of neural network used in a FPGA-based implementation, and/or the algorithm used for *on-chip learning* is a classification feature which often depends on its intended application. On-chip learning [11] occurs when the learning algorithm is implemented in hardware, or in this case, on the FPGA. *Offline learning* occurs when learning occurs on a general-purpose computing platform before the learned system is implemented in hardware.

2.3.1   Non-RTR Implementation

The digital ANN architecture implemented in this chapter is an example of a non-RTR (Run-Time Reconfiguration) reconfigurable computing application, where all stages of the algorithm reside together on the FPGA at once. A finite state machine was used to ensure proper sequential execution of each step of the back-propagation algorithm as described in Section 2.2.2, which consists of the following two states:

1 Forward state (**F**) - used to emulate the *forward pass* associated with the back-propagation algorithm. Only the ANN's input signals, synapses, and neurons should be active in this state, in order to calculate the ANN's output. All forward pass operations (i.e. Forward Computations as described by Equations 2.1, 2.2, and 2.3) should be completed by the time the Forward State (**F**) ends.

2 Backward state (**B**) - used to emulate the *backward pass* associated with the back-propagation algorithm. All the circuitry associated with helping the ANN learn (i.e. essentially all the circuitry *not* active in Forward State) should be active here. All backward pass operations (i.e. Back-

ward Computations as described by Equations 2.4, 2.5, and 2.6) should be completed by the time the Backward state ends.

It should be noted that both states of the finite state machine continually alternate, and synaptic weights are updated (as described in Equation 2.7) during the transition from Backward State to Forward State.

As far as the ANN's components (eg. neurons, synapses) were concerned, the finite state machine is generally a means of synchronizing when various sets of components should be active. The duration of each state depends on the number of clock cycles required to complete calculations in each state, the length of the system's clock period, and the propagation delay associated with each state(Note that propagation delay is platform dependent, and can only be determined after the digital VLSI design has been synthesized on a targeted FPGA. The propagation delay is then determined through a timing analysis/simulation using the platform's EDA tools). The architecture of the active ANN components associated with each state dictates the propagation delay for that state.

Each of the ANN components implemented in hardware, such as the synapse and neuron, housed a *chip select* input signal in their architecture which is driven by the finite state machine. This chip select feature ensured that only those components that were associated with a particular state were enabled or active throughout that state's duration. With regards to initialization of the circuit, a *reset* input signal was used which would fulfill two important requirements when activated:

- Ensure the finite state machine initially starts in "Forward State".

- Initialize the synaptic weights of the ANN, to some default value.

Finally, the BP algorithm calculations, Equations 2.1–2.7, are realized using a series of arithmetic components, including addition, subtraction, multiplication, and division. Standardized high-description language (HDL) libraries for digital hardware implementation can be used in synthesizing all the arithmetic calculations involved with the back-propagation algorithm, in analogous fashion of how typical math general programming language (GPL) libraries are used in software implementations of ANNs. The architecture described here is generic enough to support arithmetic HDL libraries of different amplitude-based precision, whether it be floating-point or fixed-point.

2.3.2 Arithmetic Library

The architecture was developed using *VHDL*. Unfortunately, there is currently no explicit support for fixed- and floating-point arithmetic in VHDL (according to the IEEE Design Automation Standards Committee [43], an extension of IEEE Std 1076.3 has been proposed to include support for fixed-

and floating-point numbers in VHDL, and is to be addressed in a future review of the standard). As a result, two separate arithmetic VHDL libraries were custom designed for use with the FPGA-based ANN. One of the libraries supports the IEEE-754 standard for single-precision (i.e. 32-bit) floating-point arithmetic, and is referred to as uog_fp_arith, which is an abbreviation for *University of Guelph Floating-Point Arithmetic*. The other library supports 16-bit fixed-point arithmetic, and is referred to as uog_fixed_arith, which is an abbreviation for *University of Guelph Fixed-Point Arithmetic*. These two representations were chosen based on previous results from the literature [41] that showed that 16 bit fixed point representation is the minimum needed to allow the BP algorithm to converge and the fact that 32 bit floating point precision is the standard floating point representation. We could have used a custom floating point representation (maybe with less precision) but it is very likely that any future VHDL floating point implementation will follow this standard representation. As such we specifically wanted to test the tradeoff with this standard presentation. This is also important for applications in Hardware/Software co-design using languages like SystemC and HandleC.

Fixed-point representation is signed 2's complement binary representation, which is made rational with a *virtual* decimal point. The virtual radix point location used in uog_fixed_arith is $SIII.FFFFFFFFFFFF$, where

S = sign bit

I = integer bit, as implied by location of binary point

F = fraction bit, as implied by location of binary point

The range for a 16-bit fixed-point representation of this configuration is [-8.0, 8.0), with a quantization error of 2.44140625E-4. Description of the various arithmetic VHDL design alternatives considered for use in the uog_fp_arith and uog_fixed_arith libraries are summarized in Table 2.1. All HDL designs with the word std in their name signify that one of the IEEE standardized VHDL arithmetic libraries was used to create them. For example, uog_std_multiplier was easily created using the following VHDL syntax:
$z <= x * y$;
where x and y are the input signals, and z the output signal of the circuit. Such a high level of abstract design is often associated with *behavioral* VHDL designs, where ease of design comes at the sacrifice of letting the FPGA's synthesis tools dictate the fine-grain architecture of the circuit.

On the other hand, an engineer can explicitly define the fine-grain architecture of a circuit by means of *structural* VHDL and schematic-based designs, as was done for uog_ripple_carry_adder, uog_c_l_adder (please refer to Figure 2.4 for detailed implementation) and uog_sch_adder respectively. How-

Table 2.1. Summary of alternative designs considered for use in custom arithmetic VHDL libraries

HDL Design	Description
uog_fp_add*	IEEE 32-bit single precision floating-point pipelined parallel adder
uog_ripple_carry_adder	16-bit fixed-point (bit-serial) ripple-carry adder
uog_c_l_addr	16-bit fixed-point (parallel) carry lookahead adder
uog_std_adder	16-bit fixed-point parallel adder created using standard VHDL arithmetic libraries
uog_core_adder	16-bit fixed-point parallel adder created using Xilinx LogiCORE *Adder Subtracter v5.0*
uog_sch_adder	16-bit fixed-point parallel adder created using Xilinx *ADD16* schematic-based design
uog_pipe_adder	16-bit fixed-point pipelined parallel adder created using Xilinx LogiCORE *Adder Subtractor v5.0*
uog_fp_sub*	IEEE 32-bit single precision floating-point pipelined parallel subtracter
uog_par_subtracter	16-bit fixed-point carry lookahead (parallel) subtracter, based on uog_std_adder VHDL entity
uog_std_subtracter	16-bit fixed-point parallel subtracter created with standard VHDL arithmetic libraries
uog_core_subtracter	16-bit fixed-point parallel subtracter created using Xilinx LogiCORE *Adder Subtracter v5.0*
uog_fp_mult*	IEEE 32-bit single precision floating-point pipelined parallel multiplier
uog_booth_multiplier	16-bit fixed-point shift-add multiplier based on Booth's algorithm (with carry lookahead adder)
uog_std_multiplier	16-bit fixed-point parallel multiplier created using standard VHDL arithmetic libraries
uog_core_bs_mult	16-bit fixed-point bit-serial (non-pipelined) multiplier created using Xilinx LogicCORE *Multiplier v4.0*
uog_pipe_serial_mult	16-bit fixed-point bit-serial (pipelined) multiplier created using Xilinx LogiCORE *Multiplier v4.0*
uog_core_par_multiplier	16-bit fixed-point parallel (non-pipelined) multiplier created using Xilinx LogiCORE *Multiplier v4.0*
uog_pipe_par_mult	16-bit fixed-point parallel (pipelined) multiplier created using Xilinx LogiCORE *Multiplier v4.0*
active_func_sigmoid	Logsig (i.e. sigmoid) function with IEEE 32-bit single precision floating-point
uog_logsig_rom	16-bit fixed-point parallel logsig (i.e. sigmoid) function created using Xilinx LogiCORE *Single Port Block Memory v4.0*

* Based on VHDL source code dontated by Steven Derrien (sderrien@irisa.fr) from *Institut de Recherche en Informatique et systemes aleatoires (IRISA)* in France. In turn, Steven Derrien had originally created this through the adaptation of VHDL source code found at **http://flex.ee.uec.ac.jp/ yamaoka/vhdl/index.html**.

ever, having complete control over the architecture's fine-grain design comes at the cost of additional design overhead for the engineer.

Many of the candidate arithmetic HDL designs described in Table 2.1 were created by the *Xilinx CORE Generator System*. This EDA tool helps an engineer parameterize ready-made Xilinx intellectual property (ip) designs (i.e. *LogiCOREs*), which are optimized for Xilinx FPGAs. For example, uog_core_adder was created using the Xilinx proprietary LogiCORE for an adder design.

Approximation of the logsig function in both floating-point and fixed-point precision, were implemented in hardware using separate lookup-table architectures. In particular, active_func_sigmoid was a modular HDL design, which encapsulated all the floating-point arithmetic units necessary to carry out calculation of logsig function. According to Equation 2.3, this would require the use of a multiplier, adder, divider, and exponential function. As a result, active_func_sigmoid was realized in VHDL using uog_fp_mult, uog_fp_add, a custom floating-point divider called uog_fp_div, and a *table-driven* floating-point exponential function created by Bui *et al* [44]. While this is not the most efficient implementation of the logsig, it allows implementing other transfer functions with min efforts (like Tan Hyperbolic) since it shares the basic functions with the Sigmoid. It is very common in training BP networks to test different transfer functions.

The uog_logsig_rom HDL design utilized a Xilinx LogiCORE to implement single port block memory. A lookup-table of 8192 entries was created with this memory, which was used to approximate the logsig function in fixed-point precision.

In order to maximize the processing density of the digital VLSI ANN design proposed in Section 2.3.1, only the most area-optimized arithmetic HDL designs offered in Table 2.1 should become part of the uog_fp_arith and uog_fixed_arith VHDL libraries. However, the space-area requirements of any VHDL design will vary from one FPGA architecture to the next. Therefore, all the HDL arithmetic designs found in Table 2.1 have to be implemented on the same FPGA as was targeted for implementation of the digital VLSI ANN design, in order to determine the most area-efficient arithmetic candidates.

2.4   Experiments using logical-XOR problem

The logical-XOR problem is a *classic* toy problem used to benchmark the learning ability of an ANN. It is a simple example of a non-linearly separable problem.

The minimum ANN *topology* (a *topology* includes the number of neurons, number of layers, and the layer interconnections (i.e. synapses)) required to

Table 2.2. Truth table for logical-XOR function

Inputs		Output
x_0	x_1	y
0	0	0
0	1	1
1	0	1
1	1	0

solve a non-linearly separable problem consisting of at least one hidden layer. An overview of the ANNs topology used in this particular application, which consists of only one hidden layer, is shown in Figure 2.3.

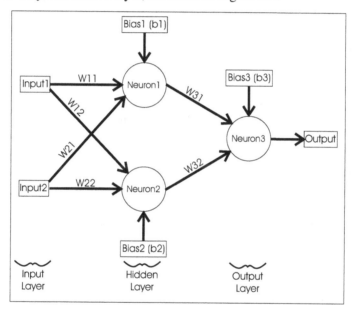

Figure 2.3. *Topology of ANN used to solve logical-XOR problem*

For each ANN implementation, a set of thirty training sessions were performed individually. Each training session lasted for a length of 5000 epoch, and used a learning rate of 0.3. Each of the training sessions in the set used slightly different initial conditions, in which all weights and biases were randomly generated with a mean of 0, and a standard deviation of ±0.3. Once generated, every BP implementation was tested using the same set of thirty training sessions. This way, the logical-XOR problem discussed acts as a common testing platform, used to benchmark the performance of all BP implementations.

Xilinx Foundation ISE 4.1i EDA tools were used to synthesize, and map (i.e. place and route) two variations of the FPGA-based ANN designs – one using uog_fp_arith library, and one using uog_fixed_arith library. All experiments and simulations were carried out on a PC workstation running Windows NT (SP6) operating system, with 1 GB of memory and Intel PIII 733MHz CPU.

These circuit designs were tested and validated in simulation only, using ModelTech's ModelSim SE v5.5. *Functional* simulations were conducted to test the syntactical and semantical correctness of HDL designs, under ideal FPGA conditions (i.e. where no propagation delay exists). *Timing* simulations were carried out to validate the HDL design under non-ideal FPGA conditions, where propagation delays associated with the implementation as targeted on a particular FPGA are taken into consideration.

Specific to VHDL designs, timing simulations are realized using an IEEE standard called VITAL (VHDL Initiative Toward ASIC Libraries). VITAL libraries contain information used for modeling accurate timing of a particular FPGA at the gate level, as determined *a priori* by the respective FPGA manufacturer. These VITAL libraries are then used by HDL simulators, such as ModelSim SE, to validate designs during timing simulations.

A software implementation of a back-propagation algorithm was created using MS Visual C++ v6.0 IDE. The software simulator was set up to solve the logical-XOR problems using the topology shown in Figure 2.3. The purpose for creating this software simulator was to generate expected results for testing and validating FPGA-based BP. To speed up development and testing, two other software utilities were created to automate numeric format conversions— one for converting real decimal to/from IEEE-754 single precision floating-point hexadecimal format, and one for converting real decimal to/from 16-bit fixed-point binary format.

2.5   Results and discussion

We implemented the previous HDL designs on Xilinx FPGAs. The resulting space-time requirements for each arithmetic HDL design are summarized in Table 2.3. In order to maximize the neuron density of the FPGA-based MLP, the area of the various arithmetic HDL designs that a neuron is comprised of should be minimized. As a result, the focus here is to determine the most area-optimized arithmetic HDL designs for use in the implementations.

2.5.1   Comparison of Digital Arithmetic Hardware

Comparison of the different adder results, shown in Table 2.3, reveals that the three carry lookahead adders (i.e. uog_std_adder, uog_core_adder, and uog_sch_adder) require the least amount of area and are the fastest among all non-pipelined adders. Note that the sophistication of today's EDA tools have allowed the VHDL-based designs for carry lookahead adders to achieve the same fine-grain efficiency of their equivalent schematic-based designs.

Since a carry lookahead adder is essentially a ripple-carry adder with additional logic as seen in Figure 2.4, it isn't immediately clear why a carry lookahead adder is shown here to use less area compared to a ripple-carry adder

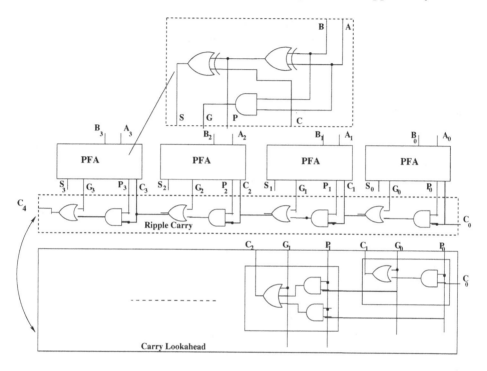

Figure 2.4. Carry Lookahead Adder

when implemented on a Xilinx Virtex-E FPGA. The carry lookahead design can be obtained [35] by a transformation of the ripple carry design in which the carry logic over fixed groups of bits of the adder is reduced to two-level logic. The carry lookahead adder would consist of (for example a 4-bit carry lookahead adder) four partial full adders (PFA) each consisting of two EXOR gates and an AND gate. The ripple carry logic (AND gate and OR gate for each bit) will be substituted with the Carry lookahead logic. Since the Virtex-II FPGA has built in (fast arithmetic functions) of look-ahead carry chains, the number of CLBS utilized by the carry lookahead adder will be equivalent or smaller than that of the ripple adder (i.e the extra logic used by the Carry Lookahead is free! since it is custom built within the FPGA). In addition, the Virtex-E CLBs dedicated fast lookahead logic enhances the performance of the adder. As a result, it's best to use HDL adder designs which take advantage of the Virtex-E's fast carry lookahead logic.

The Virtex-E's fast carry-lookahead logic is again utilized to produce the best area-optimized subtracters (i.e. uog_std_subtracter and uog_core_subtracter), as well as, the best area-optimized multiplier (i.e. uog_booth_multiplier).

Only the most area-optimized arithmetic HDL designs discussed here were used in the construction of custom arithmetic HDL libraries, as listed in Table 2.4. In the case where there was more than one choice of best area-optimized arithmetic HDL design to choose from, *behavioral* VHDL designs were preferred because they promote high-level abstract designs and portability. For example, such was the case in selecting a fixed-point adder and subtracter for the uog_fixed_arith library.

Table 2.4 also reveals how much more area-optimized the individual fixed-point arithmetic HDL designs in uog_fixed_arith were compared to the floating-point arithmetic HDL designs in uog_fp_arith. Since a floating-point adder is essentially a fixed-point adder plus additional logic, not to mention the fact that floating-point uses more precision than fixed-point arithmetic, it's no surprise to find that the 16-bit fixed-point adder is much smaller than the 32-bit floating-point adder. Similar in nature is the case for subtracter and multiplier comparisons shown in Table 2.4.

The comparison of area-optimized logsig arithmetic HDL designs reveals that the 32-bit floating-point version is over 250 times bigger than the 16-bit fixed-point version. Aside from the difference in amount of precision used, the significant size difference between logsig implementations is due to the fact that floating-point implementation encapsulates a table-lookup architecture in addition to other *area-expensive* arithmetic units, while the fixed-point version *only* encapsulates a table-lookup via memory.

Uog_fp_arith and uog_fixed_arith have been clearly defined with only the best area-optimized components, as shown in Table 2.4. This will help to ensure that 32-bit floating-point and 16-bit fixed-point FPGA-based MLP implementations achieve a processing density advantage over the software-based MLP simulations. As was shown here, the larger area requirements of floating-point precision in FPGA-based ANNs makes it not nearly as feasible as fixed-point precision.

2.5.2   Comparison of ANN Implementations

Table 2.5 summarizes logical-XOR benchmark results for each of the following implementations with identical topology:

- 32-bit floating-point FPGA-based MLP, which utilizes uog_fp_arith library.

Table 2.3. Space/Time Req'ts of alternative designs considered for use in custom arithmetic VHDL libraries

HDL Design	Area (CLB)s	Max. Clock Rate (MHz)	Pipe-lining Used?	Clock cycles per calc.	Min. Total Time per calc. (ns)
uog_fp_add	174	19.783	1-stage	2	101.096 (for first calc.)
uog_ripple_carry_adder	12	67.600	No	16	236.688
uog_c_1_addr	12	34.134	No	1	29.296
uog_std_adder	4.5	66.387	No	1	15.063
uog_core_adder	4.5	65.863	No	1	15.183
uog_sch_adder	4.5	72.119	No	1	13.866
uog_pipe_adder	96	58.624	15-stage	16	272.928
uog_fp_sub	174	19.783	1-stage	2	101.096
uog_par_subtracter	8.5	54.704	No	1	18.280
uog_std_subtracter	4.5	56.281	No	1	17.768
uog_core_subtracter	4.5	60.983	No	1	16.398
uog_fp_mult	183.5	18.069	1-stage	2	110.686 (for first calc.)
uog_booth_multiplier	28	50.992	No	34	668.474
uog_std_multiplier	72	32.831	No	1	30.459
uog_core_bs_mult	34	72.254	No	20	276.800
uog_pipe_serial_mult	39	66.397	?-stage	21	316.281 (for first calc.)
uog_core_par_multiplier	80	33.913	No	1	29.487
uog_pipe_par_mult	87.5	73.970	?-stage	2	27.038 (for first calc.)
active_func_sigmoid*	3013	1.980	No	56	29282.634
uog_logsig_rom	12	31.594	No	1	31.652

*Target platform used here was Xilinx Virtex-II FPGA (xc2v8000-5bf957)
Please note the following:

1 All fixed-point HDL designs use signed 2's complement arithmetic

2 Unless otherwise mentioned, all arithmetic functions were synthesized and implemented (i.e. place and route) under the following setup:

Target Platform: *Xilinx Virtex-E FPGA (xcv2000e–6bg560)*

Development Tool: *Xilinx Foundation ISE 4.1i (SP2)*

Synthesis Tool: *FPGA Express VHDL*

Optimization Goal: *Area (Low Effort)*

3 Max. Clock Rate is determined usig the *Xilinx Timing Analyzer* on Post-Place and Route Static Timing of HDL design. $Max.ClockRate = \min\{(Min.CombinationalPathDelay)^{-1}, [(Min.InputArrivalTimeBeforeClk) + Max.OutputRequiredTimeBeforeClk)]^{-1}\}$

Table 2.4.   Area comparison of uog_fp_arith vs. uog_fixed_arith

Arithmetic Function	uog_fixed_arith HDL Design	uog_fp_arith HDL Design	Area Optimization (CLB/CLB)
Adder	uog_std_adder	uog_fp_add	38.66x smaller
Subtracter	uog_std_subtracter	uog_fp_sub	38.66x smaller
Multiplier	uog_booth_multiplier	uog_fp_mult	6.55x smaller
Logsig Function	uog_logsig_rom	activ_func_sigmoid	251.08x smaller

- 16-bit fixed-point FPGA-based MLP, which utilizes uog_fixed_arith library.

- software-based MLP simulations using C++.

Due to the relative difference in size of arithmetic components used, the fixed-point FPGA-based MLP is over 13 times smaller than the floating-point FPGA-based MLP. It can only be assumed that the area requirements for the software-based MLP implemented on an Intel PIII CPU (i.e. general-purpose computing platform) is infinitely big in comparison to the FPGA-based MLPs.

Of concern was the fact that timing simulations via ModelSim SE v5.5 required two weeks for floating-point and six days for fixed-point runs just to complete one training session in each. In general, any VLSI design which is not area-optimized may impede the design and test productivity.

The fact that all three MLP-BP implementations converged at all is enough to validate the successful design of each. Note that a MLP is not always guaranteed to converge since it may get trapped in local minima.

What's interesting about the convergence percentages given in Table 2.5 is that they're the same for the software-based and 32-bit FPGA-based MLPs, but not for the 16-bit FPGA-based MLPs. The software-based MLP and FPGA-based MLP that used uog_fp_arith achieved the same convergence percentages because they both use 32-bit floating-point calculations, and will follow identical paths of gradient descent when given the same initial MLP parameters. Due to the quantization errors found in 16-bit fixed-point calculations, its respective FPGA-based MLP will follow down a slightly different path of gradient descent when exposed to the same initial MLP parameters as the other two implementations.

In the context of MLP applications, reconfigurable computing looks to increase the neuron density above and beyond that of general-purpose computing. Due to the fact that three neurons exist in the MLP topology used to solve the logical-XOR problem, and based on the benchmarked speeds of back-propagation iteration for each particular MLP implementation, the processing density can be calculated for each. For MLP applications, processing density is realized as the number of weight updates per unit of space-time. As shown in

Table 2.5. Summary of logical-XOR ANN benchmarks on various platforms

XOR ANN Architecture	Precision	Total Area (CLBs, [Slices])*	% of Convergence in thirty trials**	Max. Clock Rate (MHz)
Xilinx Virtex-E xcv2000e FPGA	16-bit fixed-pt	1239 [2478]	100%	10
Xilinx Virtex-II xc2v8000 FPGA	32-bit floating-pt	8334.75 [33339]	73.3%	1.25
Intel Pentium III CPU	32-bit floating-pt	NA	73.3%	733

	Total Clock Cycles per Backprop Iteration	Backprop Iteration Period (μs)	Weight Updates per Sec (WUPS)	Processing Density (per Slice)
Xilinx Virtex-E xcv2000e FPGA	478	47.8	62762	25.33
Xilinx Virtex-II xc2v8000 FPGA	464	580	5172	0.1551
Intel Pentium III CPU	N/A	2045.15***	1466.89	NA

* Note Virtex-II CLB is over twice the size of Virtex-E CLB. Virtex-II CLB consists of 4 slices, whereas Virtex-E CLB consists of 2 slices.

** Convergence is defined here as less than 10% error in the ANN's output, after it has been trained.

*** This is an average based on time taken to complete 200,000,000 iterations of the backpropagation algorithm for the software-based ANN. Microsoft Platform SDK multimedia timers were used, which had a resolution of 1ms.

Table 2.5, the relative processing density of the 16-bit fixed-point implementation is significantly higher than that of the 32-bit floating-point one. This reveals how a combination of minimum allowable range-precision and greater degree of area-optimization results in a direct impact on the processing density in implementation.

In addition to infinitely large area requirements, the software-based MLP was shown to be over 40x slower in comparison to the 16-bit fixed-point FPGA-based implementation. Therefore, it can only be assumed that the relative processing density of the software-based MLP is infinitely small in comparison to the other two implementations.

2.6 Conclusions

In general, we have shown that the choice of range-precision and arithmetic hardware architecture used in reconfigurable computing applications has a direct impact on the processing density achieved. A minimal allowable range-precision of 16-bit fixed-point continues to provide the most optimal *range-precision vs. area trade-off* for MLP-BP implemented on today's FPGAs.

The *classic* logical-XOR problem was used as a common benchmark for comparing the performance of a software-based MLP, and two FPGA-based MLPs – one with 16-bit fixed-point precision, and the other with 32-bit floating-point precision. Despite the limited range-precision, the MLP with area-optimized fixed-point arithmetic managed to maintain the same quality of performance (i.e. in terms of the MLPs ability to learn) as demonstrated with floating-point arithmetic. Results showed that the fixed-point MLP-BP implementation was over 12x greater in speed, over 13x smaller in area, and achieved far greater processing density compared to the floating-point FPGA-based MLP-BP. Also, the processing density achieved by the FPGA-based MLP-BP with 16-bit fixed-point precision compared to the software-based MLP-BP best demonstrates the processing density advantage of reconfigurable computing over general-purpose computing for this particular application. As a result, floating-point precision is not as feasible as fixed-point in this type of application.

One disadvantage of using 16-bit fixed-pt, is that its limited range poses risk of saturation. Saturation adds error to a system, the extent of which is application dependent. The logical-XOR example demonstrated in this chapter still managed to achieve convergence, despite the saturation error caused by 16-bit fixed-pt with range [-8.0,8.0). Another important lesson to learn from this study is that the area savings of using 16-bit fixed-point rather than floating-point precision in a FPGA-based ANN help minimize simulation durations when validating HDL designs. The current performance rate of digital HDL simulators, like *ModelSim SE 5.5*, is an ongoing concern. Not only does the duration of timing simulations increase proportionally with the size of the circuit being simulated, but the magnitude of duration is in the order of 'days' and even 'weeks' for large VLSI HDL designs.

References

[1] Xin Yao. Evolutionary Artificial Neural Networks, In: *Encylopedia of Computer Science and Technology*, A. Kent and J. G. Williams, Eds., Vol. 33, Marcel Dekker Inc., New York, NY 10016, pp. 137-170, 1995

[2] K. Balakrishnan and V. Honavar, Evolutionary Design of Neural Architectures – A Preliminary Taxonomy and Guide to Literature, Tech. Report no. CS TR95-01, Artificial Intelligence Research Group, Iowa State University, pp. January, 1995.

[3] J. G. Eldredge, FPGA Density Enhancement of a Neural Network Through Run-Time Reconfiguration, Department of Electrical and Computer Engineering, Brigham Young University, pp. May, 1994.

[4] J. G. Eldridge and B. L. Hutchings, Density Enhancement of a Neural Network using FPGAs and Run-Time Reconfiguration, In: *IEEE Workshop on FPGAs for Custom Computing Machines*, pp. 180-188, 1994.

[5] J. G. Eldridge and B. L. Hutchings, RRANN: A Hardware Implementation of the Backpropagation Algorithm Using Reconfigurable FPGAs, In: *Proceedings, IEEE International Conference on Neural Networks*, Orlando, FL, 1994.

[6] J. D. Hadley and B. L. Hutchings. Design Methodologies for Partially Reconfigured Systems, In: *Proceedings, IEEE Workshop on FPGAs for Custom Computing Machines*, pp. 78-84, 1995.

[7] Hugo de Garis and Michael Korkin . The CAM-BRAIN MACHINE (CBM) An FPGA Based Hardware Tool which Evolves a 1000 Neuron Net Circuit Module in Seconds and Updates a 75 Million Neuron Artificial Brain for Real Time Robot Control, *Neurocomputing journal*, Vol. 42, Issue 1-4, 2002.

[8] Amanda J. C. Sharkey, (Ed.). *Combining Artificial Neural Nets – Ensemble and Modular Multi-Net Systems*, Perspectives in Neural Computing, Springer-Verlag London Publishing, 1999.

[9] Eric Ronco and Peter Gawthrop. Modular Neural Networks: a state of the art, Tech. Report, no. CSC-95026, Center for System and Control, University of Glasgow, Glasgow, UK, May 12, 1999.

[10] Hugo de Garis and Felix Gers and Michael Korkin. CoDi-1Bit: A Simplified Cellular Automata Based Neuron Model, *Artificial Evolution Conference (AE97)*, Nimes, France, 1997

[11] Andres Perez-Uribe. Structure-Adaptable Digital Neural Networks, Ph.D. Thesis, Logic Systems Laboratory, Computer Science Department, Swiss Federal Institute of Technology-Lausanne, 1999.

[12] H. F. Restrepo and R. Hoffman and A. Perez-Uribe and C. Teuscher and E. Sanchez . A Networked FPGA-Based Hardware Implementation of a Neural Network Application. In: *Proceedings of the IEEE Symposium on Field Programmable Custom Computing Machines (FCCM'00)*, pp. 337-338, 2000.

[13] J.-L. Beuchat and J.-O. Haenni and E. Sanchez. Hardware Reconfigurable Neural Networks, 5th Reconfigurable Architectures Workshop (RAW'98), Orlando, Florida, USA, pp. March 30, 1998.

[14] Aaron Ferrucci. ACME: A Field-Programmable Gate Array Implementation of a Self-Adapting and Scalable Connectionist Network, University of California, Santa Cruz, January, 1994.

[15] Marcelo H. Martin. A Reconfigurable Hardware Accelerator for Back-Propagation Connectionist Classifiers, University of California, Santa Cruz, 1994.

[16] Tomas Nordstrom. Highly Parallel Computers for Artificial Neural Networks, Ph.D. Thesis,Division of Computer Science and Engineering, Lulea University of Technology, Sweden, 1995.

[17] T. Nordstrom and E. W. Davis and B. Svensson. Issues and Applications Driving Research in Non-Conforming Massively Parallel Processors, book. In: *Proceedings of the New Frontiers, a Workshop of Future Direction of Massively Parallel Processing*, 1992

[18] T. Nordstrom and B. Svensson. Using and Designing Massively Parallel Computers for Artificial Neural Networks, *Journal of Parallel and Distributed Computing*, Vol. 14, 1992.

[19] B. Svensson and T. Nordstrom and K. Nilsson and P.-A. Wiberg. Towards Modular, Massively Parallel Neural Computers, In: *Connectionism in a Broad Perspective: Selected Papers from the Swedish Conference on Connectionism - 1992*, L. F. Niklasson and M. B. Boden, Eds., Ellis Harwood, pp. 213-226, 1994.

[20] Arne Linde and Tomas Nordstrom and Mikael Taveniku. Using FPGAs to implement a reconfigurable highly parallel computer. In: *Field-Programmable Gate Arrays: Architectures and Tools for Rapid Prototyping*, Springer-Verlag, Berlin, 1992.

[21] T. Nordstrom. On-line Localized Learning Systems Part 1 - Model Description, Research Report, Division of Computer Science and Engineering, Lulea University of Technology, Sweden, 1995.

[22] Tomas Nordstrom. On-line Localized Learning Systems Part II - Parallel Computer Implementation, Research Report, no. TULEA 1995:02, Division of Computer Science and Engineering, Lulea University of Technology, Sweden, 1995.

[23] Tomas Nordstrom. Sparse distributed memory simulation on REMAP3, Research Report, no. TULEA 1991:16, Division of Computer Science and Engineering, Lulea University of Technology, Sweden, 1991.

[24] T. Nordstrom. Designing Parallel Computers for Self Organizing Maps, In: *Proceedings of the 4th Swedish Workshop on Computer System Architecture (DSA-92)*, Linkoping, Sweden, January 13-15, 1992.

[25] B. Svensson and T. Nordstrom. Execution of neural network algorithms on an array of bit-serial processors. In: *Proceedings, 10th International Conference on Pattern Recognition, Computer Architectures for Vision and Pattern Recognition*, Vol. II, pp. 501-505, 1990.

[26] M. Skrbek. Fast Neural Network Implementation. In: *Neural Network World*, Elsevier, Vol. 9, no. No. 5, pp. 375-391, 1999.

[27] Introducing the XC6200 FPGA Architecture: The First FPGA Architecture Optimized for Coprocessing in Embedded System Applications, *Xcell*, Xilinx Inc., No. 18 : Third Quarter, pp. 22-23, 1995, url = http://www.xilinx.com/apps/6200.htm.

[28] Xilinx. XC6200 Field Programmable Gate Arrays, Data Sheet, Version 1.7, 1996.

[29] Mikael Taveniku and Arne Linde. A reconfigurable SIMD computer for artificial neural networks, Licentiate Thesis, Department of Computer Engineering, Chalmers University of Technology, Goteborg, Sweden, 1995.

[30] Gate Count Capacity Metrics for FPGAs, Application Note, no. XAPP 059 - Version 1.1, Xilinx, Inc., Feb. 1, 1997, URL = http://www.xilinx.com/xapp/xapp059.pdf.

[31] FLEX 10K Embedded Programmable Logic Device Family, Data Sheet, Version 4.1, Altera, Inc., March, 2001, URL = http://www.altera.com/literature/ds/dsf10k.pdf

[32] XC3000 Series Field Programmable Gate Arrays, Product Description, Version 3.1, Xilinx, Inc., November 9, 1998, URL = http://www.xilinx.com/partinfo/3000.pdf

[33] XC4000XLA/XV Field Programmable Gate Arrays, Product Specification, no. DS015 - Version 1.3, Xilinx, Inc., October 18, 1999, URL = http://www.xilinx.com/partinfo/ds015.pdf

[34] Andres Perez-Uribe and Eduardo Sanchez. Speeding-Up Adaptive Heuristic Critic Learning with FPGA-Based Unsupervised Clustering, In: *Proceedings of the IEEE International Conference on Evolutionary Computation ICEC'97*, pp. 685-689, 1997.

[35] M. Morris Mano and Charles R. Kime. *Logic And Computer Design Fundamentals*, Prentice Hall Inc., New Jersey, USA, 2000.

[36] Andre Dehon. The Density Advantage of Configurable Computing, *IEEE Computer*, vol. 33, no. 5, pp. 41–49, 2000.

[37] David E Rumelhart and James L McClelland and PDP Research Group. *Parallel Distrubuted Processing: Explorations in the Microstructure of Cognition*, vol. Volume 1: Foundations, MIT Press, Cambridge, Massachusetts, 1986.

[38] Simon Haykin, *Neural Networks: A Comprehensive Foundation*, Prentice-Hall, Englewood Cliffs, New Jersey, 1999.

[39] Stephen D. Brown and Robert J. Francis and Jonathan Rose and Zvonko G. Vranesic, Field-Programmable Gate Arrays, Kluwer Academic Publishers, USA, 1992

[40] Virtex-E 1.8 V Field Programmable Gate Arrays, Perliminary Product Specification, no. DS022-2 (v2.3), Xilinx, Inc., November 9, 2001, URL = http://www.xilinx.com/partinfo/ds022-2.pdf

[41] Jordan L Holt and Thomas E Baker. Backpropagation simulations using limited precision calculations. In: *Proceedings, International Joint Conference on Neural Networks (IJCNN-91)*, vol. 2, Seattle, WA, USA, pp. 121 - 126, 1991.

[42] Hikawa Hiroomi. Frequency-Based Multilayer Neural Network with On-chip Learning and Enhanced Neuron Characterisitcs. *IEEE Transactions on Neural Networks, vol. 10, no. 3, pp. 545-553, May*, 1999.

[43] Peter J. Ashenden. VHDL Standards. *IEEE Design & Test of Computers*, vol. 18, no. 6, pp. 122-123, September–October, 2001.

[44] Hung Tien Bui and Bashar Khalaf and Sofiene Tahar. Table-Driven Floating-Point Exponential Function, Technical Report, Concordia University, Department of Computer Engineering, October, 1998, URL = http://www.ece.concordia.ca/ tahar/pub/FPE-TR98.ps

[45] W.B. Ligon III and S. McMillan and G. Monn and K. Schoonover and F. Stivers and K.D. Underwood. A Re-evaluation of the Practicality of Floating Point Operations on FPGAs. In: *Proceedings, IEEE Symposium on FPGAs for Custom Computing Machines*, pp. 206–215, 1998.

[46] Pete Hardee. System C: a realistic SoC debug strategy, *EETimes*, 2001.

[47] G.M. Amdahl. Validity of the single-processor approach to achieving large scale computing capabilities, In: *AFIPS Conference Proceedings*, vol. 30, AFIPS Press, Reston, Va., pp. 483–485, 1967.

[48] Stephen Chappell and Chris Sullivan, Celoxica Ltd. Oxford UK. Handel-C for co-processing & co-design of Field Programmable System on Chip FPSoC, 2002, url = www.celoxica.com/technical-library/

[49] Synopsys, Inc. Describing Synthesizable RTL in SystemC v1.1, Synopsys, Inc., January, 2002.

[50] Martyn Edwards. Software Acceleration Using Coprocessors: Is it Worth the Effort?, In: *Proceedings, 5th International Workshop on Hardware/Software Co-design Codes/CASHE'97*, Braunschneig, Germany, pp. 135–139, March 24-26, 1997.

[51] Giovanni De Micheli and Rajesh K. Gupta. Hardware/Software Co-design. *Proceedings of the IEEE*, vol. 85, no. 3, pp. 349-365, March, 1997.

[52] John Sanguinetti and David Pursley. High-Level Modeling and Hardware Implementation with General- Purpose Languages and High-level Synthesis, White Paper, Forte Design Systems, 2002.

[53] Don Davis. Architectural Synthesis: Unleasing the Power of FPGA System-Level Design. *Xcell Journal*, Xilinx Inc., no. 44, vol. 2, pp. 30-34, pp. Winter, 2002.

[54] Xilinx, XAPP467 Using Embedded Multipliers in Spartan-3 FPGAs, Xilinx Application Note, May 13, 2003, http://www.xilinx.com.

[55] Nestor Inc. Neural Network Chips, http://www.nestor.com.

Chapter 3

FPNA: CONCEPTS AND PROPERTIES

Bernard Girau

LORIA INRIA-Lorraine
Nancy France
girau@loria.fr

Abstract   Neural networks are usually considered as naturally parallel computing models. But the number of operators and the complex connection graph of standard neural models can not be handled by digital hardware devices. Though programmable digital hardware now stand as a real opportunity for flexible hardware implementations of neural networks, many area and topology problems arise when standard neural models are implemented onto programmable circuits such as FPGAs, so that the fast FPGA technology improvements can not be fully exploited. The theoretical and practical framework first introduced in [21] reconciles simple hardware topologies with complex neural architectures, thanks to some configurable hardware principles applied to neural computation: *Field Programmable Neural Arrays* (FPNA) lead to powerful neural architectures that are easy to map onto FPGAs, by means of a simplified topology and an original data exchange scheme. This two-chapter study gathers the different results that have been published about the FPNA concept, as well as some unpublished ones. This first part focuses on definitions and theoretical aspects. Starting from a general two-level definition of FPNAs, all proposed computation schemes are together described and compared. Their correctness and partial equivalence is justified. The computational power of FPNA-based neural networks is characterized through the concept of *underparameterized convolutions*.

Keywords: neural networks, fine-grain parallelism, digital hardware, FPGA

3.1   Introduction

Various fast implementations of neural networks have been developed, because of the huge amount of computations required by neural network appli-

A. R. Omondi and J. C. Rajapakse (eds.), FPGA Implementations of Neural Networks, 63–101.

cations, especially during the learning phase.[1] A broad introduction to parallel neural computing may be found in [50]. Several kinds of parallel devices have been used for these fast implementations, such as massively parallel computers, neuro-computers, analog or digital ASICs (application specific integrated circuits). Programmable digital circuits (such as FPGAs, field programmable gate arrays) mix the efficiency and parallelism grain of hardware devices with the flexibility of software implementations, so that they appear as particularly well-adapted to the usual needs of neural implementations (see [26]).

However, the 2D-topology of FPGAs is not able to handle the connection complexity of standard neural models. What is more, FPGAs still implement a limited number of logic gates, whereas neural computations require area-greedy operators (multipliers, activation functions). Usual solutions handle sequentialized computations with an FPGA used as a small neuroprocessor, or they implement very small low-precision neural networks without on-chip learning. Connectivity problems are not solved even by the use of several re-configurable FPGAs with a bit-serial arithmetic, or by the use of small-area operators (stochastic bit-stream or frequency-based). Subsection 3.2.3 precisely describes the specific problems raised by neural network implementations on FPGAs.

In [29], we propose an implementation method for multilayer perceptrons of any size on a single FPGA, with on-chip learning and adaptable precision. This work takes advantage of an area-saving on-line arithmetic that is well-adapted to neural computations. It also uses an original parallelization of the internal computations of each neuron. Yet, this implementation method barely exploits the massive fine-grain parallelism that is inherent to neural network architectures.

In [21] and following publications, I have defined and studied a set of neural models called FPNAs (*Field Programmable Neural Arrays*). These models share a computation method whose local data processing is more complex than in the standard neural computation scheme. It makes it possible to handle simplified neural network topologies, though such neural networks may be functionally equivalent to standard neural models (multilayer perceptron, Hopfield network, etc). Therefore, FPNAs lead to the definition of large and efficient neural networks that are adapted to hardware topological constraints. They reconcile the high connection density of neural architectures with the need of a limited interconnection scheme in hardware implementations.

---

[1] A neural network has got some learnable parameters. The learning phase consists in determining parameter values such that the neural network computes a function that performs an expected task (e.g. classification of a set of patterns). When the parameters have been determined, the neural network is ready to be applied to unknown patterns. This is the *generalization* phase.

FPNAs are based on an FPGA-like approach: a set of resources with freely configurable interactions. These resources are defined to perform computations of standard neurons, but they behave in an *autonomous* way. As a consequence, numerous *virtual* connections are achieved thanks to the application of a simple protocol to the resources of a sparse neural network. A standard but topologically complex neural network may be replaced by a simplified neural network that uses this new neural computation concept.

The aim of this two-chapter study is to gather the different results, published or unpublished, about the FPNA concept ([21, 20, 27, 25, 24, 22, 23, 28]), so as to give an exhaustive overview of this hardware-oriented neural paradigm. This first chapter focuses on definitions and theoretical aspects. Section 3.2 first proposes a brief survey of the use of parallel devices for neural implementations. Then it describes the specific advantages and problems of the use of programmable hardware for such implementations. Section 3.3 starts from a general two-level definition of FPNAs. Then it describes and compares all proposed computation schemes for FPNA-based neural networks. Their correctness and partial equivalence is justified in section 3.4. Section 3.5 characterizes the computational power of FPNA-based neural networks through the concept of *underparameterized convolutions*. Chapter 4 focuses on implementations and applications.

3.2   Choosing FPGAs

This section shows how FPGAs stand as appropriate hardware devices for neural implementations:

- §3.2.1 is a rapid survey of existing devices for parallel implementations of neural networks

- §3.2.2 describes some specific advantages of the use of FPGAs to implement neural applications

- §3.2.3 points out the main issues of such implementations on FPGAs

- §3.2.4 is a short survey of existing solutions, from the simplest to the most well-built

3.2.1   Parallel implementations of neural networks

Fast implementations of neural network applications are useful because of the very high number of required arithmetic operations. Such implementations might use massively parallel computers as well as digital or analog hardware designs. This subsection rapidly discusses the use of the various possible parallel devices.

3.2.1.1   General purpose parallel computers.   Fine-grain parallel implementations on massively parallel computers (either SIMD, as in [72, 64], or MIMD, as in [42, 55, 51, 7]) suffer from the connectivity of standard neural models that results in costly information exchanges ([63, 51, 7, 62]). Coarse-grain parallel implementations are mainly applied to neural learning, so that their efficiency suffers from the sequentiality of standard learning algorithms such as stochastic gradient descent (see [53, 55, 14]). Furthermore, massively parallel computers are expensive resources and they cannot be used in embedded applications. Such solutions are usually preferable for huge neural structures and complex neural computations or learning methods ([45, 30, 7, 62]).

3.2.1.2   Dedicated parallel computers.   *Neuro-computers* are parallel systems dedicated to neural computing. They are based on computing devices such as DSPs (*digital signal processors*) as in [39, 49, 46, 62], or neuro-processors ([3, 68, 62]). Their use suffers from their cost and their development time: they rapidly become out-of-date, compared to the most recent sequential processors. Most well-known neurocomputers are described in [50, 62].

3.2.1.3   Analog ASICs.   Many analog hardware implementations have been realized. They are very fast, dense and low-power, but they introduce specific problems (see [41]), such as precision, data storage, robustness. On-chip learning is difficult ([9]). It is an expensive and not flexible solution, as any ASIC (even if some design methods tend to be flexible, [48]). And their very long development is very tricky for users who are not analog technology experts.

3.2.1.4   Digital ASICs.   Many digital integrated circuits have also been designed for neural networks. Compared to analog chips, they provide more accuracy, they are more robust, and they can handle any standard neural computation. Yet their design requires a strong effort to obtain working chips and it is very expensive when only a few chips are needed (standard drawbacks of ASICs). They usually implement limited parts of neural networks, so as to be included in neuro- computer systems ([69, 13]). Sometimes they implement a whole specific neural network, nevertheless the architecture of this neural network is not directly mapped onto the chip (e.g. [38]).

3.2.1.5   The FPGA solution.   To summarize the above survey, hardware implementations are more likely to fit the parallelism grain of neural computations, and both digital and analog ASICs may efficiently implement regular structures, but they require important development times. This major drawback may be avoided with the help of programmable integrated circuits

Table 3.1. Appropriate/inappropriate devices for neural network implementations

	analog ASIC	digital ASIC	FPGA	processor based	parallel computer
speed	+++	++	+	−	+
area	+++	++	+	−	−−
cost	−−	−−	++	++	−−
design time	−−	−−	++	+++	+
reliability	−−	+	++	++	++

−−: very unfavourable, −: unfavourable,
+: favourable, ++: very favourable, +++: highly favourable

such as FPGAs: since the appearance of programmable hardware devices, algorithms may be implemented on very fast integrated circuits with software-like design principles, whereas usual VLSI designs lead to very high performances at the price of very long production times (up to 6 months).

FPGAs, such as Xilinx FPGA ([70]), are based on a matrix of *configurable logic blocks* (CLBs). Each CLB contains several logic cells that are able to implement small logical functions (4 or 5 inputs) with a few elementary memory devices (flip-flops or latches) and some multiplexors. CLBs can be connected thanks to a configurable routing structure. In Xilinx FPGAs, CLBs can be efficiently connected to neighbouring CLBs as well as CLBs in the same row or column. The configurable communication structure can connect external CLBs to *input/output blocks* (IOBs) that drive the input/output pads of the chip.

An FPGA approach simply adapts to the handled application, whereas a usual VLSI implementation requires costly rebuildings of the whole circuit when changing some characteristics. A design on FPGAs requires the description of several operating blocks. Then the control and the communication schemes are added to the description, and an automatic "compiling" tool maps the described circuit onto the chip. Therefore configurable hardware appears as well-adapted to obtain efficient and flexible neural network implementations.

Table 3.1 roughly summarizes the main advantages and drawbacks of the most common solutions (regardless of the implemented computations, neural or not). Each solution is roughly estimated with respect to each implementation aspect.

3.2.2 Neural networks on FPGAs: specific assets

As stated above, FPGAs offer a cheap, easy and flexible choice for hardware implementations. They also have several specific advantages for neural implementations:

- Reprogrammable FPGAs permit prototyping: in most applications, several neural architectures must be tested so as to find the most efficient one. This may be directly performed with the hardware efficiency of an FPGA-based implementation, without any additional cost. Moreover a good architecture that has been designed and implemented may be replaced later by a better one without having to design a new chip.

- On-chip learning is often considered as difficult and useless. Indeed it is used very seldom. But on-chip learning usually results in a loss of efficiency in a hardware implementation, since it requires some specific operators, a higher precision, ... etc. Therefore off-chip learning is naturally chosen when no dynamic learning is necessary. In a reconfigurable FPGA, on-chip learning may be performed prior to a specific optimized implementation of the learned neural network on the same chip.

- FPGAs may be used for embedded applications, when the robustness and the simplicity of neural computations is most needed, even for low-scale productions.

- FPGA-based implementations may be mapped onto new improved FPGAs, which is a major advantage, considering that FPGA speeds and areas approximately double each year. Even large neural networks may soon be implemented on single FPGAs, provided that the implementation method is scalable enough. The FPNA concept is a major advance to ensure the scalability of direct hardware mappings of neural networks (the neural architecture itself is mapped by the compiler onto the FPGA).

3.2.3   Implementation issues

Neural network implementations on FPGAs have to deal with the usual issues of digital hardware implementations of neural network applications. But these issues become more acute because of the specific constraints of FPGAs: it is particularly obvious when it is a matter of area consumption. Even if the software-like use of FPGAs is simple, the user should not forget that the compiler has to deal with these strong constraints.

3.2.3.1   Topology-related issues.

- Hardware devices have 2-D topologies, that do not easily fit the regular but complex connection graph of most standard neural models (e.g. multilayer ones). Furthermore, in the case of FPGAs, the user must deal with an already specified connection frame, even if it is programmable. Routing problems may not be solved by means of a shrewd use of the different metal layers.

- Neurons have a large number of parallel input links in most neural models, whereas hardware implementations of operators can handle only a limited fan-in. In the case of FPGAs, part of the available CLBs may be used just to increase the routing capacities, which results in a loss of computation resources.

3.2.3.2   Area-related issues.

- Multipliers and elementary function operators (e.g. the sigmoidal activation function *tanh*) are area-greedy. This problem strongly depends on the required precision. Though neural networks do not usually require very precise computations, the required precision varies with the application ([2]). Moreover, learning is not as tolerant as generalization ([34, 58, 47, 48]), so that on-chip learning does not only introduce algorithmic needs, but also makes low-precision implementations impossible.

- Weight storage can not be included in the FPGA for large neural networks. When efficient implementations are obtained, a careful management of the memory accesses must be taken into account.

- Neural network applications have a notable drawback: there is an imbalance between the overall computation time and the implementation area of the different operators involved in neural computing. In most standard neural models, the main part of the computation time is due to matrix-vector products that require multipliers and adders. But other area-greedy operators, such as activation functions, must be implemented. The area of an FPGA is strictly bounded. A significant part of the available area is required to implement these other operators, so that only a reduced part of the whole FPGA is available to implement almost 100 % of the overall computation time of the neural application.

3.2.4   Existing solutions

The easiest way to bypass these issues is to consider very small neural networks (as in [8, 54, 57, 44]). More generally, standard solutions include:

- Implementation of the generalization phase only ([4, 8, 60, 10, 37, 66]).

- Implementation of a small neuro-processor on the FPGA, and sequential (and therefore slower) use of this computing device ([10, 37, 15]).

- Implementation of a small neuro-processor on each one of several FPGAs with data communications ([11, 54–1, 33]) that raise bandwidth problems and that need to minimize the number of required I/O pads, which is difficult when densely connected neural architectures are handled.

- Simplified computations, discretized activation function, very low precision ([8, 10, 31]). Such solutions only adapt to specific applications with very weak precision requirements.

More advanced methods have been considered:

3.2.4.1   Area-saving solutions.

- Optimized implementation of the activation function (e.g. CORDIC-like algorithm in [2], or piecewise polynomial approximation controled by look-up tables in [29]), or fast multipliers ([60]). Such solutions may be combined with other advanced method. Their choice depends on some application characteristics, such as required precision, or required on-chip learning.

- Bit-serial arithmetic (standard: [11], or optimized: [66]) or on-line serial arithmetic ([29]). The choice of the best arithmetic for digital hardware implementations of neural networks is discussed in [61], but the specificity of FPGAs is not taken into account, and on-line arithmetic is not considered. Any serial arithmetic requires more clock cycles, and should be used with pipelined operators.

- Pulse-based arithmetic (bit-stream neurons in [4, 59, 43, 67, 40, 56], or some frequency-based solution in [32, 44]). It provides tiny operators, so that an average-sized neural network can be fully implemented on a single FPGA, but the routing constraints of FPGAs still do not permit large neural networks to be implemented. What is more, such solutions need to be applied to the development of the neural application from the beginning (arithmetical operations are not equivalent to standard ones).

3.2.4.2   Solutions to reduce CLB idleness.

- Run-time reconfiguration of the FPGA ([11, 6]). Such methods still clash with significant dynamic reconfiguration times.

- Pipelined architecture (e.g. orthogonal systolic array, [17]). Strong constraints are put on the kind of neural architectures to which such methods apply.

- Sub-neural parallelism level and computation overlappings ([29]): this work makes it possible to implement multilayer perceptrons of any size on a single FPGA, with on-chip learning and adaptable precision, but it offers a limited parallelism level (this solution is a compromise between a neuro-processor and a direct parallel mapping of neural operators).

3.2.4.3   Topologically hardware-friendly solutions.

- Partition of the neural graph into equivalent subgraphs, so that a repeated use of such a simple subgraph covers the whole graph ([31]).

- Neural computation paradigm tolerant of constrained topologies: it corresponds to the FPNA concept that is described in this paper.

3.2.4.4   Adapted frameworks of neural computation.   As shown above, many neural implementations on FPGAs handle simplified neural computations. Furthermore, many efficient implementation methods (on ASICs, neuro-computers, ... etc) have to limit themselves to few well-fitted neural architectures. An upstream work is preferable: neural computation paradigms may be defined to counterbalance the main implementation problems, and the use of such paradigms naturally leads to neural models that are more tolerant of hardware constraints, without any additional limitation. Since the main implementation issues are linked to area-greedy operators and complex topologies, two kinds of (complementary) hardware-adapted neural computation paradigms may appear: area-saving neural computation frameworks on one hand, and paradigms that may handle simplified topologies on the other hand. The bit-stream technology mentioned above has led to an example of area-saving neural paradigm, while the definition of the FPNA framework makes it possible to simplify the architectures of standard neural models without significant loss of approximation capability.

Next section introduces the FPNA concept, before §3.3.5 explains how it makes it possible to overcome most implementation issues mentioned above.

3.3   FPNAs, FPNNs

This section describes the FPNA/FPNN concept. These models appear as parameterized task graphs that are specialized so as to perform neural computations, but they differ from standard neural models that are graphs of non-linear regressors.

The description below directly illustrates the links of the FPNA concept with both implementation choices and standard neural theory.

The distinction between FPNAs and FPNNs is mainly linked to implementation properties. An FPNA is a given set of neural resources that are organized according to specific neighbourhood relations. An FPNN is a way to use this set of resources: it only requires local configuration choices. Therefore, the implementations of two different FPNNs exactly require the same mapping on FPGA, as long as they are based on the same FPNA.

This section is organized as follows:

- §3.3.1 briefly describes where this concept comes from, then it defines an FPNA as a set of computing resources which architecture is fully specified, but in which local computations still need to be parameterized

- §3.3.2 describes the parameterization step that is required to get to a functional neural model, called FPNN,

- §3.3.3 describes the basic computation algorithm that has been defined for FPNNs, and that sequentially handles a list of tasks (node-level computation); an example is described in detail and it illustrates the main properties of FPNN computation, such as how virtual connections are obtained as composite links, §

- 3.3.4 describes a parallel version of the above algorithm, in order to show that it only requires local computations; this computation algorithm is the basis of a direct parallel mapping onto FPGAs in 4.5; an example is described in detail and it illustrates how virtual connections generate a functional equivalence between an FPNN and a standard neural network with a completely different architecture,

- §1.3.5 gets back to the discussion of §3.2.3 and it explains how most neural implementation issues may be overcome by the use of FPNA architectures combined with the virtual connections that are created by FPNN computation.

3.3.1   FPNAs

3.3.1.1   From FPGAs to FPNAs.   The first aim of the FPNA concept is to develop neural structures that are easy to map directly onto digital hardware, thanks to a simplified and flexible topology. The structure of an FPNA is based on FPGA principles: complex functions realized by means of a set of simple programmable resources. The nature and the relations of these FPNA resources are derived from the mathematical processing FPNAs have to perform.

To summarize, in a standard neural model, each neuron computes a function applied to a weighted sum of its inputs: if $\vec{x}_i$ is the input vector of neuron i, and $\vec{w}_i$ is its weight vector, it computes $f_i(\vec{w}_i.\vec{x}_i)$. See [18, 19] for a unified theoretical approach of the computation of standard neural networks. The input vector $\vec{x}_i$ may contain neural network inputs as well as outputs of other neurons, depending of the *connection graph* of the neural network. In such a standard model, each link is a connection between the output of a neuron and an input of another neuron. Therefore the number of inputs of each neuron is its fan-in in the connection graph. On the contrary, neural resources become *autonomous* in an FPNA: their dependencies are freely set, and the resulting processing is more complex than in standard neural models. As in

FPGAs, FPNA configuration handles both resource interconnections and resource functionalities.

3.3.1.2 FPNA resources.

An FPNA is a programmable set of neural resources that are defined to compute partial convolutions for non-linear regression (see section 3.5), as standard multilayer neural networks do. Two kinds of autonomous FPNA resources naturally appear: *activators* that apply standard neural functions to a set of input values on one hand, and *links* that behave as independent affine operators on the other hand.

These resources might be handled in different ways. The easiest scheme would allocate any link to any activator, with the help of an underlying programmable interconnection network. This would lead to massively pruned standard neural networks, or to multiple weight sharing connections. Topological problems would still appear (such as high fan-ins), and weight sharing would lead to few different $\vec{w}_i$ weight vectors. Therefore, another principle has been chosen for FPNAs: any link may be *connected* to any *local resource*. The aim of locality is to reduce topological problems, whereas connected links result in more various weight vectors.

More precisely, the links connect the nodes of a directed graph, each node contains one activator. The specificity of FPNAs is that the relations between *any* local resources of each node may be freely set. A link may be connected or not to the local activator *and to other local links*. Direct connections between affine links appear, so that the FPNA computes numerous composite affine transforms. These compositions create numerous *virtual neural connections*, so that different convolution terms may be obtained with a reduced number of connection weights.

3.3.1.3 Definition of FPNAs.

The following definition specifies the structure of an FPNA (directed graph), as well as the functional nature of each individual neural resource.

An FPNA is defined by means of:

- A directed graph $(\mathcal{N}, \mathcal{E})$, where $\mathcal{N}$ is an ordered finite set of nodes, and $\mathcal{E}$ is a set of directed edges without loop. $\mathcal{E}$ may be seen as a subset of $\mathcal{N}^2$.

  For each node n, the set of the direct predecessors (resp. successors) of n is defined by $Pred(n) = \{p \in \mathcal{N} \mid (p, n) \in \mathcal{E}\}$ (resp. $Succ(n) = \{s \in \mathcal{N} \mid (n, s) \in \mathcal{E}\}$). The set of the input nodes is $\mathcal{N}_i = \{n \in \mathcal{N} \mid Pred(n) = \emptyset\}$.

- A set of affine operators $\alpha_{(p,n)}$ for each (p, n) in $\mathcal{E}$.

- A set of activators (i_n, f_n), for each n in $\mathcal{N} - \mathcal{N}_i$: i_n is an iteration operator (a function from $I\!\!R^2$ to $I\!\!R$), and f_n is an activation function (from $I\!\!R$ to $I\!\!R$).

To simplify, (p, n) and (n) now stand for the corresponding links and activators.

3.3.1.4   Interpretation.   Resources are associated with the nodes, whereas locality is defined by the edges. For each node $n \in \mathcal{N}$, there is one activator and as many communication links as this node has got predecessors. Each link is associated with an affine operator. An activator is defined by (i_n, f_n), since it will handle any neuron computation as in a sequential program. Indeed, any standard neuron computation may be performed by means of a loop that updates a variable with respect to the neuron inputs, and a final computation that maps this variable to the neuron output. The iteration function i_n stands for the updating function inside the loop. The neuron output is finally computed with f_n. See [21] for the definition of (i_n, f_n) so as to obtain most standard neurons, e.g.:

- A d-input sigmoidal neuron of a multilayer perceptron computes

$$\sigma \left(\theta + \sum_{j=1}^{d} w_j x_j \right) \tag{3.1}$$

    where σ is a sigmoid function (bounded, monotonous), θ is a threshold value.

- A d-input radial basis function of a RBF network computes

$$\gamma \left(\sqrt{\sum_{j=1}^{d} (w_j(x_j - t_j))^2} \right)$$

    where γ is a gaussian function and $\vec{t}$ is a translation vector.

- A d-input wavelet neuron of a wavelet network ([71, 5]) computes

$$\prod_{j=1}^{d} \psi(w_j(x_j - t_j))$$

    where ψ is a wavelet function (localized in both space and frequency domains).

When the FPNA graph and the different operators have been defined, a general implementation can be given: each resource corresponds to a basic block, and these blocks are organized according to the graph. This implementation may be used by any FPNN derived from this FPNA. Some of these FPNNs compute very complex functions (equivalent to standard neural networks), though the FPNA graph is made simple (reduced number of edges, limited node fan-ins and fan-outs, so that the FPNA is easily mapped by the compiler onto the hardware device).

3.3.2 FPNNs

An FPNN (field programmed neural network) is a configured FPNA: an FPNA whose resources have been connected in a specific way (furthermore, some parameters must be given to specify the computation of each resource).

In other words, the underlying FPNA of an FPNN is a topologically fully specified architecture of neural resources, whereas this FPNN is a given way to specify how these resources will interact to define a functional behaviour.

3.3.2.1   Definition of FPNNs.   An FPNA configuration requires local connections (from link to link, link to activator or activator to link) as well as precisions about the iterative process performed by each activator (initial value, number of iterations). Therefore an FPNN is specified by:

- an FPNA (available neural resources),

- for each node n in $\mathcal{N} - \mathcal{N}_i$,

  - a real value θ_n (initial value of the variable updated by iteration function i_n)

  - a positive integer a_n (number of iterations before an activator applies its activation function)

  - for each p in $Pred(n)$, two real value $W_n(p)$ and $T_n(p)$ (coefficients of affine operator $\alpha_{(p,n)}(x) = W_n(p)x + T_n(p)$),

  - for each p in $Pred(n)$, a binary value $r_n(p)$ (set to 1 *iff* link (p, n) and activator (n) are connected),

  - for each s in $Succ(n)$, a binary value $S_n(s)$ (set to 1 *iff* activator (n) and link (n, s) are connected),

  - for each p in $Pred(n)$ and each s in $Succ(n)$, a binary value $R_n(p, s)$ (set to 1 *iff* links (p, n) and (n, s) are connected),

- for each input node n in $\mathcal{N}_i$,

  - a positive integer c_n (number of inputs sent to this node),

  - for each s in $Succ(n)$, a binary value $S_n(s)$ (see above).

3.3.2.2   Computing in an FPNN.   Several computation methods have been defined for the FPNNs in [21]. Their common principles may be described as follows:

- All resources behave independently.

- A resource receives values. For each value,

    - the resource applies its local operator(s),
    - the result is sent to all neighbouring resources to which the resource is locally connected (an activator waits for a_n values before sending any result to its neighbours).

The main differences with the standard neural computation are:

- A resource may or may not be connected to a neighbouring resource. These local connections are set by the $r_n(p)$, $S_n(s)$ and $R_n(p,s)$ values.

- A link may directly send values to other communication links.

- A resource (even a link) may handle several values during a single FPNN computation process.

A sequential version of the most general FPNN computing method is first described below. This method clearly illustrates the above principles. Moreover, it stands as a reference computation scheme when establishing theoretical properties.

Then a parallel version is described. This computation scheme is aimed at being directly implemented onto FPGAs.

3.3.3   Sequential computation

This computation method handles a list of tasks $\mathcal{L}$ that are processed according to a FIFO scheduling. Each task $[(p,n),x]$ corresponds to a value x sent on a link (p,n). Task handling corresponds to the various steps that must be performed at node-level to deal with such an input:

- Input nodes just have to send global input values to their connected neighbours.

- An incoming value is first processed in an affine way.

- Then it is directly forwarded to neighbouring nodes when direct connections exist between links.

- It is also processed by the local activator when a connection exists between the incoming link and this activator. The latter generates an output when it has received enough inputs. This output is sent to all connected neighbours.

The algorithm is as follows:

Initialization:

INPUTS

For each input node n in $\mathcal{N}_i$, c_n values $\left(x_n^{(i)}\right)_{i=1..c_n}$ are given (outer inputs of the FPNN), and the corresponding tasks $[(n, s), x_n^{(i)}]$ are put in $\mathcal{L}$ for all s in $Succ(n)$ such that $S_n(s) = 1$. The order of task creation corresponds to a lexicographical order (n, i, s) (according to the order in $\mathcal{N}$).

VARIABLES

For each node $n \in \mathcal{N} - \mathcal{N}_i$, two local variables are used: c_n is the number of values received by the local activator, whereas x_n is the value that is updated by the iteration function i_n. Initially $c_n = 0$ and $x_n = \theta_n$.

Sequential processing:

While $\mathcal{L}$ is not empty

Let $[(p, n), x]$ be the first element in $\mathcal{L}$.

1 suppress this element in $\mathcal{L}$

2 $x' = W_n(p)x + T_n(p)$

3 for all $s \in Succ(n)$ such that $R_n(p, s) = 1$, create $[(n, s), x']$ in $\mathcal{L}$ according to the order of successors s in $\mathcal{N}$

4 if $r_n(p) = 1$

  – increment c_n and update x_n: $x_n = i_n(x_n, x')$

  – if $c_n = a_n$

  (a) $y = f_n(x_n)$

  (b) reset local variables ($c_n = 0$, $x_n = \theta_n$)

  (c) for all $s \in Succ(n)$ such that $S_n(s) = 1$, create $[(n, s), y]$ in $\mathcal{L}$ according to the order of successors s in $\mathcal{N}$

If $r_n(p) = 1$, activator (n) is said to be receiving the value of task $[(p, n), x]$.

3.3.3.1   Example.   The following simple example is not functionally useful: there is no clear global output. Yet it illustrates some specific characteristics of FPNN computing:

- a value that is processed by a link is not always received by the corresponding activator

- some links and some local connections may be useless, if no value is sent through them,

- some activators may receive values without sending outputs

Above all it illustrates the major aspect of FPNN computing: some values processed by links are *directly sent to other links*, so that *virtual links* are created.

These characteristics are outlined just after the detailed description of this example.

Let Φ be the FPNA defined by:

- $\mathcal{N} = (n_1, n_2, n_3, n_4, n_5)$

- $\mathcal{E} = \{(n_1, n_3), (n_2, n_3), (n_2, n_4), (n_3, n_4), (n_3, n_5), (n_4, n_5), (n_5, n_3)\}$

- $i_{n_3} = i_{n_4} = i_{n_5} = ((x, x') \mapsto x + x')$

- $f_{n_3} = f_{n_4} = (x \mapsto \tanh(x))$ and $f_{n_5} = (x \mapsto x)$

Figure 3.1 shows the activators, the links and the configured local connections of an FPNN ϕ derived from Φ. This FPNN is more precisely defined by:

- $\theta_3 = 2.1, \theta_4 = -1.9, \theta_5 = 0$

- $a_3 = 3, a_4 = 2, a_5 = 2$

- $\forall (n_i, n_j) \in \mathcal{E} \ \ W_{(n_i, n_j)} = i * j \ \ T_{(n_i, n_j)} = i + j$

- $r_{n_3}(n_1) = 0, r_{n_3}(n_2) = 1, r_{n_3}(n_5) = 1,$
  $r_{n_4}(n_2) = r_{n_4}(n_3) = 1, r_{n_5}(n_3) = r_{n_5}(n_4) = 1$

- $S_{n_1}(n_3) = 1, S_{n_2}(n_3) = S_{n_2}(n_4) = 1,$
  $S_{n_3}(n_4) = S_{n_5}(n_3) = 0, S_{n_3}(n_5) = S_{n_4}(n_5) = 1$

- $R_{n_3}(n_1, n_4) = 1, R_{n_3}(n_1, n_5) = 0, R_{n_3}(n_2, n_4) = R_{n_3}(n_2, n_5) = 0,$
  $R_{n_4}(n_2, n_5) = R_{n_4}(n_3, n_5) = 0, R_{n_5}(n_3, n_3) = R_{n_5}(n_4, n_5) = 0$

- $c_1 = 2, c_2 = 1$

The sequential computation method applies to ϕ as follows:

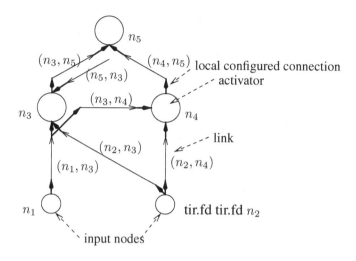

Figure 3.1. FPNN ϕ (resources and local connections)

Initialization: Some FPNN inputs are chosen: $x_{n_1}^{(1)} = 1.5$, $x_{n_1}^{(2)} = -0.8$, $x_{n_2}^{(1)} = 1.1$. The first tasks are $[(n_1, n_3), 1.5]$, $[(n_1, n_3), -0.8]$, $[(n_2, n_3), 1.1]$ and $[(n_2, n_4), 1.1]$.
Finally, $c_{n_3} = c_{n_4} = c_{n_5} = 0$, $x_{n_3} = \theta_3 = 2.1$, $x_{n_4} = \theta_4 = -1.9$ and $x_{n_5} = \theta_5 = 0$.

Sequential processing:

1 Task $[(n_1, n_3), 1.5]$ is first processed:
$x' = 1.5\,W_{n_3}(n_1) + T_{n_3}(n_1) = 8.5$, task $[(n_3, n_4), 8.5]$ is created since $R_{n_3}(n_1, n_4) = 1$, and the value 8.5 is not received by activator (n_3) since $r_{n_3}(n_1) = 0$.

2 Task $[(n_1, n_3), -0.8]$ is processed in the same way:
task $[(n_3, n_4), 1.6]$ is created.

3 Task $[(n_2, n_3), 1.1]$ is processed: $x' = 11.6$ is received by activator (n_3), $c_{n_3} = 1$ (therefore $c_{n_3} < a_{n_3}$), $x_{n_3} = 13.7$.

4 Task $[(n_2, n_4), 1.1]$ is processed: $x' = 14.8$ is received by activator (n_4), $c_{n_4} = 1$, $x_{n_4} = 12.9$.

5 Task $[(n_3, n_4), 8.5]$ is processed: $x' = 109$ is received by activator (n_4), $c_{n_4} = 2$, $x_{n_4} = 121.9$. Now $c_{n_4} = a_{n_4}$: reset ($c_{n_4} = 0$, $x_{n_4} = -1.9$), $y = \tanh(109) \simeq 1$, task $[(n_4, n_5), 1]$ is created.

6 Task $[(n_3, n_4), 1.6]$ is processed: $x' = 26.2$, $c_{n_4} = 1$, $x_{n_4} = 24.3$.

7 Task $[(n_4, n_5), 1]$ is processed: $x' = 29$, $c_{n_5} = 1$, $x_{n_5} = 29$.

As mentioned above, this simple example is not functionally useful: one would expect ϕ to map input values to some output values (for example computed by n_5). Yet it illustrates the specific characteristics of FPNN computing that have been mentioned: the values processed by the links towards n_3 are not always received by activator (n_3), link (n_5, n_3) is not used, activators (n_3) and (n_5) receive values without sending outputs, ... etc.

Most of all this example illustrates the creation of virtual links thanks to direct connections between links. In this simple example there is a direct connection from (n_1, n_3) towards (n_3, n_4), so that a virtual link (n_1, n_4) is created. Its virtual affine coefficients are $W_{n_4}(n_1) = W_{n_4}(n_3)W_{n_3}(n_1)$ and $T_{n_4}(n_1) = W_{n_4}(n_3)T_{n_3}(n_1) + T_{n_4}(n_3)$. Of course far more complex virtual links are required to obtain significant topology simplifications (see section 4.3).

3.3.4   Asynchronous parallel computation

The asynchronous parallel computation scheme illustrates best the resource autonomy. It is based on an asynchronous local handling of requests: there is no global list of requests. A request $req[\varrho_1, \varrho_2, x]$ is created when a value x is exchanged by two connected resources ϱ_1 and ϱ_2. Therefore, this parallel computation is performed at resource-level instead of node-level. The same local computations are performed as in the above sequential algorithm, except that resources behave independantly, and this behaviour depends on the type of resource (link or activator).

The corresponding algorithm is as follows:

Initialization:

- For each input node n in $\mathcal{N}_i$, c_n values $\left(x_n^{(i)}\right)_{i=1..c_n}$ are given (FPNN inputs), and the corresponding requests $req[(n), (n, s), x_n^{(i)}]$ are created for all s in $Succ(n)$ such that $S_n(s) = 1$.

- Each node n in $\mathcal{N} - \mathcal{N}_i$ has got a local counter c_n and a local variable x_n, initially set as $c_n = 0$ and $x_n = \theta_n$.

Parallel processing:

All resources work concurrently. Each one of them sequentially handles all requests it receives. Resource ϱ_2 chooses a request $req[\varrho_1, \varrho_2, x]$ among the unprocessed requests that have already been received (with a fair choice policy). This request is processed by ϱ_2 as follows:

1 an acknowledgement is sent to ϱ_1

2 if ϱ_2 is activator (n)

- c_n and x_n are updated: $c_n = c_n + 1$, $x_n = i_n(x_n, x)$
- if $c_n = a_n$ (the local activator computes its output)
  - for all s in $Succ(n)$ such that $S_n(s) = 1$, create $req[(n), (n, s), f_n(x_n)]$
  - wait for all acknowledgements
  - reset : $c_n = 0$ $\quad x_n = \theta_n$

else (resource ϱ_2 is link (p, n))

- compute $x' = W_n(p)x + T_n(p)$
- for all $s \in Succ(n)$ such that $R_n(p, s) = 1$, create $req[(p, n), (n, s), x']$
- if $r_n(p) = 1$ then create $req[(p, n), (n, n), x']$
- wait for all acknowledgements

A general implementation architecture has been directly derived from this parallel computation (see section 4.5).

3.3.4.1   Example.   The following example is a detailed description of a whole parallel computation process. It illustrates how requests are handled at a resource-level that is purely local and that does not require any global scheduling. Most of all it shows that the FPNN computation paradigm permits a given set of neural resources to behave as a standard neural network with a very different architecture (here a multilayer perceptron).

Let Ψ be the FPNA defined by:

- $\mathcal{N} = (n_x, n_y, n_1, n_2, n_3, n_4)$
- $\mathcal{E} = \{(n_x, n_1), (n_y, n_4), (n_1, n_2), (n_2, n_1), (n_1, n_4), (n_4, n_1), (n_4, n_3)$
  $, (n_3, n_4), (n_2, n_3)\}$
- $i_{n_1} = i_{n_2} = i_{n_3} = i_{n_4} = ((x, x') \mapsto x + x')$
- $f_{n_1} = f_{n_2} = f_{n_3} = f_{n_4} = (x \mapsto \tanh(x))$

Figure 3.2 shows the activators, the links and the configured local connections of a Ψ-derived FPNN ψ. The binary parameters of ψ are equal to 0, except:

- for input nodes n_x and n_y: $S_{n_x}(n_1)$, $S_{n_y}(n_2)$,
- for node n_1: $r_{n_1}(n_x)$, $r_{n_1}(n_4)$, $S_{n_1}(n_4)$, $R_{n_1}(n_x, n_2)$, $R_{n_1}(n_2, n_4)$, $R_{n_1}(n_4, n_2)$,
- for node n_2: $r_{n_2}(n_1)$, $S_{n_2}(n_1)$, $R_{n_2}(n_1, n_3)$,

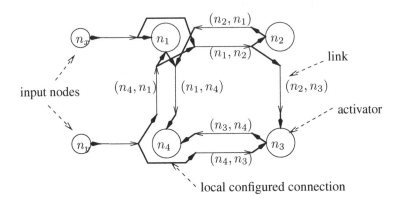

Figure 3.2. FPNN ψ (resources and local connections)

- for node n_3: $r_{n_3}(n_2)$, $r_{n_3}(n_4)$, $S_{n_3}(n_4)$,

- for node n_4: $r_{n_4}(n_1)$, $r_{n_4}(n_3)$, $R_{n_4}(n_y, n_1)$, $R_{n_4}(n_y, n_3)$.

Moreover: $a_1 = 2$, $a_2 = 2$, $a_3 = 3$, $a_4 = 3$, and $c_{n_x} = c_{n_y} = 1$.

The asynchronous parallel computation method may apply to ψ in *different ways* that depend on the scheduling policy of each neural resource. Figure 3.3 sketches the nine steps of a possible parallel computation. Processing resources are grey filled. Processed requests are easily identified thanks to the configured connection (dark grey thick arrows) between their sender and their receiver. The number of created requests becomes apparent with the corresponding configured connections (light grey thick arrows). The main aspects of this parallel processing are:

Initialization: FPNN inputs are assigned to input nodes: $x_{n_x}^{(1)} = x$, $x_{n_y}^{(1)} = y$.
The initial set of requests is
$$\{req[(n_x), (n_x, n_1), x], req[(n_y), (n_y, n_4), y]\}.$$
Moreover $\forall i \in \{1, 2, 3, 4\}$ $c_{n_i} = 0$ and $x_{n_i} = \theta_i$.

Parallel processing progress:

1 Resources (n_x, n_1) and (n_y, n_4) may work concurrently, so that both initial requests are simultaneously processed:

- Request $req[(n_x), (n_x, n_1), x]$ is processed: an acknowledgement is sent to (n_x), request
  $$req[(n_x, n_1), (n_1, n_2), \alpha_{(n_x, n_1)}(x)]$$
  is created since $R_{n_1}(n_x, n_2) = 1$, and $\alpha_{(n_x, n_1)}(x)$ is also sent to activator (n_1) since $r_{n_1}(n_x) = 1$, i.e.

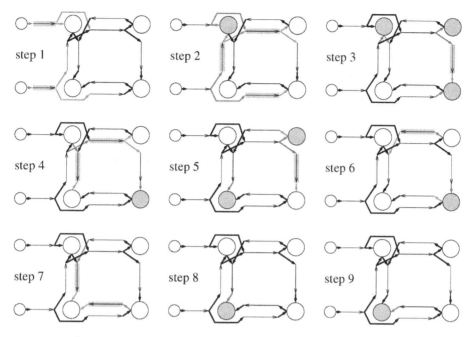

▩ resource that processes a request

← connection from the sender of the processed request

← connection(s) towards the resources to which data are sent (created requests)

Figure 3.3. A parallel computation of FPNN ψ

$req[(n_x, n_1), (n_1), \alpha_{(n_x,n_1)}(x)]$ is created.

- Request $req[(n_y), (n_y, n_4), y]$ is processed: an acknowledgement is sent to (n_y), requests

$$req[(n_y, n_4), (n_4, n_1), \alpha_{(n_y,n_4)}(y)]$$
and $req[(n_y, n_4), (n_4, n_3), \alpha_{(n_y,n_4)}(y)]$

are created since $R_{n_4}(n_y, n_1) = R_{n_4}(n_y, n_3) = 1$.

2 Resources (n_1), (n_1, n_2), (n_4, n_1) and (n_4, n_3) work concurrently to process all available requests. Links proceed as above. Activator (n_1) processes $req[(n_x, n_1), (n_1), \alpha_{(n_x,n_1)}(x)]$ as follows:

- An acknowledgement is sent to (n_x, n_1). Local updates are: $c_{n_1} = 1$, $x_{n_1} = \theta_1 + \alpha_{(n_x,n_1)}(x)$. No request is created, since $c_{n_1} < a_{n_1}$.

3 Resources (n_1), (n_2), (n_2, n_3) and (n_3) work concurrently. Link (n_2, n_3) and activators (n_2) and (n_3) proceed as above, whereas:

- $req[(n_4, n_1), (n_1), \alpha_{(n_4,n_1)}(\alpha_{(n_y,n_4)}(y))]$ is processed: an acknowledgement is sent to (n_4, n_1). Local updates are: $c_{n_1} = 2$, $x_{n_1} = \theta_1 + \alpha_{(n_x,n_1)}(x) + \alpha_{(n_4,n_1)}(\alpha_{(n_y,n_4)}(y))$. Now $c_{n_1} = a_{n_1}$, so that $req[(n_1), (n_1, n_4), tanh(x_{n_1})]$ is created since $S_{n_1}(n_4) = 1$. Then $c_{n_1} = 0$ and $x_{n_1} = \theta_1$.

  $req[(n_4, n_1), (n_1, n_2), \alpha_{(n_4,n_1)}(\alpha_{(n_y,n_4)}(y))]$ is another available request, but resource (n_1, n_2) still waits for the acknowledgement of the two requests that it created at step 2.

4 Resource (n_1, n_2) is now able to process its waiting request. Resources (n_1, n_4) and (n_3) also work concurrently.

  $\ldots$

8. Activator (n_4) chooses a request among the two ones it has received.

9. Activator (n_4) processes the last request, so that $c_{n_4} = a_{n_4}$. Therefore (n_4) computes its output, but no request is created, since all $S_{n_4}(.)$ binary values are '0'. This output is

$$
\begin{aligned}
\tanh \big(\theta_4 &+ \alpha_{(n_1,n_4)}(\tanh(\theta_1 + \alpha_{(n_x,n_1)}(x) + \alpha_{(n_4,n_1)}(\alpha_{(n_y,n_4)}(y)))) \\
&+ \alpha_{(n_1,n_4)}(\alpha_{(n_2,n_1)}(\tanh(\theta_2 + \alpha_{(n_1,n_2)}(\alpha_{(n_x,n_1)}(x)) \\
&\qquad\qquad + \alpha_{(n_1,n_2)}(\alpha_{(n_4,n_1)}(\alpha_{(n_y,n_4)}(y)))))) \\
&+ \alpha_{(n_3,n_4)}(\tanh(\theta_3 + \alpha_{(n_2,n_3)}(\alpha_{(n_1,n_2)}(\alpha_{(n_x,n_1)}(x))) \\
&\qquad + \alpha_{(n_2,n_3)}(\alpha_{(n_1,n_2)}(\alpha_{(n_4,n_1)}(\alpha_{(n_y,n_4)}(y)))) \\
&\qquad + \alpha_{(n_4,n_3)}(\alpha_{(n_y,n_4)}(y)))))
\end{aligned}
$$

The above result is *exactly* the same as with the MLP in Figure 3.4, provided that:

- all coefficients $T_n(p)$ are equal to 0,

- the compositions of link weights (virtual synaptic weights) are equal to the weights of Figure 3.4, which implies for example that $w_{2,2} = W_{n_2}(n_1)W_{n_1}(n_4)W_{n_4}(n_y)$
  and $w_{3,2} = (W_{n_3}(n_2)W_{n_2}(n_1)W_{n_1}(n_4) + W_{n_3}(n_4)) W_{n_4}(n_y)$.

As mentioned above, this example shows that the FPNA computation paradigm permits a given set of neural resources to behave as a standard neural network with a very different architecture. Nevertheless, this FPNN is not useful: the architecture of Figure 3.2 does not simplify the MLP architecture of Figure 3.4. Indeed, this MLP is so simple that it is neither possible nor necessary to expect any architecture simplification. Some examples of significant simplifications of larger neural networks are described in 4.3.

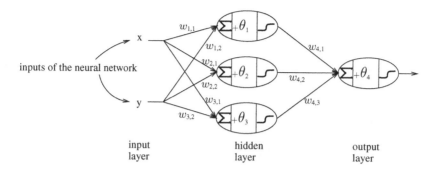

Figure 3.4. A multilayer perceptron

3.3.5 Implementation issues: FPNA answer

As explained in §3.2.4, the FPNA paradigm is a hardware-adapted framework of neural computation, instead of an implementation method that either handles simplified neural computations or uses shrewd ways to counterbalance some of the problems mentioned in §3.2.3.

In many cases, designing an FPNA consists in choosing a standard neural architecture that fits the application, then in determining an FPNA that both fits the implementation device and can be configured[2] so as to be functionally equivalent[3] to the chosen neural network. Therefore, the implementation problems are not solved by some limits put on either neural architectures or neural computations: the original FPNA computing scheme creates a bridge between complex neural networks and available hardware devices. Since this bridge is based on a general computation paradigm, it may adapt to many implementation choices (arithmetic, pipeline, etc). Finally, the obtained FPNA is directly mapped onto the hardware device thanks to a completely modular implementation: configurable predefined blocks are given for each neural resource, and they are assembled according to the hardware-friendly architecture of the FPNA.

This approach answers to the various challenges of neural implementations on FPGAs (or more generally on digital hardware) as follows:

3.3.5.1 2D topologies and fan-in. The FPNA computation scheme counterbalances a simplified neural network topology by an increased local complexity of the interactions between neural resources. This complexity may

---

[2]Section 4.3 shows that a good FPNA configuration is not searched blindly: an FPNA design and configuration is built from several basic layouts of neural resources.
[3]The function type (regression) is equivalent. Then a learning algorithm is used to adapt the FPNA weights to the application. See Chapter 4 for FPNA learning algorithms.

be tuned so as to make the simplified topology fit the hardware: a compromise must be found, since FPNA learning becomes more difficult when the complexity of the local configured connections increases (see Chapter 4.

3.3.5.2   Area-greedy operators.   FPNAs use the same operators as standard neural networks, but a simplified topology is related to a reduced number of such operators, so that area and interconnection problems are simultaneously solved.

3.3.5.3   Weight storage and time/area imbalance.   The FPNA computation paradigm creates numerous virtual connections despite a reduced number of links. Therefore, topology simplifications mostly result in less links, i.e. less weights and multipliers. Moreover each weight is simultaneously used in different ways to create the virtual connections, so that FPNA computations result in a reduced share of the overall computation time for the input-weight multiplications.

3.4   Correctness

The aim of this section is to mention the various properties that have been established about FPNN computations so as to justify their correctness.

3.4.1   Feedforward FPNNs

A FPNN is feedforward if the local configured connections between resources do not infer any cyclical dependency. Therefore an FPNN may be feedforward even if its underlying FPNA uses a cyclical graph. Two formal definitions express this. The fisrt one defines the dependency graph of an FPNN: the nodes are the neural resources (links and activators) and the edges correspond to the configured local connections.

DEFINITION 1 *Let* $\phi = \{\mathcal{N}, \mathcal{E}, (\theta_n, i_n, f_n, W_n, T_n, a_n, R_n, r_n, S_n)_{n \in \mathcal{N}}\}$ *be an FPNN. Its dependency graph* $\mathcal{G}_D(\phi) = (\mathcal{N}', \mathcal{E}')$ *is defined by:*

- $\mathcal{N}' = \mathcal{E} \cup \{(n) \mid n \in \mathcal{N}\}$

- $\mathcal{E}' = \{((p,n),(n,s)) \mid R_n(p,s) = 1\} \cup \{((p,n),(n)) \mid r_n(p) = 1\}$
  $\cup \{((n),(n,s)) \mid S_n(s) = 1\}$

Indeed, the computation scheme of an FPNN follows the oriented edges of its dependency graph. Therefore:

DEFINITION 2 *Let* ϕ *be an FPNN. It is feedforward iff* $\mathcal{G}_D(\phi)$ *does not include any cycle.*

3.4.1.1   Correctness of the asynchronous sequential computation.

The above computation algorithms (sequential and parallel) appear as quite difefrent, and they are based on local processings. The first question is whether they define a global behaviour for the whole FPNN. This correctness problem may be expressed as follows: do the asynchronous sequential and parallel computations halt ?

A first simple result answers this question as far as the asynchronous sequential computation is concerned:

THEOREM 3.1 *If ϕ is a feedforward FPNN, then its asynchronous sequential computation halts.*

The proof is straightforward, since the task generation process is locally finite, and follows the oriented edges of $\mathcal{G}_D(\phi)$. See [21] for more details.

3.4.1.2   Deadlock in the asynchronous parallel computation.   The

case of the asynchronous parallel computation is a bit more complicated. It is linked with, but not equivalent to, the notion of deadlock. Though this notion is rather standard when studying distributed communication protocols, it must be first defined for the above parallel algorithm. The following definition describes a deadlock situation as a "cycle" of waiting requests.

DEFINITION 3 *A FPNN is said to be in deadlock conditions during its asynchronous parallel computation if there exist resources $\varrho_0, \ldots, \varrho_p$ such that each ϱ_i is processing a request, and if this processing may not halt unless ϱ_{i+1} terminates the request processing it is performing (considering that $\varrho_{p+1} = \varrho_0$).*

Then a rather simple result ensures that the asynchronous parallel computation applies to feedforward FPNNs without deadlock.

THEOREM 3.2 *If ϕ is a feedforward FPNN, then it may not be in deadlock conditions during its asynchronous parallel computation.*

The proof is easy. It is based on an equivalence between deadlock conditions and the existence of a cycle in a subgraph of $\mathcal{G}_D(\phi)$ (graph inferred by simultaneously processed requests). See [21] for more details.

3.4.1.3   Conditions of parallel non-termination.   The main result is

the relation between deadlock conditions and non-termination. It ensures that deadlock is the only possible cause of non-termination.

THEOREM 3.3 *If ϕ is an FPNN whose asynchronous sequential computation halts, then its asynchronous parallel computation does not halt if and only if it is in deadlock conditions.*

<u>sufficient condition</u>: *obvious (see definition 3)*
<u>necessary condition</u> *(proof outline, see [21] for full details):*
1. The set of requests may be partially ordered.
2. The number of requests is finite (there exists an injection towards the set of sequential tasks).
3. There is a request whose processing does not halt.
4. A deadlock cycle may be built in the set of direct and indirect successors of a non-terminating request.

A straightforward consequence is that both asynchronous sequential and parallel computations of a feedforward FPNN halt. Moreover the study of [21] shows how simple conditions ensure that all defined computation methods are equivalent and deterministic (see Chapter 4).

3.4.2   Recurrent FPNNs: synchronized computations

The above computation schemes can not satisfactorily handle recurrent FPNNs. Nevertheles only slight changes are required. In a synchronized FPNN computation, synchronisation barriers are used in order to separate the strict sending of neural output values from the simple direct value transmissions between connected links. It provides a correct recurrent computation, provided that there is no loop in the direct connections between the links.

At a given moment, the requests are processed as above, except for any value sent by an activator. Such a request is put in a waiting list, without being processed. When a new synchronization barrier is raised, all waiting requests are set active, so that their processing may start.

Similar correctness results are available (Theorem 3.2 may be extended to recurrent FPNNs under some obvious conditions).

3.5   Underparameterized convolutions by FPNNs

The aim of this section is to study the computational power of FPNNs.

As shown in [21], it is *always* possible to define an FPNN that exactly computes the same function as any standard neural network. But this FPNN may be as complex as the desired neural network. Therefore, it should be known which functions may be computed by *hardware-friendly* FPNNs. This question leads to the idea of *underparameterized convolutions*.

This section is organized as follows:

- §3.5.1 is a necessary reminder of some links between neural computation and convolution,

- §3.5.2 briefly explains how FPNN computation leads to *underparameterized* convolutions,

- §3.5.3 gives two opposite examples to show that this underparameterization may be fully overcome in some cases, whereas it is a strict theoretical limit in some other cases; the special case of *grid-based layered FPNNs* is studied, since underparameterization may be put in equations in this case,

- §3.5.4 briefly explains that the influence of underparameterization on FPNN practical utility deserves an experimental study (that is provided in sections §4.3 and §4.4),

- §3.5.5 describes how the above study of the underparameterization convolution that is performed by FPNNs leads to a theoretically justified method for FPNN weight initialization.

3.5.1   Partial convolutions by neural networks

Feedforward neural models are often justified by their approximation capabilities as non-linear regression tools, since [16, 36, 35]. This regression is performed by means of discrete frequential convolutions based on neuron transfer functions. Such convolutions clearly appear in several works that study models with localized neural functions: for example [52] for RBF networks (radial basis function), or [71] for wavelet networks (which directly derive from the reversibility of convolution with wavelets). Convolution is more hidden for the most standard feedforward neural model, called multilayer perceptrons (MLP). Each neuron in a layer of a MLP applies a sigmoidal transfer function σ to a weighted sum of the outputs of all neurons in the previous layer:

- In [16], the universal approximation property is built with two main arguments:

    1 Let $\Psi(x) = \sigma\left(\frac{x}{\delta} + \alpha\right) - \sigma\left(\frac{x}{\delta} - \alpha\right)$, where $(\delta, \alpha) \in I\!\!R^2$. A function f may be approximated by the reverse transform of a discrete frequential convolution function, in which a vectorial form of Ψ convolutes Fourier transform $\tilde{f}$. This convolution function must be estimated at several convolution points, so as to build the reverse transform. And each estimation uses multiple convolution terms that are computed at each point of a fine-grain covering of a compact frequency domain.

    2 The hidden weights give a set of points that cover this frequency domain, whereas the output weights are set by means of the values of $\tilde{f}$ at each of these points.

The second argument directly derives from the multilayer architecture and its ability to map a fine n-dimensional mesh onto the frequency domain. It explains why the work lies in the proof that a reversible transform may be built on a sigmoidal function.

■ The convolution is still more hidden in [36], which is often taken as the basic reference for MLP universal approximation. The property is a consequence of the Stone-Weierstrass Theorem (sigmoid properties make it possible to go without some of its hypotheses). Yet it may be noted that this standard theorem refers to a covering by low-diameter balls. And the strong hypotheses of Stone-Weierstrass Theorem are bypassed thanks to the approximation of trigonometric polynomials by sigmoids. Therefore discrete frequential convolutions re-emerge.

To simplify, an expected function in a space domain is obtained thanks to a reverse transform of its frequential form. Neuron transfer functions create a base of localized functions in a compact frequency domain. Convolutions are performed to get the projection. Let $\vec{w}$ and $\vec{t}$ be the multiplying parameters and the additive parameters of a hidden neuron (in a MLP, $\vec{w}$ is the weight vector, whereas $\vec{t}$ is reduced to the threshold). Then $\vec{w}$ may be seen as a frequency vector involved in one discrete frequential convolution, whereas $\vec{t}$ expresses the corresponding convolution point (indirectly, since the convoluting function depends on the neural network input vector). The output weight that applies to this neuron depends on both $\vec{t}$ and the frequential value of f at $\vec{w}$.

Concrete neural applications only require a limited precision approximation, so that only few convolution terms are used (several ones instead of v^d for each convolution point, where d is the input space dimension and v is inversely proportional to the expected precision).

3.5.2   Underparameterized partial convolutions

FPNNs are also defined to compute such partial convolutions for non- linear regression, though their architecture is simplified with respect to standard neural models. The FPNN definition permits to connect any link to any *local resource*. Locality permits simplified topologies, whereas connected links result in more various frequency vectors: direct connections between affine links appear, so that the FPNN may compute numerous composite affine transforms (*virtual neural connections*). Therefore different convolution terms may be obtained with a reduced number of connection weights (underparameterization).

Though various, the weights of the virtual connections are not independent of each other. The complex induced dependencies are studied in [21]. They imply that the computational power of an FPNN with k neural resources and K virtual connections is lying between the computational power of a standard neural network with k neural operators and the computational power of a standard neural network with $k + K$ neural operators. Next subsection illustrates this intermediate situation through two opposite results: a first example shows that there are FPNNs for which the computational power takes advantage of all virtual connections, whereas Theorem 3.4 shows that there are FPNNs that are strictly less powerful than the corresponding standard neural model with $k + K$ operators.

3.5.3 Underparameterized exact computing

3.5.3.1 Towards hardware-targetted simplified topologies: example.
FPNNs make it possible to obtain complex neural network behaviours with simple 2D topologies. Such FPNNs have been studied for the standard parity problem (see Chapter 4): the d-dimensional parity problem consists in classifying vectors of $\{0, 1\}^d$ as *odd* or *even*, according to the number of non zero values among the d coordinates.

This problem may be solved by d-input multilayer perceptron (MLP) or shortcut perceptron. The search for *optimal* two-hidden layer shortcut perceptrons in [65] has solved the d-dimensional parity problem with only $\sqrt{d}(2 + o(1))$ neurons[4]. This neural network uses $d(\sqrt{d} + 1 + o(1))$ weights. For all d, an FPNN with the same number of neurons (activators), but only $\frac{15}{2}\sqrt{d} + o(\sqrt{d})$ weights *exactly* performs the same computation (see Chapter 4).

3.5.3.2 Grid-based layered FPNNs : weight dependencies.
A FPNA graph may use full-duplex links with a grid topology. Let λ_n be the index of the row of node n. A *layered* FPNN derived from such an FPNA is an FPNN where:

- $r_n(p) = 1$ *iff* $\lambda_p \leq \lambda_n$

- $R_n(p, s) = 1$ *iff* $\lambda_p \leq \lambda_n$ and $\lambda_n = \lambda_s$

- $S_n(s) = 1$ *iff* $\lambda_n < \lambda_s$

Figure 3.5 shows the local structure of such a *grid-based layered FPNN*.

The virtual connections of a layered FPNN are such that consecutive rows are *virtually fully connected* as in a multilayer perceptron. But there are strong dependencies between the weights of different virtual connections.

PROPERTY 1 *Let $n_1, \ldots, n_d$ be the nodes of a single row. Let p_i be the predecessor of n_i in the previous row ($\lambda_{p_i} = \lambda_{n_i} - 1$). Let y_{n_i} (resp. y_{p_i}) be the output computed by the neuron of node n_i (resp. p_i).*

---

[4]For some functions f and g, $f = o(g)$ means that $\frac{f(x)}{g(x)} \xrightarrow{x \to +\infty} 0$

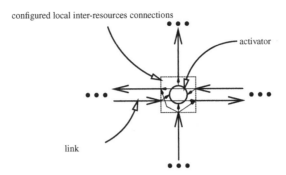

configured local inter-resources connections

activator

link

Figure 3.5. Node of a grid-based layered FPNN

Computing virtual connection weights shows that if $1 \leq i < j < k \leq d$ *then:*

$$\frac{\partial y_{n_i}}{\partial y_{p_k}} = \frac{W_{n_k}(p_k)}{W_{n_j}(p_j)} \prod_{k=j}^{k-1} W_{n_k}(n_{k+1}) \left(\frac{\partial y_{n_i}}{\partial y_{p_j}} \right)$$

Therefore, if all the FPNN weights are fixed, then for all i, $1 \leq i < d$ *and for all real values* $y_{p_1}, \ldots, y_{p_i}$, *the function* $(y_{p_{i+1}}, \ldots, y_{p_d}) \mapsto (y_{n_1}, \ldots, y_{n_i})$ *defines an arc from* $\mathbb{R}^{d-i}$ *into* $\mathbb{R}^i$.

This result is an example of the strong dependencies that may exist between the different points where an FPNN virtually estimates the terms of a partial convolution (each y_{n_i} is a convolution term computed at $(y_{p_1}, \ldots, y_{p_d})$). Section 4.3 shows in what way these dependencies influence FPNNs, and how this unfavourable result has led to the definition of a specific way to set FPNN weights.

3.5.3.3   Shortcut FPNNs.   Standard neural models may be used as FP-NAs (so that each neuron and its input and output connections become freely connectable). And any standard neural network can be exactly simulated by an FPNN based on the FPNA form of this neural network. For example, one can use the multilayer topology of a one hidden-layer MLP to build an FPNA, and then set a derived FPNN that computes the same function as the original MLP. To obtain this, there must be no direct configured connection between two different links, that is $R_n(p, s) = 0$. If some $R_n(p, s)$ values are set to 1, virtual shortcut connections are obtained between the input and output neural layers. Such FPNNs may then be called *one-hidden layer shortcut FPNNs*. Examples can be given in which these FPNNs can solve specific tasks that no one-hidden layer MLP can solve with the same number of neurons. No conclusive result has been obtained when comparing a one-hidden layer shortcut

FPNN with a one-hidden layer MLP having more hidden neurons. The only general result addresses the issue of the respective computational powers of such FPNNs and of real *one- hidden layer shortcut perceptrons* (MLP plus shortcut connections).

Exact learning of a set of N_p patterns in general position in $\mathbb{R}^{d_i}$ is studied in [12] for standard one-hidden layer MLPs. An intermediate theorem of this study can be first extended to one-hidden layer shortcut perceptrons:

PROPERTY 2 *Let $\mathcal{F}$ be a vectorial function which maps neural network weights onto the values computed by the d_o output neurons when all input patterns are presented.*

Let $N_h = \dfrac{d_o(N_p - d_i)}{d_i + d_o}$. If there are N_h hidden neurons in a one-hidden layer shortcut perceptron, then there is a vector of neural weights at which $\mathcal{F}$ is a local diffeomorphism.

This property can not be reached by any one-hidden layer shortcut FPNN with the same hidden layer size:

THEOREM 3.4 *Let the above function $\mathcal{F}$ be computed by a one-hidden layer shortcut FPNN, with N_h hidden nodes. Sard's Theorem proves that $\mathcal{F}$ can not be locally surjective, since its input space dimension (number of weights) is lower than its output space dimension $N_p d_o$.*

In an FPNN, the weights of the virtual shortcut connections depend on the weights between its consecutive layers. This *underparameterization* phenomenon results in a computation power weaker than the one of the simulated shortcut perceptron. This unfavourable result is softened by the weak influence of exact learning capabilities in concrete neural applications.

3.5.4   Underparameterized approximate computing

Usual neural network applications only use approximation capabilities. Therefore, the most useful question is whether hardware-friendly FPNNs are able to approximately compute the same function as standard neural networks that are too large for direct hardware implementation. This problem is studied in [21] through the determination of hardware-friendly FPNNs such that their virtual connections are the connections of standard neural networks used in several applications. The parameters of these FPNNs are then learned so as to achieve the application task, and the results are compared to the ones of the initial standard neural networks.

Standard neural benchmarks have been tested with different FPNNs. These experiments attest that FPNNs may learn classifications as well as standard models (gradient descent learning algorithm), although they use an almost 2D underlying topology and 4 to 10 times less resources than the equivalent standard neural networks.

Chapter 4 will describe such experiments (for example based on the *diabetes* classification problem). It will also show how the hardware-friendly architecture of the obtained FPNNs leads to efficient FPGA-based implementations that the equivalent standard neural models do not permit.

3.5.5   Weight initialization

As mentioned in §3.5.1, too many weights and neurons would be necessary for a neural network to perform full discrete convolutions. Choosing a good scattering of weight vectors seems quite natural, when approximation must be performed with a limited number of neurons. But such a good scattering is a basic property of the independent weights of standard neural models. Therefore evolved weight initialization methods focus on the adequacy of neural network parameters and input patterns (usual solutions listed in [5]). Additive parameters are essential in such methods (biases, translation vectors, ... etc), particularly when localized transfer functions are used (these parameters are scattered in a convex hull defined by expected inputs). FPNA constraints do not concern additive parameters, but only multiplying ones.

FPNN virtual weights can not be *individually* set. Weight dependencies described in §3.5.3 might lead to bad classification applications. To obtain low precision approximations, FPNN should first be able to approximate the low-band components of expected functions[5]. Convolution terms reduce to a satisfactory covering of a reduced frequency domain. This approach does not depend on the input patterns. It simultaneously considers all weight vectors, so that it better adapts to FPNNs.

This subsection shows how weight vector scattering can be controlled, and what influence it has on concrete applications. The first question is to know whether FPNNs are able to obtain well scattered virtual convolution terms.

3.5.5.1   Convolution point scattering for FPNNs.   In order to estimate the scattering of the convolution terms that correspond to the virtual connections of an FPNN, the frequency domain is partitioned into as many distinct zones as there are convolution terms, with a Voronoi partition based on the euclidean distance. Scattering may be measured by the maximum zone diameter (other criteria have been studied with similar conclusions).

The special case of a grid-based layered FPNN (see §3.5.3) is studied with this criterion. It corresponds to d convolution points in a d-dimensional space.

---

[5]Low-frequency components in a frequential decomposition give the main frame of a signal.

The ratio of the number of virtual connections to the number of links used by such FPNNs is $\frac{d}{3}$[6].

An optimal scattering may be estimated as the expected mean value of the chosen criterion when using a random drawing of d points uniformly distributed in the frequency domain: it corresponds to the independent weights of a standard neural network. A similar random drawing of FPNN weights leads to a bad scattering. As mentioned in §3.5.3, virtual weight dependencies in FPNNs are mostly linked to multiplications of $W_n(p)$ parameters. A random variable w is computed so that its distribution function is as stable as possible when multiplication is applied. Let v_1 and v_2 be two uniformly distributed random variables. Then w is defined as

$$h\left(\frac{1}{2} + \frac{1}{5}\sqrt{-2\ln(v_1)}\cos(2\pi v_2)\right)$$

where $h(x) = \frac{x+1}{2-x}$. The probability density of w is

$$\begin{aligned}
\rho(x)dx &= \int_x^{x+dx}\left[\mathcal{N}(\frac{1}{2},\frac{1}{5})\right](h^{-1}(u))[h^{-1}]'(u)du \\
&\simeq \frac{15}{(1+x)^2\sqrt{2\pi}}e^{-\frac{1}{2}\left(\frac{15(x-1)}{2(x+1)}\right)^2}dx
\end{aligned}$$

so that

- The fulfilled relation $x\rho(x) = \frac{1}{x}\rho(\frac{1}{x})$ implies that the expected mean value of w keeps unchanged by random variable multiplication.

- The use of the normal random variable ensures that the standard deviation of a product of w variables keeps reduced.

The FPNN weights are then set with a random drawing of w, to which an affine transform is applied to fit the desired range. Figure 3.6 shows the ratio of the criterion mean value in the optimal case to the mean value that is observed for FPNN virtual convolutions.

3.5.5.2 Scattering and learning.

The above w-based method has been used for random initialization of FPNN weights. It mostly improve convergence times, as illustrated in §4.3.1.

On the other hand, one can wonder whether the learning process increases virtual weight dependencies or not: the gradient computation in an FPNN depends on a much larger equivalent neural structure that widely uses weight

---

[6]d^2 connections between two fully connected layers of d neurons, about $3d$ links in a grid-based layered FPNA architecture.

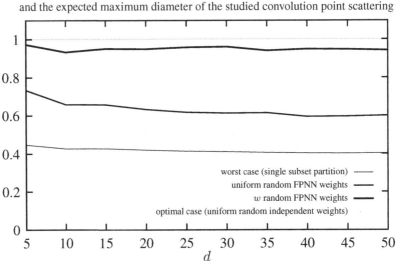

Ratio between the optimal expected maximum diameter
and the expected maximum diameter of the studied convolution point scattering

Figure 3.6.   Convolution point scattering of grid-based layered FPNNs

sharing, so that the dependencies between the partial differentials appear even
stronger (dependencies are back-propagated through the layers). And yet
§4.3.1 shows how scattering improves during learning. This improvement is
even greater with a w-based initialization method.

3.6   Conclusions

Connectionist models, such as neural networks, are the first models of parallel computing. This approach of neural computation has been penalized by its
unfitness for the characteristics of usual parallel devices. Neural network hardware implementations have to reconcile simple hardware topologies with often
complex neural architectures. Field Programmable Neural Arrays have been
defined for that. Their computation scheme creates numerous virtual neural
connections by means of a limited set of links, whatever the device, the arithmetic, and the neural structure. The FPNA computation paradigm thus makes
it possible to define neural models which performance is similar to standard
neural models despite simplified architectures.

The FPNA framework is a neural computation paradigm that has been defined in order to fit direct digital hardware implementations. It has given rise
to a complete theoretical model for topologically constrained neural computation. In this Chapter, FPNAs and FPNNs have been formally defined. Their
computation has been described and justified. Their main limitation has been
identified as an underparameterization of the frequency vectors that are used

in approximate convolutions. Possible unfavourable theoretical consequences have been pointed out. They indicate that FPNA weights deserve a specific approach. Applications take advantage of adequate FPNA weight distributions, especially in terms of convergence time. Such applications, as well as some of their implementations, are presented in Chapter 4.

References

[1] D. Abramson, K. Smith, P. Logothetis, and D. Duke. FPGA based implementation of a hopfield neural network for solving constraint satisfaction problems. In *Proc. EuroMicro*, 1998.

[2] D. Anguita, S. Bencetti, A. De Gloria, G. Parodi, D. Ricci, and S. Ridella. FPGA implementation of high precision feedforward networks. In *Proc. MicroNeuro*, pages 240–243, 1997.

[3] N. Avellana, A. Strey, R. Holgado, A. Fernandez, R. Capillas, and E. Valderrama. Design of a low-cost and high-speed neurocomputer system. In *Proc. MicroNeuro*, pages 221–226, 1996.

[4] S.L. Bade and B.L. Hutchings. FPGA-based stochastic neural networks - implementation. In *Proceedings of the IEEE Workshop on FPGAs for Custom Computing Machines*, pages 189–198, 1994.

[5] R. Baron and B. Girau. Parameterized normalization : application to wavelet networks. In *Proc. IJCNN*, volume 2, pages 1433–1437. IEEE, 1998.

[6] J.-L. Beuchat. Conception d'un neuroprocesseur reconfigurable proposant des algorithmes d'apprentissage et d'elagage: une premiere etude. In *Proc. NSI Neurosciences et Sciences de l'Ingenieur*, 1998.

[7] Y. Boniface. A parallel simulator to build distributed neural algorithms. In *International Joint Conference on Neural Networks - IJCNN'01, Washington, USA*, 2001.

[8] N.M. Botros and M. Abdul-Aziz. Hardware implementation of an artificial neural network. In *Proc. ICNN*, volume 3, pages 1252–1257, 1993.

[9] Y.K. Choi, K.H. Ahn, and S.-Y. Lee. Effects of multiplier output offsets on on-chip learning for analog neuro-chips. *Neural Processing Letters*, 4:1–8, 1996.

[10] V.F. Cimpu. Hardware FPGA implementation of a neural network. In *Proc. Int. Conf. Technical Informatics*, volume 2, pages 57–68, 1996.

[11] J.G. Eldredge and B.L. Hutchings. RRANN: a hardware implementation of the backpropagation algorithm using reconfigurable FPGAs. In *Proceedings of the IEEE World Conference on Computational Intelligence*, 1994.

[12] A. Elisseeff and H. Paugam-Moisy. Size of multilayer networks for exact learning: analytic approach. Technical Report 96-16, LIP-ENSL, 1996.

[13] W. Eppler, T. Fisher, H. Gemmeke, T. Becher, and G. Kock. High speed neural network chip on PCI-board. In *Proc. MicroNeuro*, pages 9–17, 1997.

[14] S.K. Foo, P. Saratchandran, and N. Sundararajan. Parallel implementation of backpropagation neural networks on a heterogeneous array of transputers. *IEEE Trans. on Systems, Man, and Cybernetics–Part B: Cybernetics*, 27(1):118–126, 1997.

[15] D. Franco and L. Carro. FPGA architecture comparison for non-conventional signal processing. In *Proc. IJCNN*, 2000.

[16] K.-I. Funahashi. On the approximate realization of continuous mappings by neural networks. *Neural Networks*, 2:183–192, 1989.

[17] R. Gadea, J. Cerda, F. Ballester, and A. Mocholi. Artificial neural network implementation on a single FPGA of a pipelined on-line backpropagation. In *Proc. ISSS*, pages 225–230, 2000.

[18] C. Gegout, B. Girau, and F. Rossi. A general feedforward neural network model. Technical report NC-TR-95-041, NeuroCOLT, Royal Holloway, University of London, 1995.

[19] C. Gegout, B. Girau, and F. Rossi. Generic back-propagation in arbitrary feedforward neural networks. In *Artificial Neural Nets and Genetic Algorithms – Proc. of ICANNGA*, pages 168–171. Springer-Verlag, 1995.

[20] B. Girau. Dependencies of composite connections in Field Programmable Neural Arrays. Research report NC-TR-99-047, NeuroCOLT, Royal Holloway, University of London, 1999.

[21] B. Girau. *Du parallelisme des modeles connexionnistes a leur implantation parallele*. PhD thesis n° 99ENSL0116, ENS Lyon, 1999.

[22] B. Girau. Building a 2D-compatible multilayer neural network. In *Proc. IJCNN*. IEEE, 2000.

[23] B. Girau. Conciliating connectionism and parallel digital hardware. *Parallel and Distributed Computing Practices,* special issue on *Unconventional parallel architectures*, 3(2):291–307, 2000.

[24] B. Girau. Digital hardware implementation of 2D compatible neural networks. In *Proc. IJCNN*. IEEE, 2000.

[25] B. Girau. FPNA: interaction between FPGA and neural computation. *Int. Journal on Neural Systems*, 10(3):243–259, 2000.

[26] B. Girau. Neural networks on FPGAs: a survey. In *Proc. Neural Computation*, 2000.

[27] B. Girau. Simplified neural architectures for symmetric boolean functions. In *Proc. ESANN European Symposium on Artificial Neural Networks*, pages 383–388, 2000.

[28] B. Girau. On-chip learning of FPGA-inspired neural nets. In *Proc. IJCNN*. IEEE, 2001.

[29] B. Girau and A. Tisserand. MLP computing and learning on FPGA using on-line arithmetic. *Int. Journal on System Research and Information Science,* special issue on *Parallel and Distributed Systems for Neural Computing*, 9(2-4), 2000.

[30] C. Grassmann and J.K. Anlauf. Fast digital simulation of spiking neural networks and neuromorphic integration with spikelab. *International Journal of Neural Systems*, 9(5):473–478, 1999.

[31] M. Gschwind, V. Salapura, and O. Maisch berger. A generic building block for Hopfield neural networks with on-chip learning. In *Proc. ISCAS*, 1996.

[32] H. Hikawa. Frequency-based multilayer neural network with on-chip learning and enhanced neuron characteristics. *IEEE Trans. on Neural Networks*, 10(3):545–553, 1999.

[33] R. Hoffmann, H.F. Restrepo, A. Perez-Uribe, C. Teuscher, and E. Sanchez. Implémentation d'un réseau de neurones sur un réseau de fpga. In *Proc. Sympa'6*, 2000.

[34] P.W. Hollis, J.S. Harper, and J.J. Paulos. The effects of precision constraints in a backpropagation learning algorithm. *Neural Computation*, 2:363–373, 1990.

[35] K. Hornik. Approximation capabilities of multilayer feedforward networks. *Neural Networks*, 4:251–257, 1991.

[36] K. Hornik, M. Stinchcombe, and H. White. Multilayer feedforward networks are universal approximators. *Neural Networks*, 2:359–366, 1989.

[37] N. Izeboudjen, A. Farah, S. Titri, and H. Boumeridja. Digital implementation of artificial neural networks: From VHDL description to FPGA implementation. In *Proc. IWANN*, 1999.

[38] A. Johannet, L. Personnaz, G. Dreyfus, J.D. Gascuel, and M. Weinfeld. Specification and implementation of a digital Hopfield-type associative memory with on-chip training. *IEEE Trans. on Neural Networks*, 3, 1992.

[39] J. Kennedy and J. Austin. A parallel architecture for binary neural networks. In *Proc. MicroNeuro*, pages 225–231, 1997.

[40] K. Kollmann, K. Riemschneider, and H.C. Zeidler. On-chip backpropagation training using parallel stochastic bit streams. In *Proc. MicroNeuro*, pages 149–156, 1996.

[41] A. Kramer. Array-based analog computation: principles, advantages and limitations. In *Proc. MicroNeuro*, pages 68–79, 1996.

[42] V. Kumar, S. Shekhar, and M.B. Amin. A scalable parallel formulation of the back-propagation algorithm for hypercubes and related architectures. *IEEE Transactions on Parallel and Distributed Systems*, 5(10):1073–1090, October 1994.

[43] P. Lysaght, J. Stockwood, J. Law, and D. Girma. Artificial neural network implementation on a fine-grained FPGA. In *Proc. FPL*, pages 421–432, 1994.

[44] Y. Maeda and T. Tada. FPGA implementation of a pulse density neural network using simultaneous perturbation. In *Proc. IJCNN*, 2000.

[45] S. McLoone and G.W. Irwin. Fast parallel off-line training of multilayer perceptrons. *IEEE Trans. on Neural Networks*, 8(3):646–653, 1997.

[46] I. Milosavlevich, B. Flower, and M. Jabri. PANNE: a parallel computing engine for connectionist simulation. In *Proc. MicroNeuro*, pages 363–368, 1996.

[47] P.D. Moerland and E. Fiesler. Hardware-friendly learning algorithms for neural networks: an overview. In *Proc. MicroNeuro*, 1996.

[48] A. Montalvo, R. Gyurcsik, and J. Paulos. Towards a general-purpose analog VLSI neural network with on-chip learning. *IEEE Trans. on Neural Networks*, 8(2):413–423, 1997.

[49] U.A. Müller, A. Gunzinger, and W. Guggenbühl. Fast neural net simulation with a DSP processor array. *IEEE Trans. on Neural Networks*, 6(1):203–213, 1995.

[50] T. Nordstrøm and B. Svensson. Using and designing massively parallel computers for artificial neural networks. *Journal of Parallel and Distributed Computing*, 14(3):260–285, 1992.

[51] R. Østermark. A flexible multicomputer algorithm for artificial neural networks. *Neural Networks*, 9(1):169–178, 1996.

[52] J. Park and I.W. Sandberg. Universal approximation using radial-basis-function networks. *Neural Computation*, 3:246–257, 1991.

[53] H. Paugam-Moisy. Optimal speedup conditions for a parallel back-propagation algorithm. In *CONPAR*, pages 719–724, 1992.

[54] A. Perez-Uribe and E. Sanchez. FPGA implementation of an adaptable-size neural network. In *Proc. ICANN*. Springer-Verlag, 1996.

[55] A. Petrowski. Choosing among several parallel implementations of the backpropagation algorithm. In *Proc. ICNN*, pages 1981–1986, 1994.

[56] M. Rossmann, A. Buhlmeier, G. Manteuffel, and K. Goser. short- and long-term dynamics in a stochastic pulse stream neuron implemented in FPGA. In *Proc. ICANN*, LNCS, 1997.

[57] M. Rossmann, T. Jost, A. Goser, K. Bɥhlmeier, and G. Manteuffel. Exponential hebbian on-line lerarning implemented in FPGAs. In *Proc. ICANN*, 1996.

[58] S. Sakaue, T. Kohda, H. Yamamoto, S. Maruno, and Shimeki Y. Reduction of required precision bits for back-propagation applied to pattern recognition. *IEEE Trans. on Neural Networks*, 4(2):270–275, 1993.

[59] V. Salapura. Neural networks using bit-stream arithmetic: a space efficient implementation. In *Proc. IEEE Int. Conf. on Circuits and Systems*, 1994.

[60] V. Salapura, M. Gschwind, and O. Maisch berger. A fast FPGA implementation of a general purpose neuron. In *Proc. FPL*, 1994.

[61] K.M. Sammut and S.R. Jones. Arithmetic unit design for neural accelerators: cost performance issues. *IEEE Trans. on Computers*, 44(10), 1995.

[62] M. Schaefer, T. Schoenauer, C. Wolff, G. Hartmann, H. Klar, and U. Ruckert. Simulation of spiking neural networks - architectures and implementations. *Neurocomputing*, (48):647–679, 2002.

[63] S. Shams and J.-L. Gaudiot. Parallel implementations of neural networks. *Int. J. on Artificial Intelligence*, 2(4):557–581, 1993.

[64] S. Shams and J.-L. Gaudiot. Implementing regularly structured neural networks on the DREAM machine. *IEEE Trans. on Neural Networks*, 6(2):407–421, 1995.

[65] K. Siu, V. Roychowdhury, and T. Kailath. Depth-size tradeoffs for neural computation. *IEEE Trans. on Computers*, 40(12):1402–1412, 1991.

[66] T. Szabo, L. Antoni, G. Horvath, and B. Feher. A full-parallel digital implementation for pre-trained NNs. In *Proc. IJCNN*, 2000.

[67] M. van Daalen, P. Jeavons, and J. Shawe-Taylor. A stochastic neural architecture that exploits dynamically reconfigurable FPGAs. In *Proc. of IEEE Workshop on FPGAs for Custom Computing Machines*, pages 202–211, 1993.

[68] M. Viredaz, C. Lehmann, F. Blayo, and P. Ienne. MANTRA: a multi-model neural network computer. In *VLSI for Neural Networks and Artificial Intelligence*, pages 93–102. Plenum Press, 1994.

[69] J. Wawrzynek, K. Asanovi℮, and N. Morgan. The design of a neuro-microprocessor. *IEEE Trans. on Neural Networks*, 4(3):394–399, 1993.

[70] Xilinx, editor. *The Programmable Logic Data Book*. Xilinx, 2002.

[71] Q. Zhang and A. Benveniste. Wavelet networks. *IEEE Trans. on Neural Networks*, 3(6):889–898, Nov. 1992.

[72] X. Zhang, M. McKenna, J.J. Mesirov, and D.L. Waltz. The backpropagation algorithm on grid and hypercube architectures. *Parallel Computing*, 14:317–327, 1990.

Chapter 4

FPNA: APPLICATIONS AND IMPLEMENTATIONS

Bernard Girau

LORIA INRIA-Lorraine
Nancy France
girau@loria.fr

Abstract   Neural networks are usually considered as naturally parallel computing models. But the number of operators and the complex connection graph of standard neural models can not be handled by digital hardware devices. The *Field Programmable Neural Arrays* framework introduced in Chapter 3 reconciles simple hardware topologies with complex neural architectures, thanks to some configurable hardware principles applied to neural computation. This two-chapter study gathers the different results that have been published about the FPNA concept, as well as some unpublished ones. This second part shows how FPNAs lead to powerful neural architectures that are easy to map onto digital hardware: applications and implementations are described, focusing on a class of *synchronous* FPNA-derived neural networks, for which on-chip learning is also available.

Keywords: neural networks, fine-grain parallelism, digital hardware, FPGA

Introduction

The very fine-grain parallelism of neural networks uses many information exchanges. Therefore it better fits hardware implementations. Configurable hardware devices such as FPGAs (Field Programmable Gate Arrays) offer a compromise between the hardware efficiency of digital ASICs and the flexibility of a simple software-like handling. Yet the FPGA implementation of standard neural models raises specific problems, mostly related to area-greedy operators and complex topologies. *Field Programmable Neural Arrays* have been defined in [13] such that they lead to complex neural processings based on simplified interconnection graphs. They are based on an FPGA-like approach ([15]): a set of resources whose interactions are freely configurable.

A. R. Omondi and J. C. Rajapakse (eds.), FPGA Implementations of Neural Networks, 103–136.
© 2006 *Springer. Printed in the Netherlands.*

Several different neural networks may be configured from a given FPNA. They are called *Field Programmed Neural Networks* (FPNN). FPNA resources are defined to perform computations of standard neurons, but they behave in an *autonomous* way. As a consequence, numerous *virtual* synaptic connections may be created thanks to the application of a multicast data exchange protocol to the resources of a sparse neural network. This new neural computation concept makes it possible to replace a standard but topologically complex neural architecture by a simplified one.

After a survey of parallel implementations of neural networks (particularly on FPGAs), Chapter 3 has focused on the definitions, computations and theoretical properties of FPNNs. Reading §3.3.1, §3.3.2 and §3.3.4 is required to handle FPNA/FPNN notations and to understand FPNN computation. Nevertheless, section 4.1 shortly recalls the main aspects of Chapter 3. This second chapter now describes how FPNNs may be used and implemented. Then it focuses on a class of FPNNs that result in simple implementations with on-chip learning, as well as efficient pipelined implementations. Section 4.2 shows how FPNNs may exactly perform the same computation as much more complex standard neural networks applied to boolean function computation, whereas sections 4.3 and 4.4 illustrate how they may approximately perform the same computation as much more complex standard neural networks applied to more concrete applications (benchmark problems as well as real applications). Their general implementation method is described in section 4.5. Then section 4.6 defines and studies the case of simplifiable FPNNs: both computation and learning of these *synchronous* FPNNs are described with a few equations. Their specific implementations (with on-chip learning or pipeline) are outlined in section 4.7. All implementation performances are finally discussed in section 4.8.

4.1   Summary of Chapter 3

Section 3.2 explains why FPGAs appear as a both flexible and rather efficient device for neural applications. Implementation issues should not be neglected though. Defining hardware-adapted frameworks of neural computation directly provides neural networks that are easy to map onto such devices without a significant loss of computational power. The FPNA concept is such a hardware-adapted neural computation paradigm. It mainly leads to topological simplifications of standard neural architectures.

A neural network may be seen as a graph of interconnected neural resources (neurons and weighting links). In a standard model, the number of inputs of each neuron is its fan-in in the connection graph. On the contrary, neural resources (links and activators) become *autonomous* in an FPNA: their dependencies are freely set.

Activators apply standard neural functions to a set of input values on one hand, and links behave as independent affine operators on the other hand. The FPNA concept permits to *connect* any link to any *local resource* (each node of the FPNA graph corresponds to an activator with incoming and outgoing links). Direct connections between affine links appear, so that the FPNA computes numerous composite affine transforms. These compositions create numerous *virtual neural connections*.

The activator of a node n performs any neuron computation by means of a loop that updates a variable with respect to the neuron inputs (i_n denotes the updating function), and a final computation (function f_n) that maps this variable to the neuron output. A link between two nodes p and n performs an affine transform $W_n(p)x + T_n(p)$.

An FPNA is only an unconfigured set of such resources. An FPNN is an FPNA whose resources have been connected in a specific way. The main configuration parameters are binary values: $r_n(p) = 1$ to connect link (p, n) and activator (n), $S_n(s) = 1$ to connect activator (n) and link (n, s), and $R_n(p, s) = 1$ to connect links (p, n) and (n, s): such direct connections between links create virtual composite links.

During an FPNN computation, all resources (activators as well as links) behave independently. When a resource receives a value, it applies its local operator(s), and then it sends the result to all neighbouring resources to which it is locally connected.

Section 3.4 has shown that FPNN computation is coherent and well-defined, whereas section 3.5 has estimated the computational power of FPNNs. It is linked to complex dependency relations between the weights of the virtual connections that are created by direct local connections between links. A dedicated weight initialization method has been derived so as to minimize the influence of these complex dependencies.

4.2   Towards simplified architectures: symmetric boolean functions by FPNAs

This section shows how FPNAs may be used so as to replace standard neural networks with simpler 2D-compatible architectures that *exactly* perform the same computations. Nevertheless, this way of using FPNAs is limited to few well-identified applications. The case of symmetric boolean functions is discussed here.

This section is organized as follows:

- §4.2.1 describes the initial work of [24] to find an optimal multilayer neural network with shortcut links to compute symmetric boolean functions,

- §4.2.2 shows how the FPNA concept makes it possible to get rid of all shortcut links in the above multilayer network (other links and activators being the same as in the original optimal multilayer neural network,

- §4.2.3 finally shows how an FPNN that uses an underlying 2D structure may be defined to exactly perform the same computation as the original optimal multilayer neural network; this FPNN is both easy to map onto an FPGA and optimal in terms of number of links (the number of activators being unchanged).

4.2.1 An optimal standard neural network

The neural computation of symmetric boolean functions has been a widely discussed problem. The quasi-optimal results of [24] answer a question that was posed as early as in [19]: what is the minimum neural architecture to compute symmetric boolean functions.

A boolean function $f : \{0,1\}^d \to \{0,1\}$ is symmetric if $f(x_1,\dots,x_d) = f(x_{\sigma(1)},\dots,x_{\sigma(d)})$ for any permutation σ of $\{1,\dots,d\}$. An example is the *d-dimensional parity problem*: it consists in classifying vectors of $\{0,1\}^d$ as *odd* or *even*, according to the number of non zero values among the d coordinates. This problem may be solved by d-input multilayer perceptrons (MLP) or shortcut perceptrons. A MLP consists of several ordered layers of sigmoidal neurons. Two *consecutive* layers are fully connected. A layer which is not the input layer nor the output layer is said to be *hidden*. A shortcut perceptron also consists of several ordered layers of sigmoidal neurons. But a neuron in a layer may receive the output of any neuron in *any* previous layer.

The search for optimal two-hidden layer shortcut perceptrons in [24] has solved the d-dimensional parity problem with only $\sqrt{d}(2 + o(1))$ neurons[1], thanks to an iterated use of a method introduced in [20]. The shortcut links and the second hidden layer are essential in this work, though there is no shortcut link towards the output neuron. This neural network uses $d(2\sqrt{d} + 1 + o(1))$ weights. The results of [24] apply to any symmetric boolean function. Figure 4.1(a) shows the topology of the optimal shortcut network of [24] for the 15-dimensional parity problem. In such a neural network, the first hidden layer may contain $\left\lceil \sqrt{d} \right\rceil$ neurons[2] such that the i-th neuron of this layer computes $y_{i,1} = \sigma\left(\sum_{j=1}^{d} x_j + \Theta_i \right)$, where $\sigma(x) = \begin{cases} 0 & \text{if } x < 0 \\ 1 & \text{if } x \geq 0 \end{cases}$. The sec-

---

[1]For a function f, $f(d) = o(1)$ means that $f(d) \xrightarrow{d \to +\infty} 0$
[2]$\lfloor x \rfloor$ denotes floor, and $\lceil x \rceil$ denotes ceiling

ond hidden layer contains at most $\left\lceil \frac{d+1}{\lceil\sqrt{d}\rceil} \right\rceil$: its i-th neuron computes $y_{i,2} =$

$$\sigma\left(\sum_{j=1}^{\lfloor\sqrt{d}\rfloor} w_{i,j,2} y_{j,1} + (-1)^i \sum_{j=1}^{d} x_j \right). \quad \text{Then } y = \sigma\left(\sum_{j=1}^{\sqrt{d}} w_{i,j,3} y_{j,2} + \Theta \right) \text{ is}$$

computed by the only output neuron.

4.2.2   FPNAs to remove shortcut links

The construction in [24] implies that for any (i, j), $(-1)^i$ and $w_{i,j,2}$ have opposite signs. This property may be used so that all shortcut links (between the input and the second hidden layer) are virtually replaced by some direct connections between incoming and outgoing links in the first hidden layer of an FPNA. This FPNA has got $d\sqrt{d}(1+o(1))$ weights, instead of $d\sqrt{d}(2+o(1))$ weights in [24].

More precisely, the architecture of the FPNA is the same as the shortcut perceptron of [24], *without all shortcut links*. The weights of the links between both hidden layers are as in [24]. When configured as an FPNN, each neuron is fully connected to all incoming and outgoing links ($\forall (p, n)\ r_n(p) = 1$ and $\forall (n, s)\ S_n(s) = 1$). If n is the i-th node of the first hidden layer, and if s is the i-th node of the second hidden layer, then for any $p \in Pred(n)$, there is a direct connection between (p, n) and (n, s) (i.e. $R_n(p, s) = 1$). If n is the i-th node of the first hidden layer, then for any $p \in Pred(n)$, $W_n(p) = -\frac{1}{w_{i,i,2}}$ and $T_n(p) = 0$. If n is the i-th node of the first hidden layer, then $\theta_n = -\frac{\Theta_i}{w_{i,i,2}}$. Figure 4.1(b) sketches the architecture of such an FPNN for a 15-dimensional symmetric boolean function.

4.2.3   Towards a simplified 2D architecture

Even without shortcut links, the underlying FPNA of Figure 4.1(b) still does not have a hardware-friendly architecture. A more drastic simplification of the architecture is expected. The full connection between consecutive layers may be virtually replaced by the use of sparse inter-layer and intra-layer FPNA links.

The construction of an FPNN in §4.2.2 does not depend on the symmetric boolean function. On the contrary, the determination of a hardware-friendly FPNN for symmetric boolean functions takes advantage of several function-dependent weight similarities in [24]. Such successful determinations have been performed for various symmetric boolean functions and input dimensions: it appears that for any d and for any symmetric boolean function f, an FPNN with the same number of neurons as in [24], but with only $\mathcal{O}(\sqrt{d})$

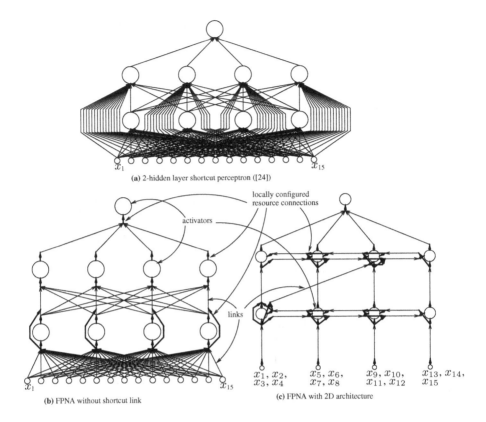

(a) 2-hidden layer shortcut perceptron ([24])

(b) FPNA without shortcut link

(c) FPNA with 2D architecture

Figure 4.1.   Neural architectures for the parity problem ($d = 15$)

weights[3] computes f *exactly* as in [24]. The parity problem may be taken as an example: in [24], the weight of the link between the i-th neuron of the first hidden layer and the j-th neuron of the second hidden layer only depends on $(-1)^j$ when $i \neq 1$. This property may be used so as to build a hardware-friendly FPNN as follows.

- Activators are globally unchanged. Therefore, the number of nodes in each hidden layer is the same as the number of neurons in [24].

- The inputs of the boolean function are gathered in groups so that the connectivity from input nodes is simplified. Therefore the number of input nodes is the number of nodes in the first hidden layer. Each input node sends up to $\left\lceil \sqrt{d} \right\rceil$ inputs.

---

[3]i.e. at most proportional to $\sqrt{d}$, indeed $\frac{15}{2}\sqrt{d} + o(\sqrt{d})$ here

- Links between consecutive layers are simplified so that a grid- structure appears. Therefore the main inter-layer links are: for any i, one link from the i-th input node to the i-th node of the first hidden layer, and one link from the i-th node of the first hidden layer to the i-th node of the second hidden layer, and one link from the i-th node of the second hidden layer to the output node.

- Moreover, a few inter-layer links are added to ensure that the special handling of the first node of the first hidden layer in [24] is reproduced. Therefore, for any $j > 1$, there is a link between the first node of the first hidden layer and the $(2j - 1)$-th node of the second hidden layer.

- Links inside each layer are chosen in a standard way that corresponds to the grid-based layered FPNNs of §3.5.3. Therefore the intra-layer links are: in both hidden layers, for any i, one link from the i-th node to the $(i + 1)$-th node, another one to the $(i - 1)$-th node.

- The $S_n(s)$, $r_n(p)$ and $R_n(p, s)$ parameters are set so as to ensure a virtual full connection scheme between consecutive layers. Moreover the $R_n(p, s)$ parameters are set so that any virtual shortcut link involves the first node of the first hidden layer. See [13] for more details and for the weight determination.

This FPNN (for any number d of inputs) is easy to map onto a 2D hardware topology, whereas the equivalent shortcut perceptron in [24] rapidly becomes too complex to be directly implemented when d increases. Figure 4.1(c) shows the architecture of the FPNN for the 15-dimensional parity problem.

Moreover, the above FPNNs satisfy several conditions that ensure a computation time proportional to the number of weights as in standard multilayer models (synchronous FPNNs, see following sections). Therefore, a $\mathcal{O}(\sqrt{d})$ computation time is achieved thanks to the topological simplifications of the underlying FPNAs. Since the arguments of [24] still hold for an FPNN, it shows that the FPNNs defined above are optimal in terms of number of neurons, and therefore also optimal in terms of number of weights as well as in terms of computation time (since it is proportional to the number of neurons).

4.3   Benchmark applications

In the previous section, FPNNs are applied to a problem of little practical interest. This section and the next one describe their application to more concrete problems.

This section focuses on benchmark problems. It shows a general way of using FPNAs to replace standard neural networks with simpler 2D-compatible architectures that *approximately* perform the same computations. More specific applications are discussed in the next section.

FPNNs have been applied to various benchmark neural applications. In this section, their application to the Proben 1 benchmarking project of [22], and to Breiman's waves of [6] are described in §4.3.1 and §4.3.2 respectively. The benchmarking rules of [22] have been respected[4].

4.3.1   FPNNs applied to Proben 1 benchmark problems

4.3.1.1   Proben 1: description.   The Proben 1 project includes many different problems to be solved by neural networks in order to compare different neural models or learning algorithms. FPNNs have been applied to all real vector classification problems of the Proben 1 project.. For each of these problems, an optimal shortcut perceptron has been determined in [22]. The computational power of such a neural network depends on the number of layers and on the number of neurons in each layer.

FPNNs have been applied in the following way. A FPNN is defined in order to compute the same kind of global function as the standard architecture of the optimal shortcut perceptron found in [22]. This FPNN derives from an FPNA with as many activators as in the optimal shortcut perceptron. The FPNA topology must be hardware-friendly. The FPNN parameters are initialized and learned so as to solve the classification problem. The aim is to obtain a classification rate equivalent to the one of the optimal shortcut perceptron, but with a far simpler architecture permitted by the FPNA concept.

The results of the different tests are given in [13]. Several initialization methods (see [12]) and gradient-based learning algorithms have been tested: their choice has an influence on the learning time (number of learning iterations), but the final performance is not significantly modified. The classification performances of the defined FPNNs are similar to the optimal classification rates obtained in [22]. The corresponding FPNAs are easy to map onto configurable hardware with the help of the implementation architectures described in section 4.5. They use 5 to 7 times less neural resources than the equivalent optimal shortcut perceptrons.

The *diabetes2* problem of the Proben 1 project may be taken as an example. Vectors of $I\!R^8$ must be classified into two classes (diabetes diagnosis with respect to quantified medical parameters). The optimal shortcut perceptron found in [22] uses two hidden layers of 16 and 8 sigmoidal neurons. All possible shortcut connections are used. Its average performance is 74.08 % correctly

---

[4]Evaluation of neural models is still a difficult problem, because an universal set of benchmark applications is lacking. Several ones have been proposed. The work of [22] is an interesting one since it offers various kinds of problems, it clearly defines the benchmarking rules that its users have to follow to make fair comparisons, and it proposes a first reference work based on the search of optimal shortcut multilayer perceptrons for each problem.

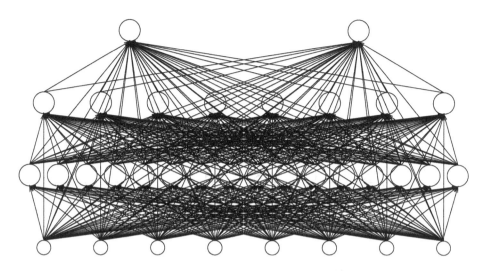

Figure 4.2. Optimal shortcut perceptron for the *diabetes2* problem (layers fully connected to all previous layers)

classified patterns (average on different runs of the learning process). A synchronous FPNN with 5 times less links reaches 73.37 % correctly classified patterns.

4.3.1.2   FPNN determination.   The optimal shortcut perceptron for the *diabetes2* problem is shown in Figure 4.2. The synchronous FPNN which has been determined so as to compute the same kind of global function is shown in Figure 4.3. It also uses $16 + 8 + 2$ activators, since the same number of nonlinear operators is required so as to compute the same kind of global function. The number of links is greatly reduced with respect to the shortcut perceptron. This reduction is made possible by the fact that the FPNA concept shares the 78 links among the various combinations of successive affine operators. Each combination corresponds to one of the 384 connections of the shortcut perceptron. Two kinds of links are used: "lateral" links connecting the activators of each layer with a bidirectional ring, and layer-to-layer links connecting each activator to one activator in the next layer. Then the local connections of the links are configured so as to broadcast the activator outputs as in a standard multilayer architecture: virtual links correspond to both fully connected layers and shortcut links.

4.3.1.3   Weight initialization.   Both uniform random and w-based random initializations of low-band weight vectors have been used for all FPNNs (see §3.5.5). Final performances do not significantly depend on the initial-

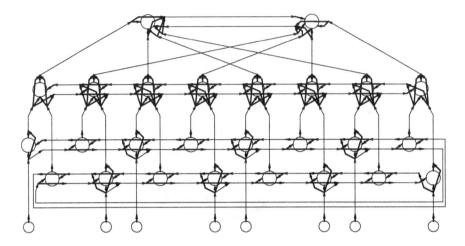

Figure 4.3. Equivalent FPNN architecture

ization method, even if the w-based method proves slightly better on average. But the convergence times (early stopping method) are greatly reduced for well scattered initializations: uniform random initializations require 55 % to 85 % more learning iterations than w-based ones.

Despite a back-propagation process that might increase weight dependencies as mentioned in §3.5.5, it appears that virtual weight scattering improves during learning. This phenomenon is strengthened when a w-based initialization is used: the average lessening of the partition maximum diameter criterion is 2.9 % with uniform random initializations (3.4 % for the best FPNNs), 4.75 % with w-based ones (5.6 % for the best FPNNs).

4.3.2   A FPNN that replaces a wavelet network: Breiman's waveforms

Wavelet networks ([26]) are far from being as famous as MLPs. Nevertheless, they have also proved efficient for approximation and classification problems. They use a standard multilayer architecture as MLPs, but their computation is less robust with respect to weight variations. This is an interesting feature to study the practical influence of weight dependencies.

As an example, wavelet networks have been successfully applied to the problem of Breiman's waveforms ([6]) in [3], with one hidden layer of 5 neurons. This problem deals with the classification of three kinds of waveforms. Signals to be classified are 21-dimensional real vectors. Therefore 120 synaptic connections have been used in wavelet networks. A FPNN with 21 input nodes, 5 hidden nodes and 3 output nodes may use 38 links to virtually create

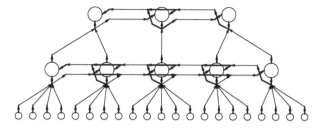

Figure 4.4. 21-5-3 FPNN (virtual multilayer architecture)

these 120 connections (see Figure 4.4). Activators (i_n, f_n) have been chosen such that they compute wavelet functions. The classification rates are the same as in [3] for the best learning runs. Nevertheless a greater variance is observed again, so that the average rate is 80 % correctly classified patterns instead of 84 %.

4.4 Other applications

FPNNs have been applied in less standard cases. Three examples are given here:

- an example of recurrent FPNN that performs Hopfield computations to illustrate the case of recurrent FPNNs,

- a first example of real application, which size problems illustrate some limits of the FPNA concept,

- a second example of real application, where FPNA topological simplifications are used on an embedded system.

A FPNN that replaces a Hopfield network Hopfield networks may be replaced by grid-based FPNNs. In such FPNNs, each node of the grid is associated with an input node, and its binary parameters configure its local connections to perform a multiple broadcast: the virtual connection graph is a clique. Figure 4.5 shows such an FPNN and details of its local resource interconnections.

Such FPNNs have been tested in [13] to replace Hopfield networks as associative memories for 7pt character recognition. They have proved efficient, though less stable than actual Hopfield networks: the diameter of their attraction basins is about one third smaller than with Hopfield networks.

FPNN to replace very large MLPs: FPNA limits A huge MLP with 240 inputs, 50 to 100 neurons in the hidden layer, and 39 outputs is used in [7] for

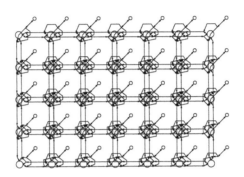

Figure 4.5. "Hopfield" FPNN

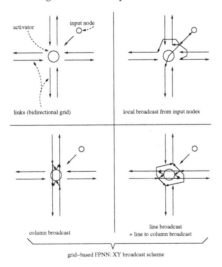

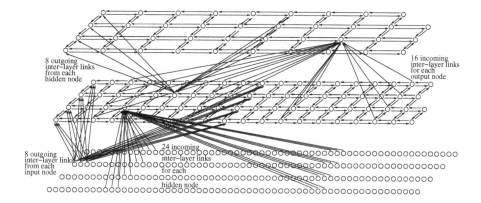

Figure 4.6.   FPNA for speech recognition

multi-band speech recognition. An FPNN version has been studied to reduce the number of parameters (and therefore the very long learning time), as well as to move towards real-time hardware implementations.

A first 2D-compatible FPNN structure has been proposed that uses the same topology simplification principles as in Figure 4.4 : the underlying multilayer FPNA contains one hidden layer of 78 activators, each of these nodes having incoming links from 3 input nodes, as well as intra-layer incoming links, and the output layer contains 39 activators, each of these nodes having incoming links from 2 hidden nodes, as well as intra-layer incoming links. It uses 548 links to virtually create the 21762 connections of the MLP. This FPNN illustrates some limits of the FPNA paradigm: its topological simplifications are too drastic for this FPNN to learn the speech recognition task as well as the MLP.

To overcome these very poor performances, a more complex FPNN has been proposed. It is based on the idea of having 8 outgoing links from each node towards the next layer (instead of only one). Figure 4.6 partially shows the underlying FPNA. It uses 1976 links. Though powerful enough, its topology is still far too complex to permit direct parallel mappings onto FPGAs.

FPNN for embedded vigilance detection   Another FPNA application is currently developed. It is based on [4], where a 23-10-9 MLP classifies physiological signals to discriminate vigilance states in humans. A simple multilayer FPNN that also uses the same topology simplification principles as in Figure 4.4 reaches the same performance, with only 67 links that virtually create the 320 connections of the MLP. A very low-power real-time implementation on a Xilinx Virtex XCV1000E FPGA is under development.

Previous sections have discussed the application of FPNNs to different problems, they have focused on the topological simplifications, and they have studied the obtained performances with respect to standard neural models. Next sections focus on implementation methods. A general one is first described in Section 4.5. It may be used for any FPNN computation. Nevertheless, more efficient solutions exist, but they only apply to specific FPNNs, called synchronous. This kind of FPNN is defined and studied in section 4.6, whereas Section 4.7 describes their specific implementation methods: a pipelined version and an implementation with on-chip learning.

Several implementation results (mostly areas) are given in sections 4.5 and 4.7. Nevertheless, technical details have to be found in Section 4.8.

4.5   General FPGA implementation

This Section describes a very general implementation method that is directly derived from the asynchronous parallel computation scheme (see §3.3.2). It is organized as follows:

- §4.5.1 explains that this implementation only requires to define basic building blocks since a modular assembling is made possible thanks to FPNA properties,

- §4.5.2 describes the first building block: the implementation of a link,

- §4.5.3 describes the second building block: the implementation of an activator.

4.5.1   Modular implementation of FPNNs

An FPNN implementation only requires configurable predefined blocks that are simply assembled as in the underlying FPNA architecture (direct parallel mapping of the neural architecture onto the FPGA). Its simplified topology make it possible for the compiling tool to perform well.

The resources (links and activators) of the FPNA are implemented: each resource corresponds to a predefined block that performs both resource computations and asynchronous protocol handling, and all blocks are assembled according to the FPNA graph. This implementation may be used by any FPNN derived from this FPNA: some elements just have to be correctly configured inside each block, such as multiplexers, registers, etc. Such FPNNs may compute complex functions as in Section 4.3, though their FPNA architecture is made simple (reduced number of edges, limited fan-ins and fan-outs). Therefore, this implementation method is flexible, straightforward and compatible with hardware constraints. Moreover this implementation shows how time may be inherent to FPNN computations.

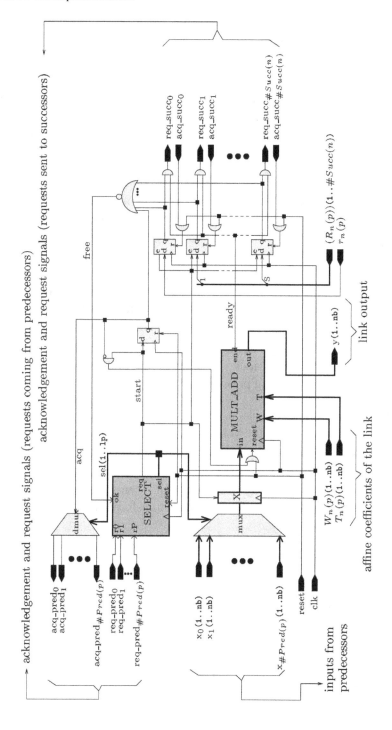

Figure 4.7. Architecture of link (p, n)

Next subsections describe the implementation of the asynchronous parallel computation by means of two predefined blocks: link and activator resources.

4.5.1.1   Recurrent FPNN.

4.5.2   Links

Figure 4.7 shows a possible implementation of a link (p, n). All asynchronous reset signals for flip-flops are active high.

SELECT :

>This block receives all request signals that may be sent by the predecessors of link (p, n) according to the FPNA topology. It also receives a signal free that is set high when link (p, n) handles no request.

>Signal start is an output that is set high during one clock cycle when a request processing begins. Signal acq is an acknowledgement signal that is routed towards the resource that sent the current request. Signal sel(1..lp) codes the number of the selected request signal. It controls the input mux and the acknowledgement dmux.

>The architecture of a SELECT block for 4 predecessors is shown on Figure 4.8. The PRIO block is a simple truth table-based component that implements a rotating priority policy.

MULT_ADD :

>This block receives the value stored in register X (value of the selected request). It also receives $W_n(p)$ and $T_n(p)$, and it computes the affine transform $W_n(p)X + T_n(p)$ of the link.

>It outputs signal ready that is set high when signal y(1..nb) contains the output value of link (p, n) (affine transform of X).

free :

>When register X stores the selected input value, a set of flip-flops stores the request signals that will have to be sent to the successors of link (p, n). These flip-flops are reset when the corresponding acknowledgements have been received. Signal free is reset when all expected acknowledgements have been received. Yet the effective output request signals remain low as long as ready is not active.

The chronogram of Figure 4.9 gives a possible example of successive requests processed by a link (p, n), with 4 preceding resources and 4 successors, and with $r_n(p) = 1$, $R_n(p, s_0) = R_n(p, s_2) = 1$ and $R_n(p, s_1) = R_n(p, s_3) = 0$. Block MULT_ADD is assumed to compute its result within 4 clock cycles: it corresponds to the use of a semiparallel multiplier (each operand is split in 2,

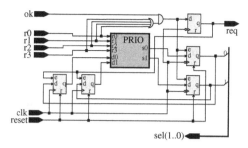

Figure 4.8. Request selection (4 predecessors)

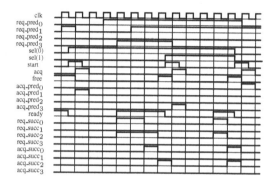

Figure 4.9. Chronogram: asynchronous parallel computation of a link

so that such a multiplier performs 4 products of $\frac{b}{2}$-bit operands sequentially computed by a a $\frac{b}{2}$-bit parallel multiplier, and then correctly added to obtain the product of b-bit values).

When all successors are free (i.e. are able to send immediate acknowledgements), a request processing requires 5 clock cycles, which means that the protocol involved in FPNN computation only costs one clock cycle. The blocks required to handle this protocol only use 5.5 CLBs on a Xilinx Virtex-E FPGA[5], whereas the affine transform requires at least 50 CLBs for a 16-bit precision (even with an area-saving semiparallel multiplier) Implementation performances will be discussed in Section 4.8.

4.5.3   Activators

4.5.3.1   Feedforward FPNN.   The main changes with respect to a link are limited to arithmetic operators (see Figure 4.10), so that i_n and f_n are com-

---

[5]In such an FPGA, each CLB contains 4 configurable logic cells.

puted instead of the affine transform of a communication link. Operators ITER and OUTPUT are used instead of MULT_ADD. These blocks perform the iterated computation of the activator output (see §3.3.2). ITER is the i_n function that is applied to the input and to the internal variable so as to update the latter. OUTPUT is the f_n function that is applied to the final state of the internal variable so as to obtain the activator output. The computation of OUTPUT only occurs when a_n values have been processed by ITER.

A linear feedback shift register is required for this arity handling, so that the FPNN asynchronous protocol requires 6 CLBs for an activator (if $i_n \leq 15$) instead of 5.5 for a link. Once again, the cost of the FPNN asynchronous protocol is negligible with respect to the arithmetic operators required by the activator itself.

§3.4.2 explains that synchronization barriers must be introduced to handle recurrent FPNNs: activator-generated requests are put in a waiting buffer until the next synchronization barrier.

Nevertheless, the theoretical study in [13] shows that an activator in a *deterministic* recurrent FPNN sends at most one value after each synchronization barrier, so that no request buffering is needed. Therefore, the general parallel implementation of recurrent FPNNs is very close to the one of feedforward FPNNs, except that a simple additional synchronization signal is received. This signal clocks a flip-flop with signal ready as input and ready_sync as output. Then ready AND ready_sync replaces signal ready in the control of the output request signals. No other change is required.

4.6   Synchronous FPNNs

This Section focuses on a specific class of FPNNs, for which both description and computation can be greatly simplified, so that an implementation with on-chip learning as well as an optimized pipelined implementation may be described (see Section 4.7). It is organized as follows:

- §4.6.1 gives the definition of synchronous FPNNs,

- §4.6.2 describes the main property of these FPNNs and it explains that such FPNNs perform computations that appear similar to usual neural computations,

- these computation simplifications lead to a backpropagated method to compute the gradient of synchronous FPNNs, so that their back-propagation-based learning is described in §4.6.3.

4.6.1   Definition of synchronous FPNNs

A synchronous FPNN is an FPNN in which any resource may *simultaneously* process all its received values. In such an FPNN, any resource can "wait"

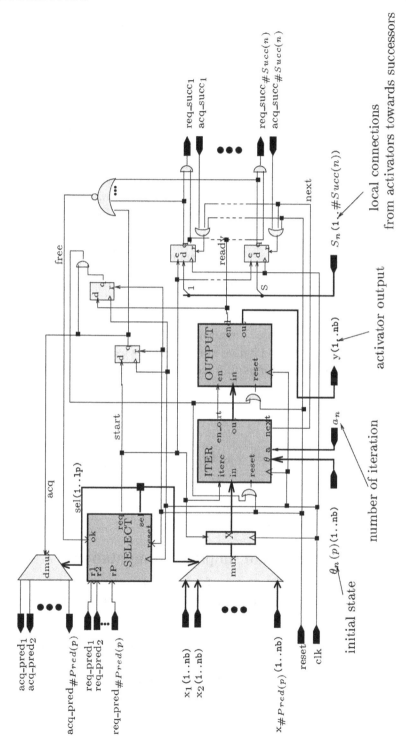

Figure 4.10. General architecture for an activator resource

for all its input values before it begins to process them, without modifying the computation of the whole FPNN.

The study of [13] defines three (rather tricky) conditions for an FPNN to be considered as synchronous. Nevertheless, these conditions are satisfied by a set of easily identified FPNNs, for which all activators use simple additions to update their variable state and for which all links apply linear functions. Therefore, only this set of synchronous FPNNs will be considered from now on. This set includes the FPNNs defined such that their activators perform the computations of standard sigmoidal neurons.

More formally, a simple definition (close to standard feedforward neural networks) may be given for synchronous FPNNs. Following the FPNA/FPNN definition given in Chapter 3, a FPNN must fulfill the following conditions to be synchronous:

- All iteration functions i_n are associative and commutative.

- Whatever computation method is chosen, the activator in node n exactly receives a_n values. In the asynchronous parallel computation method, it means that for all n, there are exactly a_n requests $req[(p,n),(n),x]$ where $p \in Pred(n)$. Therefore an activator always sends an output (except if $a_n = 0$) and it does not receive any value after having sent this output. This condition is fulfilled if the following recursive property is satisfied:

$$\forall n \quad a_n \quad = \sum_{p \in Pred(n)} d_{(p,n)} r_n(p)$$

$$\text{where} \quad d_{(p,n)} \quad = \chi_{I\!N_+^*}(a_p) S_p(n) + \sum_{q \in Pred(p)} d_{(q,p)} R_p(q,n)$$

- Links must be linear operators $x \mapsto W_n(p)x$ (for each (p,n) in $\mathcal{E}$), instead of affine ones.

4.6.2   Equivalent and simplified computations

In [13], synchronous FPNNs have been introduced when studying the determinism of both sequential and parallel computation schemes for feedforward and recurrent FPNNs. The main result[6] is:

THEOREM 4.1 *Feedforward synchronous FPNNs are deterministic: all neurons compute the same outputs with both sequential and parallel computations, and these outputs do not depend on any graph ordering nor request scheduling policy.*

---

[6]Indeed, the last condition (linearity of links) is not required here.

This is the fundamental property of synchronous FPNNs: the above conditions ensure that the order in which input requests are sent to any activator does not have any influence on its computed output.

These equivalent results may be then expressed in a very simplified way. For each input node n in $\mathcal{N}_i$, a value x_n is given (outer inputs of the FPNN). Then, activators (n) and links (n, s) compute:

$$
y_{(n)} = f_n \left(\sum_{r_n(p)=1} y_{(p,n)} + \theta_n \right)
$$

$$
y_{(n,s)} = W_s(n) \left(\sum_{R_n(p,s)=1} y_{(p,n)} + S_n(s) y_{(n)} \right)
$$

Similar results are available for recurrent synchronous FPNNs.

4.6.3   Learning of synchronous FPNNs

Both learning and generalization phases of synchronous FPNNs may be efficiently implemented on configurable hardware. This key feature shows that this kind of neural models is particularly able to take advantage of the characteristics of reconfigurable computing systems: prototyping may be efficiently performed, and then the system may be reconfigured so as to implement the learned FPNN with an outstanding generalization speed (see section 4.3).

The learning phase of many standard neural networks (such as MLPs) is often performed thanks to what is called the back-propagation learning algorithm, where an error function Err estimates the distance between the neural network output and the expected output that corresponds to the given training pattern. A positive real value ϵ is given as learning rate. Each parameter p is updated as follows:
$$
p \Leftarrow p - \epsilon \frac{\partial Err}{\partial p}
$$
.

The generalized back-propagation algorithm of [10] may be applied to synchronous FPNNs (see [13]). It shows that the gradient of a synchronous FPNN may be computed by means of simple formulae. Moreover, these computations are performed by the resources with *local* data (provided by the neighbouring resources). It provides a simple hardware implementation that is based on assembled predefined blocks (Section 4.5). The simplified architectures of FPNAs ensure that this assembling is straightforward and hardware-friendly.

A node in a synchronous FPNN is an output node if its local activator computes an output value without sending it to any other resource (i.e. $S_n(s) = 0$ for all $s \in Succ(n)$). Let Err be the quadratic distance $Err = \frac{1}{2} \sum_{n \text{ output node}} (y_{(n)} - e_n)^2$, where e_n is the corresponding expected output. The following notations are also introduced:

- $\partial Err_{(p,n)}$ is the differential of Err with respect to the output of link (p, n)

- $\partial Err_{(n)}$ is the differential of Err with respect to the output of activator (n)

Then the gradient of Err with respect to the FPNN learnable parameters $W_n(p)$ and θ_n is computed thanks to the following formulae, that are similar to a standard back-propagation process in a multilayer neural network.

Three formulae express the back-propagation of differentials.

- If n is an output node: $\partial Err_{(n)} = y_{(n)} - e_n$.

- Else: $\partial Err_{(n)} = \sum_{S_n(s)=1} W_s(n) \partial Err_{(n,s)}$

- $\partial Err_{(p,n)} = r_n(p) f'_n \left(\sum_{r_n(p)=1} y_{(p,n)} + \theta_n \right) \partial Err_{(n)}$

  $+ \sum_{R_n(p,s)=1} W_s(n) \partial Err_{(n,s)}$

Then the gradient values are locally computed.

- Differential w.r.t. θ_n:

  $$\frac{\partial Err}{\partial \theta_n} = f'_n \left(\sum_{r_n(p)=1} y_{(p,n)} + \theta_n \right) \partial Err_{(n)}$$

- Differential w.r.t. $W_s(n)$:

  $$\frac{\partial Err}{\partial W_s(n)} = \left(\sum_{R_n(p,s)=1} y_{(p,n)} + S_n(s)y_n \right) \partial Err_{(n,s)}$$

4.7   Implementations of synchronous FPNNs

The computational simplifications of synchronous FPNNs provide two specific implementation methods: on-chip learning is possible, and a fast pipeline implementation may be used (without learning). The general topological simplifications that are made possible by FPNAs are still available, so that these implementation methods still consist of an easy modular hardware mapping of basic building blocks.

4.7.1   On-chip learning

The above computations are local, so that each proposed implementation block corresponds to one neural resource in the FPNA. These basic blocks successively handle:

- the computation of the resource output

- the computations required in the back-propagation process

    - an activator (n) back-propagates

$$\partial Err_{(n)} \, f_n' \left(\sum_{r_n(p)=1} y_{(p,n)} + \theta_n \right)$$

    towards its connected predecessors

    - a link (n, s) back-propagates $W_s(n) \partial Err_{(n,s)}$

- the computations required to update the local learnable parameters.

4.7.1.1   Links.   Figure 4.11 shows a possible implementation of a link with on-chip learning. All flip-flops use asynchronous reset signals that are active high. The clock and reset signal connections are not shown in order to lighten this figure. The reset signal is active before each new iteration of the learning algorithm. There are six main sub-blocks:

- MULT is a multiplier. It is shared by the different computation steps. Signal end is active when the multiplier result is available on bus out.

- ACC_F accumulates the received values. Since the FPNA concept may handle neural architectures with reduced fan-ins, there is no need to use additional bits to store accumulation results. These results (stored in ACC_F) remain unchanged when back-propagation is performed (when signal forward is idle).

- SEL_FWD receives a request signal when any connected predecessor sends a value to the link. This block selects a request signal, so that ACC_F accumulates the corresponding value. When all expected values have been received, SEL_FWD, signal en_m is set active, so that MULT multiplies the local weight $W_n(p)$ by the accumulated value in ACC_F. When the multiplication result is available, request signals are sent to all connected successors (links or activators). Figure 4.13 shows such a SEL_FWD block, supposing that the resource fan-in is equal to 4 (as in a grid-based layered FPNA).

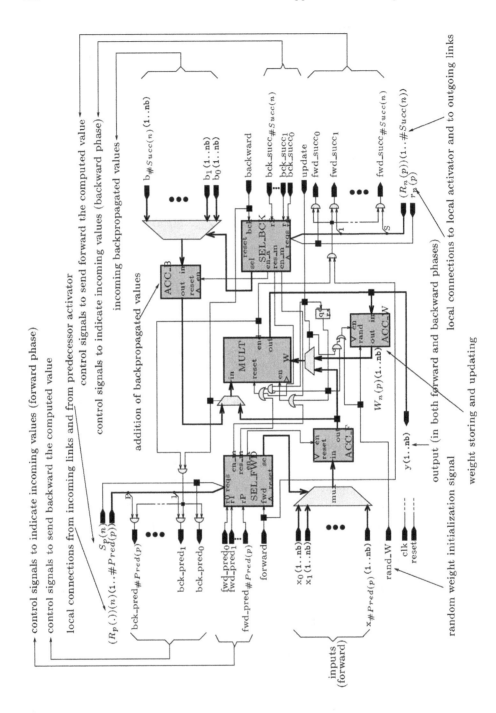

Figure 4.11.　Link architecture (learning)

- ACC_B and SEL_BCK are similar to the above ACC_F and SEL_FWD. They are used for back-propagated computations.

- ACC_W stores the local weight $W_n(p)$. When signal update is active, values stored in ACC_F and ACC_B are multiplied. The result is accumulated in ACC_W: the local weight is updated.

Various implementation choices are proposed for the accumulators and the multipliers in [13, 14]. With 16-bits[7] values (fixed point, 11 fractionary bits), 2-complement accumulators, and a semi-parallel multiplier (four 8×8 multiplications sequentially handled), this link implementation uses less than 120 CLBs on a Xilinx Virtex-E FPGA. Accumulations require one clock cycle, multiplications require 5 clock cycles (including reset), at 50 MHz.

4.7.1.2 Activators. The main changes with respect to a link are related to the computation of f_n and f'_n. When sigmoid function $f_n(x) = \tanh(x)$ is used, then $f'_n(x) = 1 - (f_n(x))^2$. It leads to the architecture of figure 4.12, with an additional part that computes a piecewise polynomial approximation[8] of tanh: a table stores the polynomial coefficients (address given by the most significant bits stored in ACC_F), a counter (CNT) gives the degree of the expected coefficient, an adder (ADD) and MULT compute the approximation with a Horner scheme.

With the above implementation choices, less than 145 CLBs of a Xilinx Virtex-E FPGA are required. The tanh function approximation lasts 11 clock cycles (a piecewise degree-2 polynomial approximation is sufficient to obtain the expected precision, provided that the coefficients are 20-bit values and MULT is a 16×20 semi-parallel multiplier).

4.7.2 Pipelined implementation

A pipeline implementation is proposed in [13] for any synchronous FPNN. It uses on-line operators (see [16] for the adequation of on-line arithmetic[9] to neural computations). In such an implementation, building blocks are (again) assembled according to the FPNA underlying structure. Thanks to simplified

---

[7]The required precisions are precisely studied in [13]. 16 bits are sufficient for most tested synchronous FPNNs. It confirms the studies of [17] despite the major differences between standard neural networks and FPNNs.

[8]Preferred to a CORDIC-like algorithm: such a method better applies to a sigmoid function such as tan^{-1} than $tanh$ (see [1]), but computing tan^{-1} derivative requires a division, whereas computing $tanh$ derivative only requires a multiplier that is already mandatory to handle back-propagated values. Moreover, this multiplier is sufficient in the forward phase for the polynomial approximation

[9]An on-line arithmetic is a bit-serial arithmetic that processes any operation or elementary function with the most significant digits first, thanks to the use of a redundant number representation system that avoids carry propagation within additions.

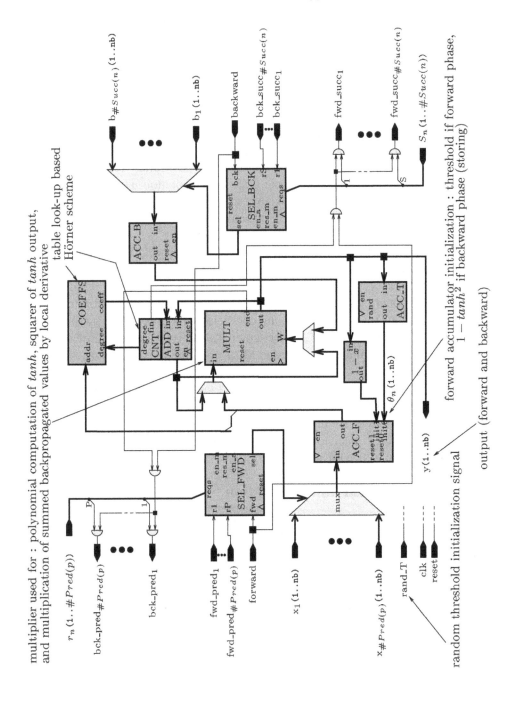

Figure 4.12. Activator architecture (learning)

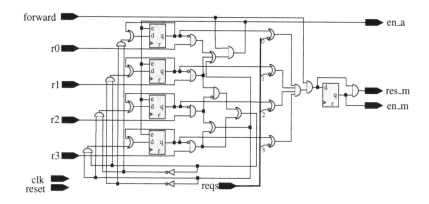

Figure 4.13. Blocks SEL_FWD and SEL_BCK

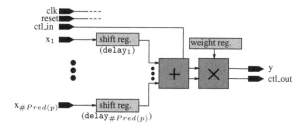

Figure 4.14. Link (pipelined synchronous FPNN)

computations (§4.6.2), these building blocks do not require any protocol handling. They handle data and computation in a serial way. Thanks to the use of on-line arithmetic, both activators and links handle data in the same way. Activators use *elementary functions* (sin, cos, tan, tanh, exp, log, arctan, . . .). The evaluation of these elementary functions can be performed using polynomial or rational approximations, shift and add algorithms or table-based algorithms. Those algorithms are presented in [21]. In this work, we have chosen the evaluation of the tanh function using polynomial approximations with a Horner scheme[10]. See [16, 13] for more details about on-line operators and delay handling.

Figure 4.14 sketches a "link" building block. It uses 37 CLBs (170 for an activator) of a Xilinx Virtex-E FPGA for a 16-bit precision. It may be clocked at 100 MHz.

4.8   Implementation performances

This section discusses the implementation performances (area and speed) of the various implementation methods that have been described in sections 4.5 and 4.7 through several examples taken from sections 4.3 and 4.4. §4.8.1 focuses on a typical synchronous FPNN, whereas §4.8.2 deals with the less common case of non- synchronous FPNNs. Finally §4.8.3 (cautiously) compares the obtained implementation performances with other works.

Our implementations are based on a Celoxica RC1000PP board with a Xilinx Virtex XCV1000E-BG560 FPGA. Xilinx FPGAs have been chosen because the fine grain parallelism of neural computations requires many elementary memory devices, and such FPGAs offer a high memory/logic ratio. We use the Virtex-E FPGA family to illustrate FPNN implementation performances, indicating the smallest one that is sufficient to implement each FPNN. Each CLB in these FPGAs corresponds to 4 configurable logic cells.

It must be pointed out that current FPGAs already outperform the capacity of the Virtex XCV1000E: such an FPGA contains 27648 logic cells, to be compared with the 73008 ones of the largest Virtex-E, as well as with the 125136 logic cells of the largest current Virtex-II Pro. Such improvements make the scalability of the FPNA concept even more useful: larger neural networks may be implemented without routing bottlenecks, or spared CLBs may be used for other computations required by the application.

We currently use Xilinx ISE 4.2 as synthesis tool. The mapping of each building block is asked to optimize the implementation area. Modular mapping is priviledged for large FPNNs. Speeds are given according to both FPGA

---

[10]Again preferred to a CORDIC-like algorithm, that would require an iterated process unfitted for a pipeline implementation

synthesis indications and memory bandwidth limits in our board. It should be pointed out that these performances have not been optimized by experts of FPGA implementation, and that they might be also improved by a combined use of area-saving neural operators (for example bit-stream neurons may be used in conjunction with the pipeline implementation of FPNNs).

4.8.1   Application to the FPNN of Figure 4.3

Several implementations are available for this FPNN.

- The implementation of the learning phase of this FPNN requires 13250 CLBs (16-bit precision, based on parallel 8×8 multipliers and a piece-wise polynomial approximation of tanh). It may use a single Xilinx Virtex XCV3200E (largest Virtex-E), or two Xilinx Virtex XCV1600E-BG560. In this case the small fan-ins of the FPNA resources make it possible to exchange all required signals with the 404 available I/O ports of the FPGAs: at most 168 ports are required for each FPGA. The shortcut perceptron of Figure 4.2 could not be directly mapped onto any number of FPGAs: these FPGAs would have to exchange too many signals, besides the topology and fan-in problems inside each FPGA. One learning iteration of the FPNN lasts 378 clock cycles, so that a 14 FPNN-MCUPS speed is reached: 14.10^6 links are updated by the learning process per second. It corresponds to 70 MCUPS (70.10^6 connections updated per second) with the optimal shortcut perceptron which outputs the same function as the FPNN.

- For the generalization phase, the asynchronous implementation of sub-sections 4.5.2 and 4.5.3 uses 7500 CLBs (a Xilinx XCV1600E is necessary). A 24 FPNN-MCPS speed is reached: 24.10^6 neural resources are processed per second. It corresponds to 120 MCPS with the equivalent optimal shortcut perceptron.

- An optimized pipeline implementation of the generalization phase of this FPNN requires only one Xilinx XCV1000E (3820 CLBs). A 250 FPNN-MCPS speed is reached, with a sufficient 11-bit precision (according to preliminary tests). It corresponds to 1.25 GCPS (giga connections per second) with the equivalent optimal shortcut perceptron.

Similar performances have been obtained for other Proben 1 applications (see [13]).

4.8.2   "Wavelet" and "Hopfield" FPNNs

The above optimized pipeline implementation is not available for these non-synchronous FPNNs, so that they require an implementation based on the blocks of 4.5.2 and 4.5.3[11].

A 13 FPNN-MCPS speed is reached for the "wavelet" FPNN of Figure 4.4 (equivalent to 42 MCPS for the corresponding wavelet network). It requires 3150 CLBs of a single XCV600E FPGA. No other hardware implementation of wavelet networks has been found (most efficient implementations are based on matrix–vector products that do not appear in such neural models). This FPNN implementation may be compared with the 5 MCPS of a sequential implementation of wavelet networks on a Intel PIII processor (800 MHz). The combined use of both FPNA paradigm and FPGA devices thus multiplies the speed–power ratio by approximately 80.

A 90 FPNN-MCPS speed is reached for the "Hopfield" FPNN of §4.5. It corresponds to 940 MCPS for the corresponding Hopfield network. Therefore this FPGA-based implementation reaches the same speed as the digital ASIC of [18], though such a comparison must be very carefully made (different implemented models, different design choices, precisions, ... etc), as explained below.

4.8.3   Performance analysis

The above speeds must be compared with the different neural network implementations discussed in Section 3.2. Nevertheless, it is difficult to make a fair comparison between the performances of the various FPGA-based methods: they implement different neural architectures, with different implementation choices (precision, on-chip learning, etc), the FPGA types vary, and a part of the performance differences may be attributed to FPGA technology improvements.

Without on-chip learning, the speeds for the implementation of standard multilayer architectures range from a few MCPS per FPGA (2 MCPS in [9], 4 MCPS in [5]) to a few tens MCPS (18 MCPS in [16], 22 MCPS in [2]). With on-chip learning, speeds range from 0.1 MCUPS in [9] to 8 MCUPS in [16].

Despite these heterogeneous previous results, it appears that the above FPNNs outperform the few previous FPGA-based implementations, with or without on-chip learning. The performances for the various FPNNs of [13] are more similar to the ones of complex neuro-computers: 5 MCUPS to 1 GCUPS and 10 MCPS to 5 GCPS (neuro-computers CNAPS, RAP, etc). But only one or a few FPGAs are required for FPNN implementations.

---

[11]Handling recurrent FPNNs just implies a few additional control signals.

Indeed, great implementation speeds are not the main advantage of the FPNA/FPNN concept: above all, this new neural framework defines neural networks that are easy to map onto some configurable hardware by means of a hardware-friendly assembling of predefined blocks, whereas such a straight-forward parallel implementation is impossible for equivalent standard neural networks.[12] As soon as a neural solution fits a chosen hardware device thanks to the FPNA paradigm, effort is put on the optimization of implementation choices (arithmetic, precision, etc) so as to reach outstanding performances.

FPNAs make it possible for current and future neural implementations to take advantage of the rapid FPGA technology improvements, whereas the previous advanced methods are limited by large neural architectures that do not fit hardware topologies.[13] Moreover the FPNA paradigm may be used in conjunction with such advanced area-saving methods.

4.9 Conclusions

FPNAs have been defined to reconcile simple hardware topologies with often complex neural architectures, thanks to a computation scheme that creates numerous virtual neural connections by means of a reduced set of links, whatever the device, the arithmetic, and the neural structure. This FPNA computation paradigm thus leads to define neural models whose computational power is similar to usual neural networks despite simplified architectures that are well suited for hardware implementations.

FPNAs and FPNNs are both a new theoretical framework for neural computation and an efficient tool to adapt most standard neural networks to digital hardware. The implementation of FPNA-derived neural networks on FPGAs has proved far more efficient than previous FPGA-based implementations of standard neural models. Moreover, the main advantage of the FPNA concept is to define neural networks that are easily mapped onto any configurable digital hardware device by means of a simple assembling of predefined blocks. This assembling is permitted by hardware-compatible FPNA architectures (the neural architecture may be directly taken as the modular design that is handled by the compiler that maps the neural computations onto the FPGA), whereas such immediate parallel implementations are impossible for equivalent standard neural networks.

The main limitation of the topology simplifications appears in FPNN learning: a too drastic reduction of the number of neural resources may result in a significant loss of computational power despite numerous distinct virtual con-

---

[12] The shortcut perceptron of Figure 4.2 cannot be directly mapped onto FPGAs, even with the help of the advanced methods of [1], [23], [9], [2], [25], or [27].

[13] A bit-stream based method leads to very small implementation areas, but the limits of the FPGA connectivity are already reached for a rather small FPGA in [2].

nection weights. Therefore topology simplifications stop when FPNN learning fails in retrieving the classification performance of the standard neural model that is functionally equivalent. Application of FPNNs to standard neural benchmarks has shown that topology may be sufficiently simplified before this limitation appears. Nevertheless, this limitation may appear for some complex real-world applications.

FPNAs still deserve further developments, such as pipeline implementations in conjunction with the bit-stream technology, additional learning algorithms, or dynamic learning of FPNN architectures. They are to be included in a neural network compiler onto FPGAs that is developped in our team (it handles implementation choices and it generates VHDL code for standard neural networks according to the user description).

FPNAs appear as an asset for FPGAs : these cheap, flexible and rapidly improving digital circuits may take advantage of the FPNA concept to stand as promising solutions for hardware neural implementations, despite their performances below ASIC speeds. Nevertheless, it is still difficult to see how FPNAs will interfere with new trends in FPGA development : mixed DSP-processor-FPGA approach of new powerful FPGAs, or routing problems in huge FPGAs ([8]).

References

[1] D. Anguita, S. Bencetti, A. De Gloria, G. Parodi, D. Ricci, and S. Ridella. FPGA implementation of high precision feedforward networks. In *Proc. MicroNeuro*, pages 240–243, 1997.

[2] S.L. Bade and B.L. Hutchings. FPGA-based stochastic neural networks - implementation. In *Proceedings of the IEEE Workshop on FPGAs for Custom Computing Machines*, pages 189–198, 1994.

[3] R. Baron and B. GIRau. Parameterized normalization : application to wavelet networks. In *Proc. IJCNN*, volume 2, pages 1433–1437. IEEE, 1998.

[4] K. Ben Khalifa, M.H. Bedoui, L. Bougrain, R. Raychev, M. Dogui, and F. Alexandre. Analyse et classification des etats de vigilance par reseaux de neurones. Technical Report RR-4714, INRIA, 2003.

[5] N.M. Botros and M. Abdul-Aziz. Hardware implementation of an artificial neural network. In *Proc. ICNN*, volume 3, pages 1252–1257, 1993.

[6] L. Breiman, J.H. Friedman, R.A. Olshen, and C.J. Stone. *Classification and regression trees*. 0-534-98054-6. Wadsworth Inc, Belmont California, 1984.

[7] C. Cerisara and D. Fohr. Multi-band automatic speech recognition. *Computer Speech and Language*, 15(2):151–174, 2001.

[8] F. de Dinechin. The price of routing in fpgas. *Journal of Universal Computer Science*, 6(2):227–239, 2000.

[9] J.G. Eldredge and B.L. Hutchings. RRANN: a hardware implementation of the backpropagation algorithm using reconfigurable FPGAs. In *Proceedings of the IEEE World Conference on Computational Intelligence*, 1994.

[10] C. Gegout, B. GIRau, and F. Rossi. Generic back-propagation in arbitrary feedforward neural networks. In *Artificial Neural Nets and Genetic Algorithms – Proc. of ICANNGA*, pages 168–171. Springer-Verlag, 1995.

[11] C. Gegout and F. Rossi. Geometrical initialization, parameterization and control of multilayer perceptron: application to function approximation. In *Proc. WCCI*, 1994.

[12] B. GIRau. Dependencies of composite connections in Field Programmable Neural Arrays. Research report NC-TR-99-047, NeuroCOLT, Royal Holloway, University of London, 1999.

[13] B. GIRau. *Du parallelisme des modeles connexionnistes a leur implantation parallele*. PhD thesis n° 99ENSL0116, ENS Lyon, 1999.

[14] B. GIRau. Digital hardware implementation of 2D compatible neural networks. In *Proc. IJCNN*. IEEE, 2000.

[15] B. GIRau. FPNA: interaction between FPGA and neural computation. *Int. Journal on Neural Systems*, 10(3):243–259, 2000.

[16] B. GIRau and A. Tisserand. MLP computing and learning on FPGA using on-line arithmetic. *Int. Journal on System Research and Information Science,* special issue on *Parallel and Distributed Systems for Neural Computing*, 9(2-4), 2000.

[17] J.L. Holt and J.-N. Hwang. Finite precision error analysis of neural network hardware implementations. *IEEE Transactions on Computers*, 42(3):281–290, March 1993.

[18] A. Johannet, L. Personnaz, G. Dreyfus, J.D. Gascuel, and M. Weinfeld. Specification and implementation of a digital Hopfield-type associative memory with on-chip training. *IEEE Trans. on Neural Networks*, 3, 1992.

[19] W. Kautz. The realization of symmetric switching functions with linear-input logical elements. *IRE Trans. Electron. Comput.*, EC-10, 1961.

[20] R. Minnick. Linear-input logic. *IEEE Trans. Electron. Comput.*, EC-10, 1961.

[21] J.M. Muller. *Elementary Functions, Algorithms and Implementation*. Birkhauser, Boston, 1997.

[22] L. Prechelt. Proben1 - a set of neural network benchmark problems and benchmarking rules. Technical Report 21/94, Fakultat fur Informatik, Universitat Karlsruhe, 1994.

[23] V. Salapura, M. Gschwind, and O. Maisch berger. A fast FPGA implementation of a general purpose neuron. In *Proc. FPL*, 1994.

[24] K. Siu, V. Roychowdhury, and T. Kailath. Depth-size tradeoffs for neural computation. *IEEE Trans. on Computers*, 40(12):1402–1412, 1991.

[25] M. van Daalen, P. Jeavons, and J. Shawe-Taylor. A stochastic neural architecture that exploits dynamically reconfigurable FPGAs. In *Proc. of IEEE Workshop on FPGAs for Custom Computing Machines*, pages 202–211, 1993.

[26] Q. Zhang and A. Benveniste. Wavelet networks. *IEEE Trans. on Neural Networks*, 3(6):889–898, Nov. 1992.

[27] J. Zhao, J. Shawe-Taylor, and M. van Daalen. Learning in stochastic bit stream neural networks. *Neural Networks*, 9(6):991–998, 1996.

Chapter 5

BACK-PROPAGATION ALGORITHM ACHIEVING 5 GOPS ON THE VIRTEX-E

Kolin Paul

Department of Computer Science & Engineering
Indian Institute of Technology, Delhi
New Delhi, INDIA
kolin@cse.iitd.ac.in

Sanjay Rajopadhye

Department of Computer Science
Colorado State University
Fort Collins, Colorado, USA
svr@CS.colostate.edu

Abstract

Back propagation is a well known technique used in the implementation of artificial neural networks. The algorithm can be described essentially as a sequence of matrix vector multiplications and outer product operations interspersed with the application of a point wise non linear function. The algorithm is compute intensive and lends itself to a high degree of parallelism. These features motivate a systolic design of hardware to implement the Back Propagation algorithm. We present in this chapter a **new** systolic architecture for the complete back propagation algorithm. For a neural network with N input neurons, P hidden layer neurons and M output neurons, the proposed architecture with P processors, has a running time of $(2N + 2M + P + max(M, P))$ for each training set vector. This is the first such implementation of the back propagation algorithm which completely parallelizes the entire computation of learning phase. The array has been implemented on an Annapolis FPGA based coprocessor and it achieves very favorable performance with range of 5 GOPS. The proposed new design targets Virtex boards.

We also describe the process of automatically deriving these high performance architectures using systolic array design tool MMALPHA. This allows us to specify our system in a very high level language (ALPHA) and perform

A. R. Omondi and J. C. Rajapakse (eds.), FPGA Implementations of Neural Networks, 137–165.

design exploration to obtain architectures whose performance is comparable to that obtained using hand optimized VHDL code.

5.1   Introduction

The design of application specific parallel architectures has become very important in many important areas that demand high performance. Field Programmable Gate Arrays (FPGAs) have become a viable alternative to full custom VLSI design in this domain. Although they offer lower performance than full custom chips, the faster development cycle make them an attractive option. Current FPGA devices offer over a million programmable gates which can, with careful design, achieve operational frequencies in excess of 100MHz. When we program a well designed highly parallel architecture, on this new breed of FPGAs , (re-programmable) performance of the order of GFLOPs become accessible to the designer at a relatively low cost. Walke et al. have presented a design [1] in the area of adaptive beam forming where they estimate a 20 GFLOPS rate of sustained computation.

In this paper, we present a new implementation of an application that is also computationally intensive. Back propagation (BP) is one of the most widely used learning algorithms for multilayer neural networks [2]. The BP algorithm is computationally intensive and often requires a lot of training time on serial machines [3]. As a result, considerable interest has been focused on parallel implementations and efficient mappings of this algorithm on parallel architectures. Many hardware implementations for neural net algorithms (either for the forward or for the complete BP learning) have been proposed [4]. Many of these designs are "paper designs" and a lot of the them suffer from inefficient use of hardware resources. One common hardware structure for neural nets is the linear systolic array model. This structure, consisting of neighbor-connected string of identical processing elements, each operating in lockstep, is accepted as an attractive solution for the rapid execution of the matrix vector and outer product computation at the core of the learning as well as the recall phase.

Our architecture, which is highly pipelined, implements the complete BP algorithm and achieves over *5 GOPS of sustained performance*. Our design exploits the inherent parallelism present in the BP algorithm. The implementation is based on a systolic design. The architecture has been designed for maximal parallelism and exploits pipelining to achieve a high frequency of operation. The architecture exhibits fine grained parallelism and is a natural candidate for implementation on an FPGA. Our design has been implemented on a PCI board which has the FPGA and can thus be considered as a coprocessor to the host computer.

This design has the best processor utilization and also the fastest reported implementation on a FPGA both in terms of the frequency of operation of the circuits as well as the number of clock cycles required. We emphasize that this architecture achieves in excess of 5 GOPS of sustained computation on a single Virtex E 1000 FPGA. The implementation of this high performance architectecture on an FPGA indicates that with careful design, we can get very high rates of sustained computation in applications that are computationally intensive.

This contribution has two parts. We discuss the implementation of our BP architecture in the first part. In the second part we show how to derive this implementation automatically using the synthesis tool MMALPHA [5] where our focus was to compare the potential performance lost in the automatic derivation to the ease in the design process. We start with equations defining the algorithm at a very high level and derive synthesizable VHDL. We also desire that this automatically generated hardware program be as efficient as possible. Towards this end, we compared its performance with the handwritten manually optimized version.

The rest of the paper is organized in the following manner. Section 5.2 describes the back propagation algorithm and explains the problem definition. In Section 5.3, we discuss the key ideas behind our proposed parallelization while in Section 5.4, we outline the design aspects of the architecture. In Section 5.5, we provide a detailed description of the architecture and its implementation on the target FPGA. This concludes the first part of the contribution. Section 5.6 introduces the MMALPHA synthesis tool. In section 5.7 we discuss some of the key transformations used. It also describes the scheduling and processor allocation process while Section 5.8 gives a high level overview of the hardware generation in the MMALPHA system. Section 5.9 describes the performance and experimental results of the implementation on the FPGA and gives the comparison between the two approaches. In Section 5.10, we review other FPGA based implementations of the back propagation algorithm. Section 5.11 discusses the key contribution proposed in the paper and concludes it by providing the directions for future work.

5.2   Problem specification

We now describe the back propagation algorithm for a neural network with one hidden layer. In the forward phase of Figure 5.2.1, the hidden layer weight matrix W_h is multiplied by the input vector $X = (x_1, x_2, \ldots, x_N)^T$ to calculate the hidden layer output where $w_{h,ij}$ is the weight connecting input unit j to unit i in the hidden layer. A non linear function f is applied to the elements of this after correction with the input bias θ yielding $y_{h,j}$ which is the output from hidden layer. This is used to calculate the output of the network, $y_{o,k}$ by a

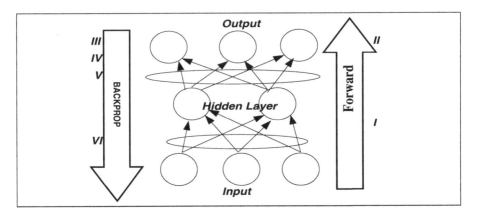

Figure 5.2.1: The Backpropagation Algorithm

Table 5.1. Equations for The Back Propagation Algorithm. Capitals denote either the matrix of the values of the weights or vectors representing the inputs and outputs. The outer product of two vectors is denoted by $\otimes$

Section	BPN Equations	Our Notation	EqNo
I	$y_{h,j} = f(\sum_{i=1}^{N_i} w_{h,ji} x_i - \theta)$	$Y_1 = W_1 X$ $Y_2 = f(Y_1)$	A B
II	$y_{o,k} = f(\sum_{j=1}^{N_h} w_{o,kj} y_{h,j} - \theta)$	$Y_3 = W_2 Y_2$ $Y_4 = f(Y_3)$	C D
III	$\delta_{o,k} = y_{o,k}(1 - y_{o,k})(d_k - y_{o,k})$	$Y_5 = Y_4(1 - Y_4)* \\ (D - Y_4)$	E
IV	$\delta_{h,j} = y_{h,j}(1 - y_{h,j}) \sum_{k=1}^{N_o} \delta_{o,k} w_{o,kj}$	$Y_6 = W_2^T Y_5$ $Y_7 = Y_2(1 - Y_2)Y_6$	F G
V	$\Delta w_{o,kj} = \eta \delta_{o,k} y_{h,j}$ $w'_{o,kj} = w_{o,kj} + \Delta w_{o,kj}$	$\Delta W_2 = Y_5 \otimes Y_2$ $W'_2 = W_2 + \Delta W_2$	I J
VI	$\Delta w_{h,ji} = \eta \delta_{h,j} x_i$ $w'_{h,ji} = w_{h,ji} + \Delta w_{h,ji}$	$\Delta W_1 = Y_7 \otimes X$ $W'_1 = W_1 + \Delta W_1$	K L

$$(5.1)$$

similar process using a different weight matrix W_o. The expected output is d_k and it is used to update the weights. A vector δ_o is computed from y_o and d (by element wise operation) and the update to the W_o matrix is simply the outer product of δ_o and y_h scaled by a learning parameter η. Similarly a vector δ_{hk} is computed by the element wise operation on y_h and the matrix vector product of W_o^T and δ_o. The update to the W_h matrix is also computed as η times the outer product of δ_h and X. The flow of computation is illustrated in Figure 5.2.1 where the roman numerals refer to the rows on Table 1. In our discussion, we ignore the learning parameter and input bias to avoid clutter. Moreover, they are scaling factors and may be trivially incorporated in the implementation.

The BP algorithm may be succinctly specified as a sequence of three Matrix Vector Multiplications or MVMs (Equations A, C and F)) and two outer product evaluations (Equations I and K). These main operations are interlaced with the application of point functions (Equations B,D, E & G). We may thus formulate our problem as follows: determine an efficient hardware implementation of the computation defined by Equations A - L.

Clearly any such efficient implementation is crucially dependent on the implementation of the MVM. Our architecture was derived by focusing on this module and by exploiting the special relationship between two dependent MVM's— one applied to the output of the first.

5.3   Systolic implementation of matrix-vector multiply

We will now outline the systolic implementation of Matrix Vector Multiply in some detail, carefully noting the two choices for the allocation strategy. The MVM can be conveniently written as

$$y_i = \sum_j w_{ij} x_j \text{ where } 1 \leq i \leq M \text{ and } 1 \leq j \leq N \qquad (5.2)$$

It is well known that the systolic implementation of this MVM can be achieved with M processors in $N + M$ clock cycles [6]. The accumulation of the summation can take place in either the vertical or the horizontal direction depending on the method we use to feed in the inputs x_j's.

In the first implementation, we propose to feed in the x_j's to the first processor in the array. In the case, the accumulation is in the horizontal direction as shown in the space-time (processor-time) diagram given below (Figure 5.3.1). The direction of accumulation is along the horizontal direction as shown in the diagram. The values of y_i's are accumulated in each processor. We note that each x_j is input to the next processor in a serial manner following one cycle delay. The accumulation takes place in the register in the processor. The output y_1 appears in processor 1 after n clock cycles and the output y_m appears in processor m after $n + m - 1$ clock pulses. The rows of the matrix W circulate within each processor (processor k having the kth row).

The other way to perform this MVM is to feed in x_j to the jth processor. This means that we feed in the data in parallel but not at the same time step— the x_j's have to fed in at the appropriate time step. Each x_j is fed into the jth processor at time step $j - 1$. This is illustrated in Figure 5.3.1, where each processor accesses a distinct column of W. The partial result is propagated to the next processor. The outputs y_i's are available in the last processor with y_1 appearing at time instant t_n and the last value y_m at time instant t_{n+m-1}.

We have two different ways of performing the MVM, just by changing the way data is fed into the systolic array. In the first case, the data was fed in a serial fashion to the first processor, and the outputs were available in parallel at all the processors in a shifted manner. In the second case, we fed in the input

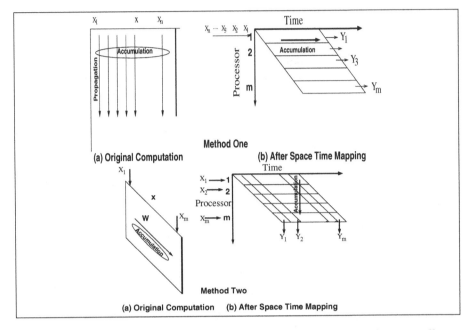

Figure 5.3.1: Two Systolic Implementations for MVM illustrating two allocations of computation processors

data in a "delayed" parallel method and we obtained the output in a serial way from the last processor. In isolation, each architecture has a drawback —in the first one, the outputs are in all the processors (one value in each processor) and are available at diffrent time instants. This would necessitate additional circuitry to "flush out" the outputs. Similarly, the second architecture has a similar problem with its inputs—it needs special circuitry and additional time to feed in the inputs before the computation can start.

5.4 Pipelined back-propagation architecture

The above property of the MVM that the *allocation function is not unique* is the basis for our new design. We use the fact that the *output pattern* for the first allocation function is **identical** to the *input pattern* of the second one. Hence there is no need for additional circuitry and no wasted cycles if one needs to perform back to back MVMs. The design is optimal with respect to processor utilization as well as time. In this section we describe the architecture in detail. We assume that X is an N-vector, W_1 is a $P \times N$ matrix and W_2 is a $M \times P$ matrix. This corresponds to N input neurons, P hidden layer neurons and M output neurons. We will use the two different allocation functions to implement the back to back MVMs. In general, to obtain the product

$$y_i = \sum_j w_{ij} x_j \text{ where } 1 \leq i \leq P \text{ and } 1 \leq j \leq N \qquad (5.3)$$

using either of the two allocation functions, we need $(N + P - 1)$ clock cycles. Therefore, if we have two consecutive MVMs where the matrices are $P \times N$ and $M \times P$, we need $(N + P - 1)$ and $(M + P - 1)$ clock pulses to have the the result. This thus amounts to a total time of $(M + N + 2P - 2)$. What we propose to do is to use the first allocation strategy to perform the first MVM. In this case, the x_j's will be input serially to the first processor and the outputs will start arriving after time N at the individual processors. We note that at time instant $N + 1$, the first processor is free and remains so for the next P cycles. Similarly, the second processor becomes free after the $N + 1$ clock pulse and remains free thereafter. Clearly, the point function can be applied to the output y_1 and this then becomes the input to the next MVM. We choose to schedule our second MVM taking advantage of the fact that the second allocation strategy allows the input to be sent to each processor but at a different instant of time as explained in Figure 5.3.1. The output of the first stage remains in one of the registers and the processor switches mode to perform the second MVM. *We note that we have taken advantage of the two different allocation strategies to ensure that the processors are not idle at any time.* Using our scheme, consecutive MVMs can be performed in $N + M + P - 1$ time steps—a significant improvement over the *"standard"* way of implementing consecutive MVMs that has been used in all previous hardware implementations. The outer products are also evaluated using this strategy with only a special artifact to configure the processors to perform only the multiply and not accumulate in any direction. Each product in the register becomes a term in the weight matrices.

Given this high level description for the basis of the proposed hardware, we now outline the complete design of "one-step" of the entire process. We emphasize that the choice of two different allocation functions have allowed us this innovative design which cuts down on the total execution time by about 40% and also has almost 100 percent processor utilization—two key areas in efficient systolic array designs. If $P = N = M$, the standard implementation would entail an execution time of $(5 * 2N)$ clock cycles, while our approach takes $(5N + N)$ clock cycles to process each training instance (which is why we claim a 40% reduction in execution time). We describe the design with reference to the Figure 5.4.1 above. In the diagram, the rhombus with the solid lines represents the first allocation strategy while the one with dashed contours represents the second allocation strategy. The diagram shows one step in the complete algorithm viz., the operation of the circuit for one training vector which is one iteration of Equations A-L. For each input vector in the "learning

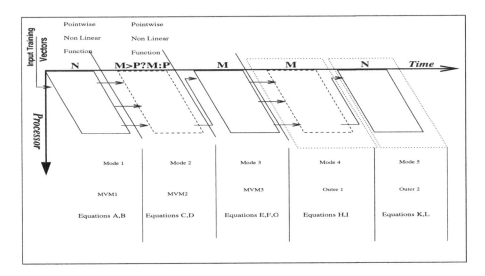

Figure 5.4.1: Schematic of Proposed Hardware (Computation Model)

phase", there are 5 modes in which each of the processors of the systolic array operate. These five modes correspond to the 3 MVMs and the 2 outer product evaluations. In the diagram we have indicated which of the defining equations are evaluated in each mode. After a vector has been processed, the updated weights remain in the registers inside the processor. This one-step is called the "macro step" in our design and the operation of the processors in the 5 different modes is refereed to as the five "mini-steps". Thus, at every macro step one vector is processed and requires an input of one training vector and the expected output for this vector. The entire architecture thus has an output only at the end of the training phase when the final values of the weight matrices are read back.

5.5   Implementation

We elaborate on the proposed architecture and its implementation on the target FPGA. Our implementation targets the AMS Starfire board [7] which is used as a coprocessor with the host computer. In Figure 5.5, we show a simplified schematic of the board. [8]. It has a single Xilinx XCV1000 VirtexE FPGA [9]. The host downloads the configuration data and the input data to the board via the PCI bus.

We now describe the processor configuration in greater detail with reference to Figure 5.5.2 which shows the details of the architecture on the FPGA. All the data that the coprocessor needs are read in from the board local memory (Right and Left Mem of Figure 5.5). The individual processors are initially supplied with small random weights. The LMSE and ITERATION registers

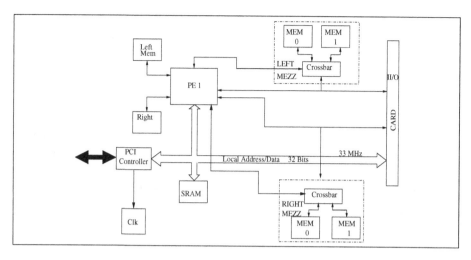

Figure 5.5.1: Simplified Diagram of the Starfire board

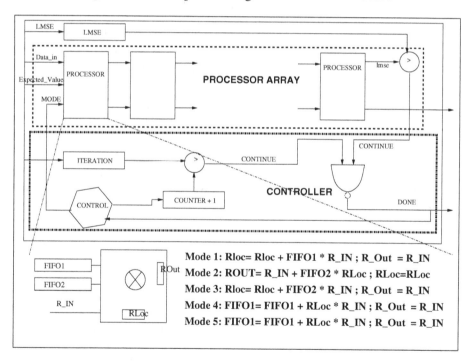

Figure 5.5.2: Details of Proposed Hardware

are control registers which determine the termination condition. The LMSE contains the (user defined) lowest permissible mean square error and the IT-ERATION contains the total number of training set vectors. The first processor reads in each training vector and the associated expected output. After either the least mean square error is lower than LMSE or the number of training set vectors are exhausted, the processor downloads the final weight vectors to the local memory and issues the '*done*' signal to the host. The host then reads out the final weight values. The entire operation is thus compute bound. In each macro step, only the first processor reads in a sequence of 2N vectors in only one of the five phases of "one-step" of the computation. The system has an output only once at the end of the entire training phase.

In Figure 5.4.1 we show the space time diagram of the computation of one step in the entire computation. The one step of the computation takes in a n-size vector and does a MVM. This takes n clock ticks on a p processor systolic array. We have a point wise function applied to the output of each processor and this is then fed as input to the processor to perform the next MVM. This second MVM continues for m or p cycles whichever is greater as indicated in the diagram. The outputs from the last processor are fed into the first processor after this time interval so that the third MVM can begin. In the final two modes, the processors actually evaluate the products and update the shift registers containing the weight values.

The operation of the processor in the five modes is described in Figure 5.5.2. We have shown the operations during each of the five modes of operation. We have abstracted away the portions of the circuit inside the processor which remain inactive during a particular mode and only show the relevant sections which are enabled. The mode signal controls this enabling/disabling of components inside the processors. For example, in mode 1 , the first shift register bank which contains the row of W_1 matrix, is enabled and the other similar bank being disabled are omitted from the diagram. Also in this mode, each processor at the end of the accumulation phase needs to evaluate the non linear function f on the final value before switching modes. We have implemented this function as a table lookup. In this mode, at any given time, one and only one of the processors will be accessing the LUT. Hence the LUT is designed as a shared resource between the individual processors of the array. We also have one additional processor following the "last" processor. This processor is active only in mode 2 and evaluates equation 14. This enables us to decrease the length of the critical path and improve the frequency of operation of the entire circuit. This is a design choice that we have made noting that the control becomes simpler if we use some of the registers to make an additional processor instead of incorporating the functionality in the last processor. The result of the multiply at each time step is added to the content of the appropriate shift register output to reflect the updated value of the weight matrices. We note in the figure, that except for mode 1, the multiply-accumulator unit in each processor in all other modes takes its input either from the shift registers or an internal register. The register, in general, contains the value computed in a previous mode. Thus in four out of five "mini steps" of each "macro step", there is no IO involved.

We decided to see if it was possible to derive a similar architecture automatically starting with the equations describing the computation. There are a lot of systems which take a high level description of the system in C/Matlab and derive synthesizable RTL level descriptions [10, 11]. However, most of these systems require that we write the description making the parallelism explicit. The motivation for trying to see if we could get a high performance VHDL description of the network automatically starting from the equations was to reduce the implementation time of the designing and coding in VHDL. We shall walk through, in detail, the automatic derivation of this architecture using our codesign research prototype "MMALPHA".

5.6 MMALPHA **design environment**

We used the systolic design prototype tool MMALPHA to systematically derive the RTL description of the proposed implementation. In this section we describe briefly the characteristics of this design environment. We also give a top level overview of the design procedure.

5.6.1 The ALPHA **Language**

We use ALPHA [5] as the language to write our algorithm. The ALPHA language was conceived in 1989 by Christophe Mauras as part of his PhD thesis. The MMALPHA system was initially developed at IRISA, and since 1995, its further development is being done at IRISA in collaboration with BYU [12] and CSU [13].

ALPHA is a single assignment, equational (functional), and inherently parallel language, based on the formalism of systems of recurrence equations. Algorithms may be represented at a very high level in ALPHA, similar to how one might specify them mathematically. The basic advantage of writing the algorithm in this functional language is that it allows us the ability to reason about the system. The ALPHA language is restrictive in the class of algorithms it can easily represent (although the language is Turing complete), but is useful for programming algorithms which have a high degree of regularity and parallelism such as those frequently occurring in linear algebra and digital signal processing applications.

Computational entities in ALPHA are called systems [5], which are parallel functions relating the system outputs to system inputs. The ALPHA system declaration specifies the types and domains of system inputs and outputs. The definition of the system function is given as a set of affine recurrence equations involving input, outpu t, and local variables.

One of ALPHA's strengths is the fact that an ALPHA program can be transformed by a series of rewriting rules into equivalent ALPHA programs suitable for:

- implementation as a regular array circuit, or

- execution on a sequential or parallel computer.

This capability is provided by the MMALPHA system. ALPHA program transformations have been independently proved to preserve the semantics of the original program, and thus derived programs are correct by construction (assuming the specification was correct). ALPHA also allows the analyzability needed to help choose what transformations to perform.

The MMALPHA **System**

The "transformation engine" interface for manipulating ALPHA programs is written in C and Mathematica. The user interface is in Mathematica which provides an interpreted language with high level built in functions for symbolic computations. MMALPHA uses these facilities for transforming ALPHA programs.

The internal representation of any ALPHA program is stored as an Abstract Syntax Tree (AST). All computations and manipulations are performed on this internal representation. Typical commands include viewing an ALPHA program, checking its correctness, generating C code to execute the program, etc.

Specific C functions are used for two purposes: various parsers and unparsers, and computations on polyhedra. All the C functions are accessed via Mathematica and extensively use the polyhedral library [14] which is the computational kernel of the MMALPHA system.

The design flow with MMALPHA is illustrated in Figure 5.6.1. A computation intensive part of a program (usually a loop) is rewritten in ALPHA and is then transformed using MMALPHA into an implementation which consists of three programs:

- The hardware part (in VHDL) which represents the implementation of the loop on the FPGA.

- The software part which replaces the loop in the original program.

- The hardware/software interface part which handles the communication of data betweeen the host and the FPGA.

MMALPHA is not a push button system, but requires the user to be "in the loop" as an *"expert"* who can take the initial specification of the program and apply the required transformations (intelligently) so that we get the three programs as mentioned above. We shall go through a detailed walkthrough of the entire process in the derivation of synthesizable RTL from the ALPHA specification of "one step" of the computation of the back propagation network to

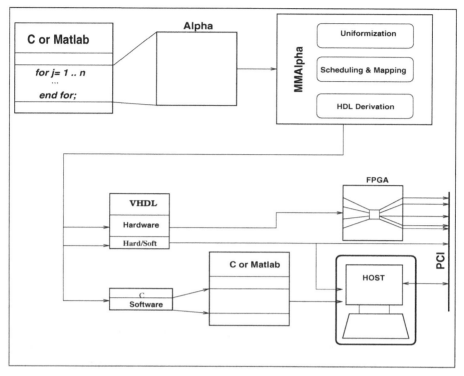

Figure 5.6.1: MMAlpha Design Flow

illustrate this. The entire ALPHA code for one step of the back propagation algorithm is shown in Figure 5.6.2. We see that the code is very similar to the equations of the BP algorithm. It has some (redundant) copies of some input and output variables which were introduced to in the design process to facilate the complete derivation of hardware with minimal amount of tweaking of the generated RTL. The three "reduce" statements refer to the 3 MVMs while the last two lines refer to the outer products. In ALPHA the order of the equations is immaterial.

We shall in the subsequent sections describe the derivation of RTL description.

5.7 Architecture derivation

We describe in this section a few of the transformations that we used in deriving the architecture. In the process, we also show the result of applying each of these transformations on the initial program of Figure 5.6.2. The progressive transformation of this code to a hardware description by careful and controlled application of these transformations is illustrated in this section.

```
--  -===================================================================--
--
-------------- Back Propagation Neural Nets
--
--  -===================================================================--
-- N Number of Input Neurons
-- P Number of hidden layer neurons
-- M Number of Output Neurons              let
system Neural :  {N,P,M |2≤M,N,P}          Y1 = reduce (+ , (p,n→p) , W1 * x.(p,n→n));
-- these are inputs to the system          Y3 = reduce(+, (m,p→p), W2copy *
(x, xcopy :  {n | 1 ≤ n ≤ N} of real;          Z4copy1.(m,p→m));
Z4copy1, Z4copy2 :  {m | 1≤m≤M} of real;   Z1 = reduce(+, (m,p→m), W2 * Y2.(m,p→p));
W1:  {p,n | 1≤p≤P; 1≤n≤N} of real;         Y2 = f(Y1);
W2, W2copy:  {m,p | 1≤m≤M; 1≤p≤P} of real; Y4 = Y3 * Y2 * (1.(p→) - Y2);
expZ :  {m | 1≤m≤M} of real)               Z2 = f(Z1);
returns (Z4 :  {m | 1≤m≤M} of real;        Z3 = expZ - Z2;
-- these are the outputs                    Z4comp = Z3 * Z2 * (1.(m→) - Z2);
DeltaW1:  {p,n | 1≤p≤P; 1≤n≤N} of real;    Z4 = Z4comp;
DeltaW2:  {m,p | 1≤m≤M; 1≤p≤P} of real);   DeltaW1[p,n] = Y4[p] * xcopy[n];
var                                        DeltaW2[m,p] = Z4copy2[m] * Y2[p];
-- output (and related vectors) of the hidden  tel;
layer
Y1, Y2, Y3, Y4 :  {p | 1≤p≤P } of real;
-- and the output layer
Z1, Z2, Z3, Z4comp :  {m | 1≤m≤M} of real;
```

Figure 5.6.2: Alpha Code

The entire script consisting of the sequence of transformations necessary to generate the entire architecture including the host code, the code for the interface as well as the target application is presented in the Appendix.

Analyze

The MMALPHA system stores the input ALPHA program in the form of an AST. This AST stored in Mathematica, is amenable to the myriad of transformations available within the development environment. One of the tools that this system provides enables us to perform a static analysis of the program stored in the AST. The routine checks the validity of the declaration of domains of ALPHA systems and variables and also performs other checks to enable us to detect dead code and unused variables, estimate certain quantities like number of calculations in a a loop nest as a function of loop bounds etc. This is very useful for the developer as it provides him with an early indication of the syntactic correctness of his program. It also provides initial estimates of the computation and the I/O requirements.

Normalize

This is a transformation that "normalizes" any ALPHA program into a predefined syntactic form which fixes the nesting order of the ALPHA constructs. Due to the mathematical properties of polyhedra and affine functions [15], the domain of any ALPHA expression is a finite union of integral polyhedra and is detremined automatically by a library for manipulating polyhedra [14]. Due to these closure properties, every expression can be simplified into a unique normal form, called the *case-restrict-dependency* form. It consists of an (optional) outer case, each of whose branches is a (possibly restricted) simple expression. A simple expression is a variable or constants composed with a single dependency function or pointwise operators applied to such expressions. This transformation consists of about 35 rewrite rules which among other things change the nesting order of incorrectly nested operations, flatten nested cases and combine restriction and affine functions whenever possible. This transformation is generally used as a followup to other transformations.

Serialization

In this transformation, we replace the reduction operations by a sequence of binary operations because the hardware elements have bounded "fan in". The MVM represents an accumulation. The summation can be evaluated in constant time on a CRCW-PRAM which supports unbounded fan in. However for any real scenario, we have a binary operator and hence the summation is performed in linear time. We note in passing that although the fastest way to add is using a logarithmic adder but that would be an overkill as we need linear time to feed in the input. In our example, we have three MVMs corresponding to the following three reductions.

We serialize them by invoking the "serializeReduce" transformation with the

```
Y1 = reduce (+ , (p,n→p) , W1 * x.(p,n→n));
Y3 = reduce (+ , (m,p→p) , W2copy * Z4copy1.(m,p→m));
Z1 = reduce (+ , (m,p→m) , W2 * Y2.(m,p→p));
```

Figure 5.7.1: Reductions

appropriate direction for the serialization. The resulting code for the first re-

```
serializeReduce[6, 1, 2, "YY1.(p,n→p,n-1)"];
serializeReduce[6, 3, 2, "YY3.(m,p→m-1,p)"];
serializeReduce[6, 5, 2, "ZZ1.(m,p→m,p-1)"];
```

Figure 5.7.2: SerializeReduce

duction is shown in Figure 5.7.3.

```
YY1acc[p,n] =
    case
    { | n=1} : 0[] + W1 * x[1];
    { | 2<=n} : YY1acc[p,n-1] + W1 * x[n];
    esac;
```

Figure 5.7.3: Serialized Code for the Reduction

Alignment

We align the computation. We do this keeping in mind the space time diagram that we want. This is a particular case of a general transformation called Change of Basis(COB).

Change of Basis

This is similar to the reindexing of loop variables and other loop transformations done in vectorizing and parallelizing compilers. Given a valid program and a change of basis variable, a new provably equivalent program is derived. We use this transformation to align our computation. If we refer to our space diagram, the computation of Y2 occurs after Y1 has been evaluated. Hence we do a COB on the equation for Y2 to align the variable Y2 at $N + 2$. The command `changeofBasis` `["Y2.(p→p,N+2)","p","n"]` aligns the computation at $N + 2$.

The following is the complete of sequence of calls to "changeOfBasis" that performs the required alignment as illustrated in the spacetime diagram (Figure 5.4.1). It may be noted that our current implementation doesnot allow us

```
changeOfBasis["Y1.(p→p,N+1)", "p", "n"];
changeOfBasis["Y2.(p→p,N+2)", "p", "n"];
changeOfBasis["ZZ1acc.(m,p→p,N+2+m)", "p", "n"];
changeOfBasis["Z1.(m→P+1,N+2+m)", "p", "n"];
changeOfBasis["Z2.(m→P+1,N+2+m)", "p", "n"];
changeOfBasis["Z3.(m→P+1,N+2+m)", "p", "n"];
changeOfBasis["Z4comp.(m→P+1,N+2+m)", "p", "n"];
changeOfBasis["YY3acc.(m,p→p,N+M+m+2)", "p", "n"];
changeOfBasis["Y3.(p→p,N+2M+3)", "p", "n"];
changeOfBasis["Y4.(p→p,N+2M+4)", "p", "n"];
addlocal["DeltaW1Loc = DeltaW1"];
addlocal["DeltaW2Loc = DeltaW2"];
normalize[];
changeOfBasis["DeltaW1Loc.(p,n→p,n+N+2M+4)"];
changeOfBasis["DeltaW2Loc.(m,p→p,m+2N+2M+4)", "p", "n"];
```

Figure 5.7.4: The transformations for performing all Change of Bases

to perform a COB on Input and Output variables of a System (since this would change the semantics of the ALPHA program). Hence we add a local variable to perform the change of bases. The resulting code after this set of transformations is shown in Figure 5.7.5. **Pipelining**
This is a very important transformation that allows us to build highly pipelined designs. This also involves making judicious choices about feeding the appropriate data at the correct time. This is a transformation that is widely used in systolic synthesis. It is also called localization or uniformization. It consists basically of replacing a broadcast by the pipeline of this value through all the computations that require it. In our example, we need to pipeline the input "X". The direction of feeding in the input in one way to perform the MVM demanded that we send one at a time to the first processor. This is performed by making a call to "pipeall" with the appropriate dependence function which indicates the direction of the pipeline. Clearly, in the equation for $YY1acc$ we want the x to be piped across the processors. The command `pipeall` `["YY1acc", "x.(p,n→ n)","Xpipe.(p,n→p+1,n)"]` does this. In this expression, the first argument is the variable whose equation is to be modified, the second argument is the expression to be pipelined and the last expression

```
system Neural :{N,P,M | 2≤N; 2≤P; 2≤M}
(x :  {n | 1≤n≤N} of real;
xcopy :  {n | 1≤n≤N} of real;
Z4copy1 :  {m | 1≤m≤M} of real;
Z4copy2 :  {m | 1≤m≤M} of real;
W1 :  {p,n | 1≤p≤P; 1≤n≤N} of real;
W2 :  {m,p | 1≤m≤M; 1≤p≤P} of real;
W2copy :  {m,p | 1≤m≤M; 1≤p≤P} of real;
expZ : {m | 1≤m≤M} of real)
returns (Z4 :  {m | 1≤m≤M} of real;
DeltaW1 :  {p,n | 1≤p≤P; 1≤n≤N} of real;
DeltaW2 :  {m,p | 1≤m≤M; 1≤p≤P} of real);
var
DeltaW2Loc :  {p,n | 1≤p≤P;
2N+2M+5≤n≤2N+3M+4} of real;
DeltaW1Loc :  {p,n | 1≤p≤P;
N+2M+5≤n≤2N+2M+4} of real;
ZZ1acc :  {p,n | 1≤p≤P; N+3≤n≤N+M+2} of
real;
YY3acc :  {p,n | 1≤p≤P; N+M+3≤n≤N+2M+2} of
real;
YY1acc :  {p,n | 1≤p≤P; 1≤n≤N} of real;
Y1 :  {p,n | 1≤p≤P; n=N+1} of real;
Y2 :  {p,n | 1≤p≤P; n=N+2} of real;
Y3 :  {p,n | 1≤p≤P; n=N+2M+3} of real;
Y4 :  {p,n | 1≤p≤P; n=N+2M+4} of real;
Z1 :  {p,n | p=P+1; N+3≤n≤N+M+2} of real;
Z2 :  {p,n | p=P+1; N+3≤n≤N+M+2} of real;
Z3 :  {p,n | p=P+1; N+3≤n≤N+M+2} of real;
Z4comp :  {p,n | p=P+1; N+3≤n≤N+M+2} of real;
let
DeltaW2[m,p] = DeltaW2Loc[p,m+2N+2M+4];
DeltaW1[p,n] = DeltaW1Loc[p,n+N+2M+4];
```

```
ZZ1acc[p,n] =
case
{ | p=1 } :  0[] + W2[n-N-2,p] * Y2[p,N+2];
{ | 2<=p } :  ZZ1acc[p-1,n] + W2[n-N-2,p] *
Y2[p,N+2];
esac;
YY3acc[p,n] =
case
{ | n=N+M+3 } :  0[] + W2copy[n-N-M-2,p] *
Z4copy1[n-N-M-2];
{ | N+M+4<=n } :  YY3acc[p,n-1] +
W2copy[n-N-M-2,p] * Z4copy1[n-N-M-2];
esac;
YY1acc[p,n] =
case
{ | n=1 } :  0[] + W1 * x[n];
{ | 2<=n } :  YY1acc[p,n-1] + W1 * x[n];
esac;
Y1[p,n] = YY1acc[p,n-1];
Y3[p,n] = YY3acc[p,n-1];
Z1[p,n] = ZZ1acc[p-1,n];
Y2[p,n] = f(Y1[p,n-1]);
Y4[p,n] = Y3[p,n-1] * Y2[p,N+2] * (1[] -
Y2[p,N+2]);
Z2[p,n] = f(Z1[p,n]);
Z3[p,n] = expZ[n-N-2] - Z2[p,n];
Z4comp[p,n] = Z3[p,n] * Z2[p,n] * (1[] -
Z2[p,n]);
Z4[m] = Z4comp[P+1,m+N+2];
DeltaW1Loc[p,n] = Y4[p,N+2M+4] *
xcopy[n-N-2M-4];
DeltaW2Loc[p,n] = Z4copy2[n-2N-2M-4] *
Y2[p,N+2];
tel;
```

Figure 5.7.5: Code after Change of Bases

```
Xpipe[p,n] =
     case
        { | p=1 } : x[n];
        { | 2<=p } : Xpipe[p-1,n];
     esac;
YY1acc[p,n] =
     case
        { | n=1 } : 0[] + W1 * Xpipe;
        { | 2<=n } : YY1acc[p,n-1] + W1 * Xpipe;
esac;
```

Figure 5.7.6: Pipelined Code

gives the direction of the pipeline as well as the name of the new variable introduced. The resulting code is shown in Figure 5.7.6 The sequence of calls to perform the pipelining of the other variables isshown in Figure 5.7.7 . The

```
pipeall["YY1acc", "x.(p,n->n)", "Xpipe.(p,n->p+1,n)"];
pipeall["YY3acc", "Z4copy1.(p,n->p-N-M-2)", "Z4pipe1.(p,n->p+1,n)"];
pipeall["DeltaW1Loc", "Y4.(p,n->p,N+2M+4)", "Y4pipe.(p,n->p,n+1)"];
```

Figure 5.7.7: Transformations for all Pipelines

ALPHA specification following a sequence of these semantic preserving transformations is shown below.

```
 system Neural :{N,P,M | 2≤N; 2≤P; 2≤M}
(x  :  {n | 1≤n≤N} of real;
xcopy :   {n | 1≤n≤N} of real;
Z4copy1 :   {m | 1≤m≤M} of real;
Z4copy2 :   {m | 1≤m≤M} of real;
W1 :  {p,n | 1≤p≤P; 1≤n≤N} of real;
W2 :  {m,p | 1≤m≤M; 1≤p≤P} of real;
W2copy :  {m,p | 1≤m≤M; 1≤p≤P} of real;
expZ :  {m | 1≤m≤M} of real)
returns (Z4 :  {m | 1≤m≤M} of real;
DeltaW1 :  {p,n | 1≤p≤P; 1≤n≤N} of real;
DeltaW2 :  {m,p | 1≤m≤M; 1≤p≤P} of real);
var
Y2pipe :  {p,n | 1≤p≤P; N+2≤n} of real;
Y4pipe :  {p,n | 1≤p≤P; N+2M+5≤n≤2N+2M+4} of
real;
Z4pipe1 :  {p,n | 1≤p≤P; N+M+3≤n≤N+2M+2} of
real;
Xpipe :  {p,n | 1≤p≤P; 1≤n≤N} of real;
DeltaW2Loc :  {p,n | 1≤p≤P;
2N+2M+5≤n≤2N+3M+4} of real;
DeltaW1Loc :  {p,n | 1≤p≤P;
N+2M+5≤n≤2N+2M+4} of real;
ZZ1acc :  {p,n | 1≤p≤P; N+3≤n≤N+M+2} of
real;
YY3acc :  {p,n | 1≤p≤P; N+M+3≤n≤N+2M+2} of
real;
YY1acc :  {p,n | 1≤p≤P; 1≤n≤N} of real;
Y1 :  {p,n | 1≤p≤P; n=N+1} of real;
Y2 :  {p,n | 1≤p≤P; n=N+2} of real;
Y3 :  {p,n | 1≤p≤P; n=N+2M+3} of real;
Y4 :  {p,n | 1≤p≤P; n=N+2M+4} of real;
Z1 :  {p,n | p=P+1; N+3≤n≤N+M+2} of real;
Z2 :  {p,n | p=P+1; N+3≤n≤N+M+2} of real;
Z3 :  {p,n | p=P+1; N+3≤n≤N+M+2} of real;
Z4comp :  {p,n | p=P+1; N+3≤n≤N+M+2} of real;
let
Y2pipe[p,n] =
case
{ | n=N+2} :  Y2[p,n];
{ | N+3<=n} :  Y2pipe[p,n-1];
esac;
```

```
Y4pipe[p,n] =
case
{ | n=N+2M+5} :  Y4[p,n-1];
{ | N+2M+6<=n} :  Y4pipe[p,n-1];
esac;
Z4pipe1[p,n] =
case
{ | p=1} :  Z4copy1[n-N-M-2];
{ | 2<=p} :  Z4pipe1[p-1,n];
esac;
Xpipe[p,n] =
case
{ | p=1} :  x[n];
{ | 2<=p} :  Xpipe[p-1,n];
esac;
DeltaW2[m,p] = DeltaW2Loc[p,m+2N+2M+4];
DeltaW1[p,n] = DeltaW1Loc[p,n+N+2M+4];
ZZ1acc[p,n] =
case
{ | p=1} :  0[] + W2[p+n-N-3,1] * Y2pipe;
{ | 2<=p} :  ZZ1acc[p-1,n] + W2[n-N-2,p] *
Y2pipe;
esac;
YY3acc[p,n] =
case
{ | n=N+M+3} :  0[] + W2copy[1,p] * Z4pipe1;
{ | N+M+4<=n} :  YY3acc[p,n-1] +
W2copy[n-N-M-2,p] * Z4pipe1;
esac;
YY1acc[p,n] =
case
{ | n=1} :  0[] + W1 * Xpipe;
{ | 2<=n} :  YY1acc[p,n-1] + W1 * Xpipe;
esac;
Y1[p,n] = YY1acc[p,n-1];
Y3[p,n] = YY3acc[p,n-1];
Z1[p,n] = ZZ1acc[p-1,n];
Y2[p,n] = f(Y1[p,n-1]);
Y4[p,n] = Y3[p,n-1] * Y2pipe * (1[] - Y2pipe);
Z2[p,n] = f(Z1[p,n]);
Z3[p,n] = expZ[p+n-N-P-3] - Z2[p,n];
Z4comp[p,n] = Z3[p,n] * Z2[p,n] * (1[] -
Z2[p,n]);
Z4[m] = Z4comp[P+1,m+N+2];
DeltaW1Loc[p,n] = Y4pipe * xcopy[n-N-2M-4];
DeltaW2Loc[p,n] = Z4copy2[n-2N-2M-4] * Y2pipe;
tel;
```

Figure 5.7.8: Before Scheduling Alpha Code

Scheduling and processor allocation

An ALPHA program does not convey any sequential ordering: an execution order is semantically valid provided it respects the data dependencies of the program. The goal of scheduling is to find a timing function for each variable which maps each point in the variable's index space to a positive integer representing a virtual execution time [16, 17]. This mapping should respect causality. The scheduling problem in the general case is undecidable. The MMALPHA system, therefore tries to find timing functions within the restricted class of affine schedules, for which necessary and sufficient conditions exist. Extra flexibility is accorded by allowing different affine schedules for each variable.

The technique used for scheduling is to formulate the causality constraints as linear integer programming program (LP) and to solve it using a software tool (like PIP [18]).

For the example of the neural nets, we required that the one step of the computation be completed in $2N + 2M + max(P, M)$ time (Figure 5.4.1). The scheduler in MMALPHA was able to compute this schedule. Also in the general case, where N, M and P are not equal, the scheduler was able to compute the optimum schedule (lowest processor idle time being the criterion for optimality). We do a reindexing of the first dimension to represent the time and the second processor. The resulting ALPHA code is shown in Figure 5.7.9. In the next section, we describe the generation of the RTL description of the architecture.

5.8 Hardware generation

One of the motivating factors in this research was to derive the RTL description of the high performance architecture implementing the back propagation algorithm. We have described the sequence of transformations which transformed the initial specification (Figure 5.6.2) into a form where an execution time and processor allocation has been assigned for each expression (Figure 5.7.9).

Control Signal Generation

To realize a systolic array in hardware, control signals are necessary to instruct each Processing Element (PE) of the array when to perform what computation. MMALPHA allows us to automatically generate the control flow information so that the right computations are done by the right PEs at the right time steps.

Alpha0 Format
After the program has been *uniformized* and *scheduled*, MMALPHA translates it to the ALPHA0 format. After the program has been scheduled, the next important step is the generation and pipelining of the control signal. Therefore control signals are generated and multiplexors added to each processors. The controls signal is then pipelined automatically by the system. The resulting ALPHA0 program contains the precise information that appears on the lifetime of each signal and on the input date of each data.
ALPHARD Format
ALPHARD is a subset of the ALPHA language and is used to describe regular circuits with local interconnections and their control and is hierarchical. At the lowest level we have *cells* consisting of combinatorial circuits, registers, multiplexors, etc. A cell may also contain other cells. We also have controllers

```
system Neural :{N,P,M | 2≤N; 2≤P; 2≤M}       Z4pipe1[t,p] =
(x :   {n | 1≤n≤N} of real;                   case
xcopy :  {n | 1≤n≤N} of real;                 { | t=p} :  Z4copy1[p-N-M-2];
Z4copy1 :  {m | 1≤m≤M} of real;               { | p+1<=t} :  Z4pipe1[t-1,p];
Z4copy2 :  {m | 1≤m≤M} of real;               esac;
W1 :  {p,n | 1≤p≤P; 1≤n≤N} of real;           Xpipe[t,p] =
W2 :  {m,p | 1≤m≤M; 1≤p≤P} of real;           case
W2copy :  {m,p | 1≤m≤M; 1≤p≤P} of real;       { | t=p} :  x[p];
expZ :  {m | 1≤m≤M} of real)                  { | p+1<=t} :  Xpipe[t-1,p];
returns (Z4 :  {m | 1≤m≤M} of real;           esac;
DeltaW1 :  {p,n | 1≤p≤P; 1≤n≤N} of real;      DeltaW2[m,p] =
DeltaW2 :  {m,p | 1≤m≤M; 1≤p≤P} of real);     DeltaW2Loc[m+p+2N+2M+6,m+2N+2M+4];
var                                           DeltaW1[p,n] = DeltaW1Loc[p+n+N+2M+7,n+N+2M+4];
Y2pipe :  {t,p | p+2≤t≤p+P+1; N+2≤p} of       ZZ1acc[t,p] =
real;                                         case
Y4pipe :  {t,p | p+3≤t≤p+P+2;                 { | t=p+3} :  0[] + W2[t-N-5,1] *
N+2M+5≤p≤2N+2M+4} of real;                    Y2pipe[t-1,p];
Z4pipe1 :  {t,p | p≤t≤p+P-1; N+M+3≤p≤N+2M+2}  { | p+4<=t} :  ZZ1acc[t-1,p] + W2[p-N-2,t-p-2]
of real;                                      * Y2pipe[t-1,p];
Xpipe :  {t,p | p≤t≤p+P-1; 1≤p≤N} of real;    esac;
DeltaW2Loc :  {t,p | p+3≤t≤p+P+2;             YY3acc[t,p] =
2N+2M+5≤p≤2N+3M+4} of real;                   case
DeltaW1Loc :  {t,p | p+4≤t≤p+P+3;             { | p=N+M+3} :  0[] + W2copy[1,t-p] *
N+2M+5≤p≤2N+2M+4} of real;                    Z4pipe1[t-1,p];
ZZ1acc :  {t,p | p+3≤t≤p+P+2; N+3≤p≤N+M+2}    { | N+M+4<=p} :  YY3acc[t-1,p-1] +
of real;                                      W2copy[p-N-M-2,t-p] * Z4pipe1[t-1,p];
YY3acc :  {t,p | p+1≤t≤p+P; N+M+3≤p≤N+2M+2}   esac;
of real;                                      YY1acc[t,p] =
YY1acc :  {t,p | p+1≤t≤p+P; 1≤p≤N} of real;   case
Y1 :  {t,p | N+2≤t≤N+P+1; p=N+1} of real;     { | p=1} :  0[] + W1[t-p,p] * Xpipe[t-1,p];
Y2 :  {t,p | N+3≤t≤N+P+2; p=N+2} of real;     { | 2<=p} :  YY1acc[t-1,p-1] + W1[t-p,p] *
Y3 :  {t,p | N+2M+4≤t≤N+P+2M+3; p=N+2M+3} of  Xpipe[t-1,p];
real;                                         esac;
Y4 :  {t,p | N+2M+7≤t≤N+P+2M+6; p=N+2M+4} of  Y1[t,p] = YY1acc[t-1,p-1];
real;                                         Y3[t,p] = YY3acc[t-1,p-1];
Z1 :  {t,p | t=p+P+3; N+3≤p≤N+M+2} of real;   Z1[t,p] = ZZ1acc[t-1,p];
Z2 :  {t,p | t=p+P+4; N+3≤p≤N+M+2} of real;   Y2[t,p] = f(Y1[t-1,p-1]);
Z3 :  {t,p | t=p+P+5; N+3≤p≤N+M+2} of real;   Y4[t,p] = Y3[t-3,p-1] * Y2pipe[t-1,p] * (1[] -
Z4comp :  {t,p | t=p+P+6; N+3≤p≤N+M+2} of     Y2pipe[t-1,p]);
real;                                         Z2[t,p] = f(Z1[t-1,p]);
let                                           Z3[t,p] = expZ[t-N-P-7] - Z2[t-1,p];
Y2pipe[t,p] =                                 Z4comp[t,p] = Z3[t-1,p] * Z2[t-2,p] * (1[] -
case                                          Z2[t-2,p]);
{ | p=N+2} :  Y2[t-1,p];                      Z4[m] = Z4comp[m+N+P+8,m+N+2];
{ | N+3<=p} :  Y2pipe[t,p-1];                 DeltaW1Loc[t,p] = Y4pipe[t-1,p] *
esac;                                         xcopy[p-N-2M-4];
Y4pipe[t,p] =                                 DeltaW2Loc[t,p] = Z4copy2[p-2N-2M-4] *
case                                          Y2pipe[t-1,p];
{ | p=N+2M+5} :  Y4[t-1,p-1];                 tel;
{ | N+2M+6<=p} :  Y4pipe[t-1,p-1];
esac;
```

Figure 5.7.9: Scheduled Alpha Code

responsible for initialization. At this point we have a description of a circuit that has no "spatial dimensions". It represents a single processing element which may be instantiated at multiple spatial locations.

The next level of the hierarchy is the *module* which specifies how the different cells are assembled regularly in one or more dimensions. This is achieved by *instantiating* previously declared cells. In addition, controllers are instantiated only once, with no spatial replication. The separation of temporal and spatial aspects is also reflected in the fact that the equations describing the behavior of cells have only one local index variable (time) and in the equations for modules, the dependencies have only spatial components indicating pure (delayless) interconnections.

The architectural description given by ALPHARD provides

- structuring, as complex designs must be hierarchical;

- genericity to allow component reuse and

- allows regularity to be described, in order to reuse hardware descriptions and simplify the design process.

Synthesizable VHDL
The translation produces synthesizable VHDL (compliant with IEEE-1076 standard [19] when the size parameters have been fixed. The hardware translator is implemented using a syntax generated meta translator which allows fast retargeting of generator depending on the type of synthesis tools used.

Interface Generation
The MMALPHA system supports the automatic generation of interfaces between the host program and the application running on the FPGA. The underlying interface of the system consists of a bus and a FIFO that interconnects the bus with the application. The bus and the FIFO comprises the *hardware interface*. The FIFO allows the data to be buffered when an interruption occurs on the bus. On top of the hardware interface, is built an application interface whose role is to rearrange as necessary, the data between the hardware interface and the application.

As described in the previous discussion on the hand derived code, we download the application data into the on board memories. This requires us to connect the FIFO to the onboard memories instead of the bus. This part of the interface is dependent to a certain degree on the hardware board being used—nevertheless MMALPHA generates generic code that can be easily customized for the different boards. The data to be downloaded to the memory is the same reordered data that is automatically generated by the system.

5.9   Performance evaluation

In this section, we discuss the performance of our architecture for the backpropagation neural network. In the first part of this paper we described the design—the VHDL specification for the entire system was written and synthesized for the target application directed for implementation on the Starfire Board. Subsequently we derived a VHDL description (automatically) of the architecture from a very high level specification using the MMALPHA research tool. This gave us two circuits implemented on hardware and we present in this section, their comparative performance figures.

The learning phase of the algorithm is the most time consuming one. We see that under the conditions where $(N = P = M)$, the processors are never idle (except the special one which operates only in one of the 5 micro steps). In the "recall" phase, the weights as obtained from the learning phase remain unchanged. Then the system is allowed to run only in modes 1 and 2. The target application for which we designed this system has 20 input neurons, 32 hidden layer neurons (one hidden layer) and one output layer neuron. The back propagation network was coded in VHDL and the circuit mapped to the XCV1000 package available on the Starfire board. The results of the place and route tool of Xilinx Foundation 2.1i is shown in Table 5.2.

The circuit operates at 35 MHz on an FPGA with a speed grade of -4. No

Table 5.2. Results for Hand Written VHDL

Number of External GCLKIOBs	:	1 out of 4	25%
Number of External IOBs	:	98 out of 404	24%
Number of Slices	:	3933 out of 12288	32%
Number of GCLKs	:	1 out of 4	25%

register retiming was performed and we are working on that to further push up the frequency of operation. The critical path delay in this circuit was $24ns$.

We could fit up to 70 processors on the XCV1000. Each processor does 2 fixed point operations. Hence the throughput of the circuits is $2 * 70 * 35MHz$ or about 5 Giga Operations per second.

In the case of the automatically derived code, the following table gives the results of the place and route tool.   The circuit operates at 30MHz which

Table 5.3. Results for Automatically generated VHDL

Number of External GCLKIOBs	:	1 out of 4	25%
Number of External IOBs	:	122 out of 404	30%
Number of Slices	:	4915 out of 12288	39%
Number of GCLKs	:	1 out of 4	25%

is about 14% slower than the hand coded one. Clearly, in comparison with the hand optimized design, the automatic implementation does take up more resources (7% more on the target FPGA). But this is to be expected and the gain is in the shortened design time. The entire design and coding cycle for the automatic derivation is a couple of days whereas the hand coded design took a couple of months to be implemented. Also we can generate C code to test functional correctness of the design after every stage of transformations.

Clearly, the number of neurons in the hidden layer determines the number of processors that the neural network needs in this design. This then becomes the major limitation of a FPGA based design because the number of processors that can fit is limited by the size of the FPGA. For the target application that we had chosen, it was not a problem. However we envision that more complex examples would need a larger number of processors and even the larger FPGA's may not be able to meet the requirement. We are working on a solution to this problem whereby we use partitioning techniques [20, 21] to partition the processor space using different methods. The underlying idea is to partition the processor array in such a way that a single physical processor sequentially emulates several virtual processors.

5.10   Related work

As mentioned in the previous sections, the focus of our research was not to come up with a new algorithm for the back propagation algorithm. Rather we intended to derive an efficient implementation of the algorithm targeted towards FPGAs. Research in the area of neural networks have been ongoing for over two decades now and hence a wealth of knowledge is available. In our study we found that the following related closely to implementations targeted towards hardware in general and FPGAs in particular. Burr [22] has a bibliography of digital neural structures. Zhang and Pal also have recently compiled a great reference of neural networks and their efficient implementations in systolic arrays [23]. Also this reference is a recent one (2002) and reviews a host of implementations of neural networks. Zhu and Sutton report a survey on FPGA implementations of neural networks in the last decade [24]. Kung and Hwang [25, 26] describe a scheme for designing special purpose systolic ring architectures to simulate neural nets. The computation that they parallelize is essentially the forward phase of the algorithm. They use the fact that neural algorithms can be rewritten as iterative Their implementation needs bidirectional communication links. It also doesn't avoid idle cycles between processors. Their architecture takes advantage of the fact that neural algorithms can be rewritten in the form of iterative matrix operations. They apply standard techniques to map iterative matrix operations onto systolic architectures. Kung et al [27] have reported results of implementing the back propagation on CMU Warp. The Warp exploited coarse grained parallelism in the problem by mapping either partitions or copies of the network onto its 10 systolic processors. Ferrucci describes a multiple chip implementation [28]. In this the author describes in fine detail of how he has implemented the basic building blocks like multipliers, adders etc of a systolic implementation of neural networks in FPGA. This work concentrates more on extracting performance out of the FPGA architecture in terms of efficient low level implementations of the elements that are used to build the neurons. The design also is reported to work at 10MHz. Elredge and Hutchings report a novel way of improving the hardware density of FPGAs using Runtime Reconfiguration(RTR) [29]. They divide the back propagation algorithm into three sequentially executed stages and configure the FPGA to execute only one stage at a time. They report significant deterioration in performance but justify the use of RTR to get density enhancement. Bade and Hutchings [30] report an implementation of stochastic neural networks based on FPGAs—their proof of concept circuit is implemented on a very small FPGA (100 CLBs). Their solution proposes the use of bit serial stochastic techniques for implementing multiplications. Their implementations achieve very high densities for stochastic neural networks. Zhang has reported an efficient implementation of the BP algorithm on the connection machine [31]. This implements only the forward phase of the network. The authors describe how to implement a "multiply-accumulate-rotate" iteration for a fully connected network using the 2D connections of CM-2.

Botros et al. [32], have presented a system for feed forward recall phase. Each node is implemented with two XC3042 FPGAs and a 1KX8 EPROM. Training is done offline on a PC. The whole system operated at 4MHz. Linde et al [33] have described REMAP which is an experimental project of building an entire computer for neural computation using only FPGAs. The suggested architecture incorporates highly parallel, communicating processing modules, each constructed as a linear SIMD (Single Instruction stream, Multiple Data stream) array, internally connected using a ring topology, but also supporting broadcast and reduction operations.This gives high performance, up to 40-50 MHz. The full scale prototype is reported to work at 10 MHz. Gadea et al. [34], report an implementation of a pipelined on-line back propagation network on a single XCV400 package. Also they have reported a maximum operating frequency of only 10MHz.

5.11 Conclusion

Our contribution in this paper is two fold. We have described a new systolic design for implementing the back propagation learning algorithm which lends itself to efficient parallelization. Our algorithm is the fastest systolic design reported as yet since it avoids the inter phase idle time from which all previous designs suffer. Secondly, we have implemented the proposed algorithm on an FPGA. Here too, we believe that we have a higher operating frequency as compared with other implementations. Although, this is an implementation issue and depends on the target boards, available memory and the FPGA itself, our operating frequency of about 35 MHz is based on the the XCV1000 FPGA which is not the most recent technology. We expect to have a faster circuit on the Virtex II FPGA. Our design is completely scalable and can accommodate a large number of processors (limited only by the size of the FPGA). The design is also scalable in the number of hidden layers. The design has been targeted for the Virtex based boards and good results have been obtained.

The main focus of this research endeavor has been two fold—one to desisgn a high performance circuit for the BP algorithm and second to use automatic synthesis of such a high performance architecture for FPGA. We implemented a new design and then used the tool to generate the RTL automatically. We have presented performance figures for both these designs. The automatically generated hardware description obviously makes greater demands on hardware resources but that is to be expected. We have been working on hardware optimizations that can be implemented in our tool so as to close the gap.

Appendix

The complete script which transforms the ALPHA specification to synthesizable VHDL.

```
load["NN.a"]; analyze[]; ashow[]
serializeReduce[{6, 1, 2}, "YY1.(p,n->p,n-1)"];
serializeReduce[{6, 3, 2}, "YY3.(m,p->m-1,p)"];
serializeReduce[{6, 5, 2}, "ZZ1.(m,p->m,p-1)"];
simplifySystem[]; ashow[]
cut["ZZ1", "{m,p | p=0}", "ZZ10", "ZZ1acc"];
cut["YY1", "{p,n | n=0}", "YY10", "YY1acc"];
cut["YY3", "{m,p | m=0}", "YY30", "YY3acc"];
Print["Simplifying, please wait"]
simplifySystem[];
substituteInDef[ZZ1acc, ZZ10];
substituteInDef[YY1acc, YY10];
substituteInDef[YY3acc, YY30];
removeAllUnusedVars[];
Print["I'm normalizing, please wait ..."]
normalize[]; simplifySystem[]; ashow[]
asave["NN-serial.a"]

load["NN-serial.a"];
changeOfBasis["Y1.(p->p,N+1)", {"p", "n"}];
changeOfBasis["Y2.(p->p,N+2)", {"p", "n"}];
changeOfBasis["ZZ1acc.(m,p->p,N+2+m)", {"p", "n"}];
changeOfBasis["Z1.(m->P+1,N+2+m)", {"p", "n"}];
changeOfBasis["Z2.(m->P+1,N+2+m)", {"p", "n"}];
changeOfBasis["Z3.(m->P+1,N+2+m)", {"p", "n"}];
changeOfBasis["Z4comp.(m->P+1,N+2+m)", {"p", "n"}];
changeOfBasis["YY3acc.(m,p->p,N+M+m+2)", {"p", "n"}];
changeOfBasis["Y3.(p->p,N+2M+3)", {"p", "n"}];
```

```
changeOfBasis["Y4.(p->p,N+2M+4)", {"p", "n"}];
addlocal["DeltaW1Loc = DeltaW1"];
addlocal["DeltaW2Loc = DeltaW2"];
normalize[];
changeOfBasis["DeltaW1Loc.(p,n->p,n+N+2M+4)"];
changeOfBasis["DeltaW2Loc.(m,p->p,m+2N+2M+4)", {"p", "n"}];
Print["and now ..."]
normalize[]; simplifySystem[]; ashow[]
asave["NN-aligned.a"]
Print["Now we need to go into NN-aligned.a and uniformize some depencences --
which are actually uniform-in-context by hand, and save the result in
NN-aligned-hand.a"]
Print["Actually, this has already been done"]
InputString["So hit return to continue > "];

load["NN-aligned-hand.a"];
pipeall["YY1acc", "x.(p,n->n)", "Xpipe.(p,n->p+1,n)"];
pipeall["YY3acc", "Z4copy1.(p,n->n-N-M-2)", "Z4pipe1.(p,n->p+1,n)"];
pipeall["DeltaW1Loc", "Y4.(p,n->p,N+2M+4)", "Y4pipe.(p,n->p,n+1)"];
addlocal["temp", "Y2.(p,n->p,N+2)"];
cut["temp", "{p,n|n>=N+2}", "Temp", "junk"];
simplifySystem[]; removeAllUnusedVars[];
pipeall["Temp", "Y2.(p,n->p,N+2)", "Y2pipe.(p,n->p,n+1)"];
substituteInDef[DeltaW2Loc, Temp];
substituteInDef[Y4, Temp];
substituteInDef[ZZ1acc, Temp];
simplifySystem[]; removeAllUnusedVars[];
convexizeAll[]; ashow[]
asave["NN-local1.a"];
Print["Apparently, convexizeAll does not work as well as we hope, so you have
to do this by hand"]
InputString["So hit return to continue > "];

load["NN-local1-hand.a"];
schedule[scheduleType -> sameLinearPart]
appSched[]; asave["NN-scheduled-app.a"];
load["NN-local1-hand.a"];
applySchedule[]; asave["NN-scheduled-apply.a"];
load["NN-scheduled-apply.a"];
toAlpha0v2[];
simplifySystem[alphaFormat -> Alpha0];
reuseCommonExpr[];
simplifySystem[alphaFormat -> Alpha0];
addAllParameterDomain[];
convexizeAll[];
simplifySystem[alphaFormat -> Alpha0];
alpha0ToAlphard[]; asaveLib["NN_Hard"]
load["NN_Hard"];
analyze[];uniformQ[];
fixParameter["N", 20];
```

```
fixParameter["P", 32];
fixParameter["M", 10];
$library = Drop[$library, -1];
$library = $library /. (real -> integer);
a2v[]
```

References

[1] R L Walke, R W M Smith, and G Lightbody. 2000. "20 GFLOPS QR processor on a Xilinx Virtex E FPGA," in *SPIE, Advanced Signal Processing Algorithms, Architectures, and Implementations X.*

[2] D. E. Rumelhart, G. E. Hinton, and R. J. Williams. 1986. "Learning Internal Representations by Error Propagation," *Nature*, vol. 323, pp. 533–536.

[3] Y. L. Cun and et al., 1989. "Backpropagation applied to Handwritten Zip Code Generation," *Neural Computation*, vol. 1, no. 4, pp. 541–551.

[4] N. Sundarajan and P. Saratchandran.1998. *Parallel Architectures for Artificial Neural Networks.* IEEE Computer Society Press, California, USA, ISBN 0-8186-8399-6.

[5] "The Alpha Homepage," Information available at http://www.irisa.fr/cosi/ALPHA.

[6] D. Lavenier, P. Quinton, and S. Rajopadhye. 1999. *Digital Signal Processing for Multimedia Systems*, ch. 23. Parhi and Nishitani editor, Marcel Dekker, New York.

[7] Annapolis Micro Systems Inc. STARFIRE Reference Manual, available at www.annapolis.com.

[8] Annapolis Micro Systems Inc. at http://www.annapmicro.com.

[9] "Xilinx Inc. 2.5V Field Programmable Gate Arrays:Preliminary Product Description," October 1999. www.xilinx.com.

[10] "Celoxica Inc," Information available at http://www.celoxica.com.

[11] "Accel Inc," Information available at http://www.accelchip.com.

[12] "The Alpha Homepage at Brigham Young University," Information available at http://www.ee.byu.edu:8080/~wilde/Alpha.

[13] "The Alpha Homepage at Colorado State University," Information available at http://www.cs.colostate.edu/~kolin/Alpha.

[14] "The polyhedral library," Information available at http://icps.u-strasbg.fr/PolyLib/.

[15] C. Mauras. 1989. *Alpha: un langage équationnel pour la conception et la programmation d'arctitectures parallèles synchrones*. PhD thesis, Thèse, Université de Rennes 1, IFSIC .

[16] P. Feautrier. 1992. "Some efficient solutions to the affine scheduling problem: I. one-dimensional time," *International Journal of Parallel Programming*, vol. 21, no. 5, pp. 313–348.

[17] P. Feautrier. 1992. "Some efficient solutions to the affine scheduling problem: part ii multidimensional time," *International Journal of Parallel Programming*, vol. 21, no. 6, pp. 389–420.

[18] "PIP/PipLib 1.2.1 'fusion'," Information available at http://www.prism.uvsq.fr/~cedb/bastools/piplib.html.

[19] "IEEE Standard VHDL Language Reference Manual," 1994. ANSI/IEEE Std 1076-1993.

[20] S Derrien and S Rajopadhye. 1991. "Loop Tiling for Reconfigurable Accelerators," in *International Workshop on Field Programmable Logic and Applications (FPL'91)*.

[21] S Derrien and S Sur Kolay and S Rajopadhye. 2000. "Optimal partitioning for FPGA Based Arrays Implementation," in *IEEE PARELEC'0*.

[22] J. Burr. 1991. *Digital Neural Network Implementations – Concepts, Applications and Implementations, Vol III*. Englewood Cliffs, New Jersey: Prentice Hall.

[23] D. Zhang and S. K. Pal. 2002. *Neural Networks and Systolic Array Design*. World Scientific Company.

[24] Jihan Zhu and Peter Sutton. 2003. "FPGA Implementation of Neural Networks - A Survey of a Decade of Progress," in *13th International Conference on Field-Programmable Logic and Applications (FPL 2003), Lisbon, Portugal.*, pp. 1062–1066, Springer-Verlag.

[25] S Y Kung. 1988. "Parallel Achitectures for Artificial Neural Networks," in *International Conference on Systolic Arrays*, pp. 163–174.

[26] S Y Kung and J N Hwang. August 1989. "A Unified Systolic Architecture for Artificial Neural Networks," *Journal of Parallel Distributed Computing*, vol. 6, pp. 358–387.

[27] H T Kung. 1988. "How We got 17 million connections per second," in *International Conference on Neural Networks*, vol. 2, pp. 143–150.

[28] A T Ferrucci. 1994. *A Field Programmable Gate Array Implementation of self adapting and Scalable Connectionist Network.* PhD thesis, Master Thesis, University of California, Santa Cruz, California.

[29] James G Elridge and Stephen L Bade. April 1994. "Density Enhancement of a Neural Network using FPGAs and Runtime Reconfiguration," in *IEEE Workshop on FPGAs for Custom Computing Machines,* pp. 180–188, IEEE.

[30] Stephen L Bade and Brad L Hutchings. April 1994. "FPGA-Based Stochaistic Neural Networks," in *IEEE Workshop on FPGAs for Custom Computing Machines,*, pp. 189–198, IEEE.

[31] X. Zhang and et al. 1990. "An Efficient Implementation of the Back-propagation Algorithm on the Connection Machine," *Advances in Neural Information Processing Systems*, vol. 2, pp. 801–809.

[32] N. M. Botros and M Abdul-Aziz. December 1994. "Hardware Implementation of an Artificial Neural Network using Field Programmable Arrays," *IEEE Transactions on Industrial Electronics,* vol. 41, pp. 665–667.

[33] A Linde, T Nordstrom, and M Taveniku, "Using FPGAs to implement a Reconfigurable Highly Parallel Computer" 1992. in *Selected papers from: Second International Workshop on Field Programmable Logic and Applications (FPL'92)*, pp. 199–210, Springer-Verlag.

[34] Rafael Gadea, Franciso Ballester, Antonio Mocholí, and Joaquín Cerdá. 2000. "Artificial Neural Network Implementation on a Single FPGA of a Pipelined On-Line Backpropagation," in *Proceedings of the 13th International Symposium on System Synthesis*, IEEE.

Chapter 6

FPGA IMPLEMENTATION OF VERY LARGE ASSOCIATIVE MEMORIES

Application to Automatic Speech Recognition

Dan Hammerstrom, Changjian Gao, Shaojuan Zhu, Mike Butts*

*OGI School of Science and Engineering, Oregon Health and Science University, *Cadence Design Systems*

Abstract   Associative networks have a number of properties, including a rapid, compute efficient best-match and intrinsic fault tolerance, that make them ideal for many applications. However, large networks can be slow to emulate because of their storage and bandwidth requirements. In this chapter we present a simple but effective model of association and then discuss a performance analysis we have done in implementing this model on a single high-end PC workstation, a PC cluster, and FPGA hardware.

Keywords: Association networks, FPGA acceleration, Inner product operations, Memory bandwidth, Benchmarks

6.1   Introduction

The goal of the work described here was to assess the implementation options for very large associative networks. Association models have been around for many years, but lately there has been a renaissance in research into these models, since they have characteristics that make them a promising solution to a number of important problems.

The term *Intelligent Signal Processing* (ISP) is used to describe algorithms and techniques that involve the creation, efficient representation, and effective utilization of complex models of semantic and syntactic relationships. In other words, ISP augments and enhances existing Digital Signal Processing (DSP) by incorporating contextual and higher level knowledge of the application domain into the data transformation process. It is these complex, "higher-order"

A. R. Omondi and J. C. Rajapakse (eds.), FPGA Implementations of Neural Networks, 167–195.

relationships that are so difficult for us to communicate to existing computers and, subsequently, for them to utilize efficiently when processing signal data.

Associative networks are neural network structures that approximate the general association areas seen in many biological neural circuits. In addition to having a strong grounding in computational neurobiology, they also show promise as an approach to ISP. This is particularly true when used in multiple module configurations where connectivity, learning, and other parameters are varied between modules, leading to complex data representations and system functionality.

This chapter is divided into 4 major sections and a Summary and Conclusions section. Section 6.2 describes the basic association memory computational model used for our analysis as well as examples of its use in real applications. Section 6.3 presents a performance analysis of a single PC and PC cluster in executing this model. These systems represent the best alternatives to the FPGA solution proposed here and constitute a standard performance benchmark for the model. Section 6.4 presents an analysis of the FPGA implementation, using a memory intensive FPGA board. And, Section 6.5 compares the FPGA analysis to the PC performance results.

6.2   Associative memory

In this section we show the basic computational model that was implemented. Subsection 6.2.1 provides some historical background, while subsection 6.2.2 presents the basic model. A theoretical model is briefly presented in 6.2.3, and two example applications in subsection 6.2.4.

In traditional computing, data are stored in memory according to a direct addressing scheme. This approach requires that the computation engine be able to provide the addresses for the data being accessed. For some applications, such as the storage of large numerical arrays, creating the necessary address is not difficult, though a fair amount of effort is still involved. However, in some applications the address is unknown, or some part of a particular set of data is known and the remaining data in the set are sought. To accomplish this, a different kind of memory function, *content addressing*, is used.

The most common kind of content addressing is *exact match*, where part of the data is known and is matched exactly to the contents of a location. There are a number of applications for exact match association, from addressing in cache memory to routing tables. There have been a number of parallel, exact match, association engines [4] created over the years. Also, there are fairly efficient algorithms for exact match association such as hash coding.

However, if the data we have are close but not exact to data stored in the memory, a *best match* content addressable memory would be useful. This memory would find the content that had the "closest" match according to some metric. And it turns out that in many kinds of ISP this is exactly what we want to do, that is, given noisy, corrupt input, we want to find the original content that is the "most likely" in some sense.

An exact match can be determined fairly easily, but a best match is a little more complicated. Unlike using hash based addressing for exact match association, there is no short cut[1] to finding the best match. For arbitrary searches it requires that every item in memory be examined. This is one reason why best match associative memory is not used extensively. However, by using distributed data representation, efficient best match associative processing can be realized. Best-match association also seems to be something that is commonly performed by neural circuitry.

Best-match association finds the "closest" match according to some metric distance defined over the data space. A simple and often used metric is Hamming distance, however more complex vector metrics can also be used. In some applications, for example, the distance between two vectors may have more to do with the meaning assigned to the words represented by the vectors than the number of letters different or the distance the letters are from each other.

A common example of best match processing is Vector Quantization, where the metric can be a Euclidean distance in a high dimensional vector space. We can be even more sophisticated and let our metric be based on probability estimates. Then under certain conditions our best match system returns the match that was the most likely to have been the source of the data being presented, based on data and error probabilities.

There is always the brute force approach which allocates a simple processor per record (or per a small number of records), all computing the match in parallel with a competitive "run-off" to see who has the best score. Though inefficient, this implementation of best-match guarantees the best results and can easily be used to generate optimal performance criteria, but it is too compute intensive for most real applications.

6.2.1   Previous Work

We cannot do justice to the rich history of associative processing in this chapter, but it is important to mention some of the key developments. It is not even clear who started using these techniques first, though Donald Hebb [10] used the term "association" in conjunction with arrays of neurons. Perhaps the first visible development was the Lernmatrix of Karl Steinbuch [23]. Stein-

---

[1] One of the best algorithms is the KD tree of [18]. *Bumptrees for efficient function, constraint, and classification learning.* Advances in Neural Information Processing Systems 3, Denver, Colorado, Morgan Kauffmann., but even these have significant limitations in many applications, for example, such as dynamically learning new data during execution.

buch's model, and similar early models by Willshaw and his group [7] and Kohonen [14], were based primarily on linear association. In *linear associa-tion*, the connection weight matrix is formed by summing the outer products (of the input vector with its output vector). Recall is done via a simple matrix / vector multiplication. If a training vector is input, the output vector associ-ated with it will be output. If a noisy version of a training vector is input, the output will be a combination of the various training vectors that are present in the noisy input vector. The components of the associated output vectors will be proportional to the degree to which the training vectors are represented in the noisy input, with the most prominent (usually the training vector before the noise was added) input vector being the most prominent in the output. To be effective, the trained vectors need to be linearly independent.

In the 80s a number of important, non-linear, variations of the linear asso-ciator were developed. Perhaps the most influential models were developed by Palm [20], Willshaw [24], and Hopfield [12]. The Hopfield model was based on the physics of spin glasses and consequently had an appealing the-oretical basis. In addition, the model required asynchronous operation. For additional models the interested reader is referred to the work of Amari [1], Hecht-Nielsen [11], Anderson [2, 3], and Lansner [15].

6.2.2   Sparse Association Networks with k-WTA Output

6.2.2.1   The Palm Model.   A simple, but effective association algo-rithm was developed by G. Palm and his associates [20, 21]. A key aspect of this model is a very sparse data representation and *k-WTA* (K Winners Take All) output filtering. The k-WTA operation increases the effectiveness of the memory and also eliminates the need for the training vectors to be linearly independent.

Although non-binary Palm models are possible, for the sake of this discus-sion we will assume that all input and output vectors are binary as is the weight matrix. In the Palm model there are certain vectors that are used to train the network. Each vector has a sparse representation, which is enforced by limit-ing the number of 1s in each vector to a fixed value, k. This limited activation enforces a certain distribution of the data representations, where each node may participate in the representation of a number of different stored elements. Sparse, distributed representations are efficient, because of the combinatoric effects of very large vector spaces. And they are massively parallel because large numbers of processing nodes operate concurrently to compute the out-put. Distributed representations also enhance fault tolerance.

The algorithm uses here stores mappings of specific input representations x_i to specific output representations y_i, such that $x_i \rightarrow y_i$. The network is constructed with an input-output training set (x_i, y_i), where $F(x_i) = y_i$. The mapping F is *approximative*, or *interpolative*, in the sense that $F(x_i + \varepsilon) = y_i + \delta$, where $x_i + \varepsilon$ is an input pattern that is close to input x^μ being stored in the network, and $|\delta| \ll \quad |\varepsilon|$ with $\delta \to 0$, where δ and ε represent noise vectors. This definition also requires a metric over both the input and output spaces.

In Palm's model, the input, X, and output, Y, vectors are binary valued (0 or 1) and have equal dimensions, n, since we are using a feedback network (output to input) to implement *auto-association*. There is also a binary valued n by n matrix, W, that contains the weights. Output computation is a two-step process, first an intermediate sum is computed for each node (there are n nodes) by thresholding

$$s_j = \sum_i w_{ji} x_i$$

where w_{ji} is the weight from input i to node j. The node outputs are then computed

$$\hat{y}_j = f(s_j - \theta_j)$$

The function, $f(x)$, is a step function ("hard max"), it is 1 if $x > 0$ and 0 if $x \leq 0$, leading to a threshold function whose output $\hat{y}_j$ is 1 or 0 depending on the value of the node's threshold, θ_j. The setting of the threshold is discussed below. In Palm's basic model, there is one global threshold, but more complex network models relax that assumption.

Programming these networks involves creating the appropriate weight matrix, W, which is developed from "training" vectors. In the case of auto-association these represent the association targets. In the Hopfield energy spin model case, for example, these are energy minima. There are m training patterns

$$S = \{(X^\mu, Y^\mu) | \mu = 1, \cdots, m\}$$

In auto-associative memories the output is fed back to the input so that $X = Y$. The weights are set according to a "clipped" Hebbian rule. That is, a weight matrix is computed by taking the outer product of each training vector with itself, and then doing a bit-wise OR of each training vector's matrix

$$\bar{w}_{ij} = \bigcup_{\mu=1}^{M} (x_i^\mu \cdot (y_j^\mu)^T)$$

An important characteristic is that only a constant number of nodes are active for any vector. The number of active nodes, k, is fixed and it is a relatively small number compared to the dimensions of the vector itself – Palm suggests $k = O(\log(n))$. This is also true for network output, where a global threshold value, θ, is adjusted to insure that only k nodes are above threshold. Although reducing theoretical capacity somewhat, small values of k lead to very sparsely activated networks and connectivity. It also creates a more effective computing structure, since the $k-$Winners Take All (k-WTA) operation acts as a filter that tends to eliminate competing representations.

Although a Palm network is similar to a Hopfield network in function and structure, the key differences of clipped Hebbian weight update and k-WTA activation rule make a big difference in the capability of the network. In theoretical studies of these networks, researchers have developed asymptotic bounds for arbitrarily scalable nets. Palm has shown, for example, that in order to have maximum memory capacity, the number of 1s and 0s in the weight matrix should be balanced, that is $p_1 = p_0 = 0.5$, where p_1 is the probability that a bit is 1 and p_0 the probability that a bit is 0. The memory capacity per synapse then is $-\ln p_0 \leq \ln 2$. In order for this relationship to hold, the training vectors need to be sparsely coded with at most $\log_2(n)$ bits set to 1, then the optimal capacity of $ln\ 2 = 0.69$ bits per synapse is reached.

6.2.2.2　　Other Model Variations.　　Palm networks are robust and scale reasonably well. However, there are still some limitations:

1　The model as currently defined does not allow for incremental, dynamic learning.

2　Even though the vectors themselves are sparsely activated, as more information is added to the network, the weight matrix approaches 50% non-zero entries, which is not very sparse. For a 1M network, that is 500Gb with each node having a fan-in (convergence) of 500K connections. Cortical pyramidal cells typically have a convergence of roughly 10K.

3　Time is not factored into the model, so that temporal data is not cleanly integrated into the association process.

4　The network requires that input data be mapped into a sparse representation, likewise output data must be mapped back to the original representation.

Concerning the last item, algorithms are available [8] for creating sparse representations ("sparsification"). Our experience has shown that for many applications getting a sparse representation is straightforward. Once a sparse representation is determined, it has not been difficult to map from an external representation into the sparse representation and back again. This is important, since much of power of these nets is due to the sparse, distributed representation of the data.

The Palm model is simple and easy to work with and it is a good start for looking at the various performance issues concerning hardware implementation of such models. However, it is likely that real applications will use more complex models. One model that we are also working with is the Bayesian Confidence Propagation Neural Network (BCPNN) developed by Anders Lansner and his group [15, 16] at the Royal Institute of Technology in Stockholm, Sweden. BCPNN addresses several of these issues and demonstrates a number of useful behaviors. For example, in addition to capturing high order Bayesian statistics, they also demonstrate the Palimpsest property, where older information is forgotten first as the network approaches full capacity. Another key characteristic of BCPNN networks is its ability to do "second match," where, when the network is allowed to habituate to its first output, it will then return the next best match, etc.

Lansner has likened a single node in the BCPNN as comparable to a cortical minicolumn and the Winner-Take-All group as a cortical hypercolumn. The model has its origin in neuroscience and the study of cortical circuits. One disadvantage of the model is the need for fairly high precision for the weight update computations. Both the Palm and BCPNN models can be reformulated as pulse or spiking based models. Pulse models have certain implementation advantages and handle the temporal dimension more efficiently [17].

It is also likely that real systems will use collections of models with different learning capabilities, thresholding systems, and connectivity. Examples of such models can be found in O'Reilly [19].

6.2.2.3   Simulation.   We have developed an association memory simulation environment, Csim (Connectionist SIMulator). Csim is object oriented and is written in C++. It uses objects that represent clusters, or "vectors" of model nodes. It can operate in a parallel processing mode and uses the Message Passing Interface (MPI) to do interprocess communication over multiple processors. We have a small set of association network models operating on the simulator. Because the simulator has a simple command line interface, and because C++ and MPI are standards, it operates under Windows 2K, Linux, and SGI Unix. We use the simulator on our Beowulf cluster (8 1GHz Pentium IIIs) and on NASA's SGI supercomputers at the Ames Research Center in Mountain View, California [25].

6.2.3   An Information Theoretic Model of Association

A communication channel is a useful way to conceptualize the operation of a simple, auto-association network. For example, in speech recognition when we say a word, we have a specific word in mind. We then encode that word into a form (speech) that is transmitted over a noisy channel (sound waves in air) to the computer, which then decodes the received wave back into the original word. The channel, Figure 6.1, is our mouth, tongue, vocal chords, the physical environment, as well as the computer's microphone and signal processing.

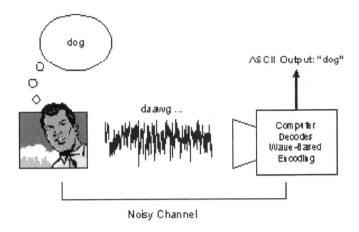

Figure 6.1. The "Decoder" Model of Association

This channel is noisy because there is significant variation in the way the word is encoded – people say the same things differently from one instant to the next, then add to that speaker variations, ambient noise in the environment, etc. The decoding process uses digital signal processing algorithms at the front end and complex "intelligent signal processing" algorithms, such as Hidden Markov Models, that model higher order dependencies at the high-end.

6.2.3.1   A Single Channel.   An auto-associative memory can be modeled as the decoder in a simple communication channel, Figure 6.2. An input generates a message y that is encoded by the transmitter as x. The message is then sent over a noisy channel and x' is received. The decoder decodes x' into, y', what is believed to be the most likely y to have been sent.

Messages are generated with probability $p(y)$. The received message, x', is the transmitted message with errors. This message is sent to the decoder. The decoder has, via many examples, learned the probabilities $p(y)$ and $p(x'|y)$. The decoder uses these data to determine the most likely y given that it received x' (based on what knowledge it has)[2]

$$p(y|x') = \frac{p(x'|y)p(y)}{p(x')} = \frac{p(x'|y)p(y)}{\sum_x p(x'|x)p(x)}$$

Message x_i is transmitted, message x' is received, for simplicity assume that all vectors are N bits in length, so the Noisy Channel only inserts substitution errors. The Hamming Distance between two bit vectors, x_i and x_j,

---

[2]Note, the probability $p(x'|y)$ reflects the channel error probabilities.

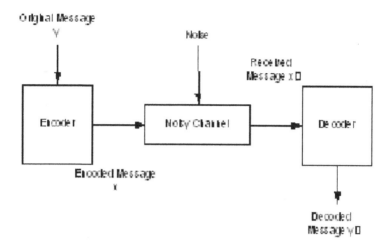

Figure 6.2. Communication Channel - The Decoder Is An Auto-Associative Memory

is $HD(x_i, x_j)$. Assume that the Noisy Channel is binary symmetric with the probability of a single bit error being ε, and the probability that a bit is transmitted intact is $(1-\varepsilon)$. The error probabilities are independent and identically distributed and ε < 0.5. Under these circumstances it can be shown that the Palm memory approximates a Bayesian inference, that is, the Palm associative memory will generally recall the most likely original message, x_i, from a noisy received message x'.

Our primary objective, therefore, is to develop an associative memory that performs such probabilistic inference in real-time over very large data sets with sophisticated metrics. We believe that a large capacity version of this memory, implemented in an inexpensive chip, has significant commercial value.

6.2.3.2   Scaling.   Association networks have much promise as a component in building systems that perform ISP. There are a number of problems that need to be solved before distributed representation, best match associative memories can enjoy widespread usage. Perhaps the most serious concerns scaling to very large networks. Even though the vectors used by the Palm memory are sparse, as you add training vectors to the memory, the Weight matrix becomes decidedly non-sparse. Palm has shown that maximum capacity occurs when there are an equal number of 1s and 0s. This is 50% connectivity. This is not very biological and it does not scale. This level of connectivity causes significant implementation problems as we scale to relatively large networks.

If you randomly delete connections from a Palm network, performance degrades gradually to a point where the network suddenly fails completely. There is no sudden appearance of emergent functionality, yet cortex has roughly 0.0001% connectivity. Clearly we're doing something wrong.

The associative networks in the cortex seem to be fairly localized. Just as in silicon, connections are expensive to biology, they take space and metabolism, increase the probability of error, and they require genetic material to specify connectivity, [6], Cortex: Statistics and Geometry of Neuronal Connectivity [6].

- "metric" (high density connections to physically local unit, based on actual two-dimensional layout) and

- "ametric" (low density point to point connections to densely connected groups throughout the array)

Braitenberg [5] used $\sqrt{n}$ hypercolumns each with $\sqrt{n}$ neurons. BCPNN uses *1-WTA* per hypercolumn. Although beyond the scope of this chapter, it is possible to create a multiple channel model of associative memory that models the hierarchical structure seen in primate neocortex. Here the memory is modelled as parallel channels that have unique inputs with some overlap. Likewise, each channel uses inputs from other parallel channels to provide information in its own decoding process. Although not required for the current models, reproducing more biological connectivity patterns will be required in the long run.

6.2.4   Example Applications

To see how best match association memories can be used, we briefly describe two applications we are working on at OGI, the first is for a sensor fusion system for the US Air Force, and the second, a robotic control system for NASA.

6.2.4.1   Enhanced Visual System for Pilot Assist.   With Max-Viz, Inc., Portland, OR, we are developing an Enhanced Vision Systems (EVS) for aircraft [13]. This system is designed to aid the pilot during approach and landing in extremely low visibility environments. The system will have visual sensors that operate at different frequencies and have different absorption spectra in natural phenomena such as fog. The EVS takes these various images and fuses them in a Bayesian most-likely manner into a single visual image that is projected on a Head-Up Display (HUD) in front of the pilot.

Computing a Bayesian most likely fused image at the necessary resolution and video rates is extremely computationally intensive. So we are developing a system that approximates Bayesian accuracy with far less computation by using an association engine.

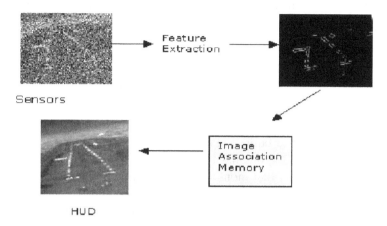

Figure 6.3.   EVS Association Application

In this system basic visual-like image pre-processing is used on each image to find key elements of the background, generally the edges of the runway. Since these features tend to be consistent across all images from the various sensors, the feature vectors are just added together, then thresholding is performed to reduce the number of features, creating a sparse representation of the strongest features. This cumulative feature vector is then input to an associative memory that returns the features of a stored image. These features are then used to do a hash (exact match association) to a database that contains the real visual image that is then projected onto the HUD. An overview of the processing steps for each video frame is shown in Figure 6.3.

6.2.4.2   Reinforcement Based Learning for Robot Vision.
We are developing a robotics control system for NASA[3] that uses biologically inspired models for bottom-up/top-down sensorimotor control with sparse representations. In this research we plan to demonstrate a system that performs a simple cognitive task, which is learned by the system via reinforcement, in much the same way a monkey would learn the task. The selected task is block copy where the animal copies a model of a block layout in one work area into another work area using a set of predefined blocks in a resource area. An example is shown in Figure 6.4. This task is known to illustrate important properties of the primate brain in terms of:

- Visual processing and representation, and visual short term memory;

[3]"Biological computing for robot navigation and control," NASA, March 1, 2001, Three years, $1,393K. PI: Marwan Jabri, Co-PIs: Chris Assad, Dan Hammerstrom, Misha Pavel, Terrence Sejnowski, and Olivier Coenen.

- Sequencing and planning, and the requirements of short-term microplans and longer term plans;

- Memory based processing, in particular, top-down information processing; and

- Spatial Representations.

Data on human and monkey experiments on various block copying tasks are available. However to build a machine that performs such a task using a biologically based model, one has to think how a monkey is trained to perform such a task. We know that monkeys learn complex sequences by maintaining their attention and their rewards as they are proceeding. To build a machine to learn to perform such a task, we need to define the set of sub-skills that the machine must have the minimum hardware capabilities for such learning to take place.

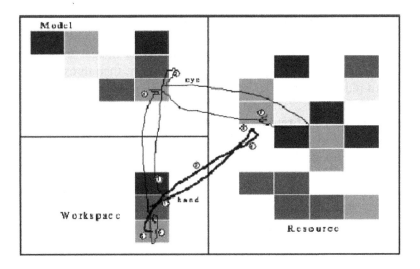

Figure 6.4.   Active Vision Block Copying Task

Our approach is to take preliminary information on the systems in the primate brain that neuroscientists believe are used in solving this task and start with highly simplified versions, but with all the required brain areas represented. The architecture of our current system is shown in Figure 6.5. Although complex, each subcomponent is basically an associative memory, though different modules have slightly different learning rules and connectivity patterns.

6.3   PC Performance Evaluation

In this section we present results and analysis from the implementation of the basic Palm model on a PC Workstation, single processor, and a multi-processor PC cluster. As we move to larger, more complex cognitive systems built from diverse association modules, the ability to execute these large models in close to real time, which would be required by most applications, becomes a problem even for high end PC Workstations. The primary purpose of this chapter is to study the use of an FGPA based emulation system for association networks. However, as part of any specialized hardware design, it is important to assess the cost performance of this hardware in comparison with high performance, commercial off-the-shelf hardware.

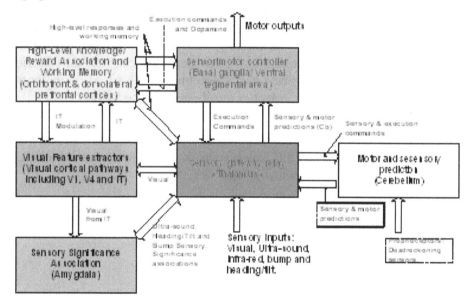

Figure 6.5.   Architecture of Block Copy System

Thanks to Moore's law, the performance-price of commercial hardware has made amazing improvements over the last 10 years. It is possible to have the same performance in a desktop machine or notebook that was only available in super computers 15 years ago. Consequently, any application that is a candidate for special purpose hardware implementation must first be measured on general-purpose hardware, since in many cases specialized hardware will not provide a sufficient performance improvement to justify the cost of the extra hardware.

The Palm algorithm has two basic operations:

1 A matrix vector multiply of a very sparse binary vector by a very large, and, for the time being, not so sparse binary matrix.

2 A *k-WTA* function on the non-sparse, non-binary vector that is the product of the matrix vector multiply, the output being another very sparse binary vector.

In our experience the matrix-vector multiply is the most time consuming and the most memory intensive. For the real applications we are contemplating, the input and output vectors contain 10s to 100s of thousands of elements. What is clear is that we need to use large, commercial memory to get the necessary capacity, density, bandwidth, and latency, which implies state of the art SDRAM. Since the matrices we use are very large and not sparse[4], consequently the entire matrix cannot be stored in any reasonably sized cache. For the inner product operation every element fetched by the cache is only used once. Since, the entire matrix must be traversed for one operation, the cache provides no performance benefit. So the matrix-vector multiply requires a long, continuous stream of data from memory with the memory bandwidth being the major factor in determining system performance. So in this situation a data caches does not contribute much performance. This is also true for the clock speed of the processor, which is mostly just waiting for data.

6.3.1   Single PC

For the single, high-end, PC experiments described here we used a DELL Dimension 8100 with the following hardware and software:

- CPU: Pentium 4 1.8GHz

- Chipset: Intel® 850

- Memory: 1024MB RDRAM

- System Bus: 400MHz

- Microsoft Windows 2000 Professional

- Microsoft Visual C++ 6.0

- Intel VTune Performance Analyzer 6.1

Using the Csim version of the Palm algorithm, we ran several experiments measuring average memory bandwidth and nodes updated per second. The algorithm that Csim implemented is not necessarily optimal. In this algorithm, both for the sparse and full representations, the weight matrix is stored by row. One row was read at a time and an inner product performed with the input

---

[4]A sparse matrix generally has fewer elements and a matrix with very large dimensions may still fit entirely into the cache.

(test) vector, which, incidentally, is stored as a set of indices. This algorithm was implemented in the same way by both the Csim and FPGA versions.

Another approach to the matrix vector inner product with very sparse data structures is to store the weight matrix by columns and then only read the columns that correspond to a non-zero element in the input vector. We are studying how best to do this in the FPGA implementation, which will be presented in future papers. In Figure 6.6, the x-axis is the vector size, and the y-axis is the *inner-product* memory bandwidth (MBytes/sec). The diamond-dotted curve is for the sparse weight matrix representation (where only the indices of the non-zero elements are stored) with the compiler set to maximum speed optimization. The square-dotted curve is for the full binary weight matrix (one bit per weight, 8 weights per byte, etc.) with the compiler set to maximum speed. Note both representations are of the same matrix, this is true in Figure 6.7 as well.

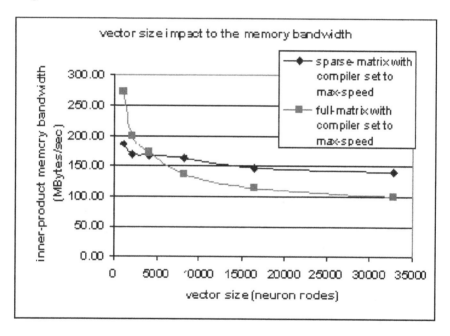

Figure 6.6.   he relationship between the number of network nodes (vector size in nodes) and the single PC inner-product memory bandwidth

In Figure 6.7 we can see that as the vector size increases (the number of training vector numbers used is 0.69 of the vector size, n, and the number of active nodes, k, is approximately $log_2(vector\ size)$), the *inner-product* memory bandwidth is reduced. For the sparse-matrix representation, the bandwidth goes to 140MB/sec when the vector size, n, is about 32K. For the full-matrix representation, the bandwidth decreases to 100MB/sec when the vector size is about 32K. As the vector size increases, L1 cache misses increase, the average memory access time increases, and the bandwidth decreases.

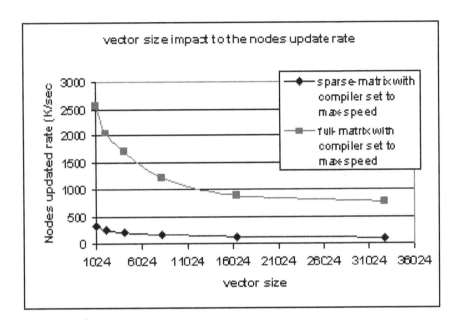

Figure 6.7.   Vector size versus node update rate

Incidentally, with the current algorithm it is not necessary for the hardware to read the entire weight matrix. If you assume that each test vector index is in a different word, which is a worst case assumption, for each test vector, we will only need to read $\lceil kn/32 \rceil$ words of weight matrix data, where for these experiments $k=log_2(n)$. The FPGA implementation does this explicitly, the PC does it by virtue of the cache only fetching the block that contains a element corresponding to a test vector index.

6.3.2   PC Cluster

One common approach to enhancing the performance of the basic PC platform is to use a parallel PC configuration, generally referred to as a PC cluster [22]. A common cluster configuration is to use Linux based PCs in what is referred to as a "Beowulf cluster," where a number of PCs are connected to a common broadband network. Software is available that allows these PCs to share tasks and communicate data and results.

In addition to systems support there are programming environments for supporting parallel programs for such clusters, the most commonly used is MPI (the Message Passing Interface) [9]. Csim has a parallel execution option that is based on MPI. In this subsection, we present the results of executing Csim on a small Beowulf cluster, where a single association network is spread across the processors as shown in Figure 6.8.

The purpose of this experiment was to understand the overhead required to execute a single association network across multiple processors. However, real implementations will probably use multiple association networks, with each module assigned to a processing node (whether it be a PC or an FPGA).

There are two components to the Palm algorithm that require different approaches to parallelization. The first is the matrix-vector multiplication. Here the weight matrix is equally divided into p groups of r rows each $(n = p \times r)$, each processor is assigned a group of r rows. The entire input vector is broadcasted to all processors so that they can perform a matrix-vector *inner-product* on those rows of the weight matrix that are assigned that processor.

The second part of the algorithm involves the k-WTA. Each processor computes a k-WTA on the r node portion of the output vector that it has (those k nodes allocated to it). These $(p - 1)$ $k-$element vectors are then sent to processor 0 (the root process), which performs another k-WTA over the total $p \times r$ elements. Since k is usually much smaller than the vector dimension, this approach is reasonably efficient and guarantees that the final k-WTA is correct. Also, the k-WTA is only performed after all the inner products are complete, which is generally the larger time component.

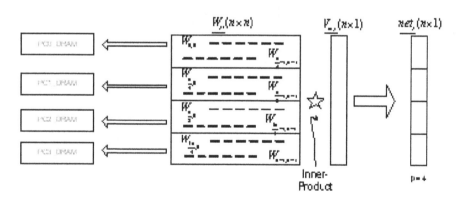

Figure 6.8. Weight Distribution (4 processor cluster)

The cluster gets its performance by dividing the computation into p equal parts that can be computed concurrently. This also increases the effective memory bandwidth by p. Ideally the speed up would be p. Unfortunately, there is no free lunch, since there is some extra overhead that parallelism creates. In our simulator this overhead has two components: 1) the broadcast of the input vector to all the processes, and 2) The broadcast of the k local winners to the root processor and the computation of the final k-WTA.

The important question then is, what is the relationship between speed up and the number of processors, p? For these experiments, we used a small PC cluster environment with 8 DELL Dimension L1000R computers. Each node was an Intel P3 1.0GHz CPU, 512MB PC133 memory. The OS is RedHat Linux 7.1. The compiler is G++ 2.96.

A summary of the results is shown in Table 6.1. The number of training vectors, m, was approximately 0.69 the vector size, n. The number of active nodes, k, was $log_2(n)$. The nodes per second per PC is smaller than for the 8100 which is most likely due to the reduced memory bandwidth of the P3 based Dimension L1000R.

The most important result is that for the larger network, going to more parallelism does not add significant overhead. For the 64K node network, only about 8% of the performance is lost in going from 2 PCs to 8 PCs, i.e., we are only 8% short of a linear performance improvement of 4x.

Table 6.1.   IPC cluster (8 processors) simulation results

Number of PCs	Vector Size, n	Knodes per second	Normalize node rate by 2 PC Number
2	4096	410	1.00
4	4096	683	0.83
8	4096	1024	0.63
2	8192	390	1.00
4	8192	745	0.95
8	8192	1170	0.75
2	16384	356	1.00
4	16384	683	0.96
8	16384	1170	0.82
2	32768	312	1.00
4	32768	607	0.97
8	32768	1130	0.91
2	65536	289	1.00
4	65536	560	0.97
8	65536	1057	0.92

6.4   FPGA Implementation

In this section we first present the architecture of a memory intensive FPGA board implementation and then second, an analysis of the performance of that board on our association memory implementation.

FPGAs are a remarkable technology, allowing a rapid, soft reconfiguration of on chip hardware to customize the hardware to a wide range of applications. As Moore's law continues, FPGAs gain in density and speed at the same rate as state of the art microprocessors. And although FPGAs pay a price in speed and density for their flexibility, for certain applications with a sufficient number of parallel operations they can offer better performance-price and power dissipation than state of the art microprocessors or DSPs.

Because of this ability to leverage parallelism, FPGAs have been used extensively in image processing applications and are beginning to be used in neural network applications, where incremental learning by on-line weight modification can be implemented concurrently and with any precision. Most of these benefits assume that the network fits entirely on a single FPGA and that on chip RAM, typically on the order of a few megabits of SRAM, is used to store the weights and other parameters. As discussed earlier in this Chapter, we believe that the network models required for many applications will be quite large, easily outstripping the capacity of the on-chip memory in a state of the art FPGA. This raises an important question about whether the FPGA is still a useful platform for the emulation of such networks. One possibility is to store the network parameters in a state of the art memory connected directly to the FPGA instead of inside the FPGA, but does this provide a sufficient performance-price advantage? In this subsection we study this question for our model.

These programs do have a fair amount of opportunity for parallel computation, especially where only a few bits of precision are required. If most of the model state is stored in external DRAM, the memory bandwidth is the primary contributor to systems performance, since, for our algorithms, even a modest sized FPGA can keep up with the memory access rate for even very fast state of the art SDRAM. We believe that an FPGA connected directly to state of the art DRAM in this manner is a useful platform for a large range of image and pattern classification problems.

6.4.1 The Relogix Accelerator Board

The Relogix Mini-Mirax Memory-Intensive Reconfigurable Accelerator Card is a stand-alone card containing one FPGA+SDRAM pair. Each accelerator has 1.6 GByte/s of dedicated memory bandwidth available for memory-intensive applications. Each accelerator card is accessed by application software running on the Unix host via an IDE/FireWire bridge. The objective of this board is to take advantage of inexpensive SDRAM PC memory, FPGAs and high speed interfaces to maximize performance-price in a reconfigurable accelerator. The Mini-Mirax is organized as an incremental brick, so that many Minis can be connected together in a single system. The basic board layout is shown in Figure 6.9.

The PQ208 pinout accepts any size Spartan-IIE chip. Spartan-IIE, XC2S50E-100, 150, 200 or 300 FPGAs may be installed at assembly. They all have the same pinout, 142 general-purpose I/Os, and four global clock inputs, and differ in only the amount of logic and memory. By using smaller FPGAs, we can stay in a QFP (Quad Flat Pack) for the early boards, since it reduces costs. And because our algorithms make such extensive use of the SDRAM, there is a point of diminishing returns as one goes to larger FPGAs.

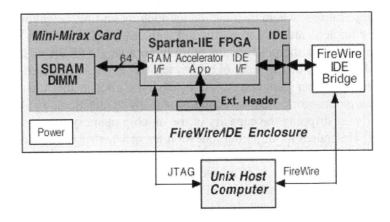

Figure 6.9. he basic FPGA card

Connected to each FPGA is a single DIMM socket, which supports a 10ns DDR (Double Data Rate) 64-bit pathway (two words are delivered in every 10ns clock) into the FPGA. With today's memory technology this allows up to 512MB (soon to be 1GB) of memory to be directly connected to each FPGA. Finally there is an IDE to FireWire chip that creates the external FireWire interface for the board. An external header uses 20 uncommitted FPGA I/Os that can be connected to other Mini cards or external devices such as sensors. A JTAG daisy chain header connects the FPGA to a Xilinx interface cable for programming by Xilinx software. Many Minis can be used in one JTAG loop. One or more Mini cards are installed in one or more FireWire/IDE enclosures, which provide an IDE to FireWire bridge, power and cooling, and a connection to a Linux, OS-X or Solaris host.

We are developing a set of IP that will be a more highly abstracted interface to the DRAM controller, the IDE controller and the header pins. The DRAM and IDE controllers are also IP that sits inside the FPGA. The user develops their own IP for the FPGA in Verilog or VHDL and then connects it to the Mini-Mirax "Operating System" IP and loads that into the FPGA. In addition, there is software on the PC Host that communicates with this "OS," in starting the user subsystem as well as providing debug services.

The primary reason for building this board is that we want to be able to place state of the art SDRAM, in the form of a DIMM, immediately next to a state of the art, FPGA. The goal is to leverage the maximum bandwidth of the largest commercially available memory. At the writing of this Chapter we are not aware of any commercial FPGA boards that provide as much directly accessible SDRAM to each FPGA.

The system will obviously be most cost-effective with larger numbers of parallel, low-precision operations, a configuration where typically there will be more computation per bit fetched from the SDRAM. The board will track the latest memory density (commercial SDRAM) and bandwidth, following Moore's law at the same rate as a general purpose microprocessor.

We do not yet have a functioning board, so we cannot provide exact performance measurements. In the next subsection we present an analysis of the expected performance of the Mini-Mirax board on the same association algorithm used throughout this paper.

6.4.2 Performance Analysis of Simulation on FPGA

A block diagram of the FPGA Palm implementation is shown in Figure 6.10. This implementation is quite similar to that of the PC, in the sense that the weight values are stored in the external SDRAM with the *inner-product* and *k-WTA* performed by the FPGA. Because the board implementation is simple and there are no complex entities such as multi-level caches, the results presented here should be a reasonably accurate representation of the performance we expect from the actual board. Also, we have implemented pieces of the Palm model and run them on other commercially available FPGA boards.

6.4.2.1 FPGA Functional Block Description. There are a number of assumptions that were made about the Palm implementation on the Mirax board:

- A single FPGA / SDRAM board is used for analysis, even though the Mini-Mirax system supports multiple FPGA configurations. Likewise, it is likely that the communication overhead for connecting multiple boards, as a percentage of the total computation time, will be much smaller than with multiple PCs in a cluster.

- Only a full binary weight matrix implementation is assumed where each bit is a binary weight. One 64-bit word at a time is fetched from the SDRAM. Only words that have at least one test vector index pointing into them are fetched.

- The input test vectors are sparsely encoded and the time transfer of data to and from the PC host is ignored. Each vector is assumed to have single bit precision and consists of a list of indices of the weight bits in

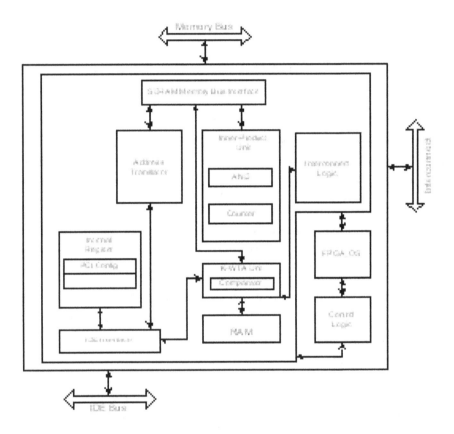

Figure 6.10.   FPGA functional block diagram

the matrix row. When a test vector is processed, this list of indices is brought from the PC into the FPGA.

- The *inner-product* produces a vector of sums. This vector is generally not sparse. The bits required to represent each sum is $\lceil \log_2 k \rceil$ bits, and there are k sums. The k-*WTA* is computed dynamically, that is, after each inner-product element is generated it is compared to the existing k largest sums, which, along with their indices, are stored in SRAM in the FPGA. The new index is then either inserted or discarded. This approach eliminates the need to move the sums off chip and then back again to form the k-*WTA*.

- The final output vector is sparse and its indices are written to the PC host, thus completing the computation of a single vector. The time to move the final result is also ignored.

- There are a number of additional optimizations that are possible which we have ignored here in favor of keeping the PC and FPGA implementation as close as possible. Future work will examine the effect of such optimizations for both platforms more closely.

Inner-product operation: The input vector $V_{in,i}$ (represented by its active nodes' indices) is transferred to the *IDE Interface*. The fetching of the weight matrix rows begins, with the weight matrix row addresses being generated by the *Address Translator*. The SDRAM transfers the data requested by the FPGA a single 64-bit word at a time through the *SDRAM Memory Bus Interface*. There are a number of ways a binary inner product can be performed. But a simple look up and count is all that is required. Because of the sparse representation of the test vector, many words will not have a matching index, and those that do will generally only have one or two matches.

k-*WTA* operation: After executing the inner-product for each output vector sum, the FPGA, which keeps a list of the k largest sums, checks to see if the resulting element is larger than the smallest of the current k largest sums, this requires a single comparison. If the new element is larger, then it needs to be inserted into the list of k largest in the proper, sorted, position.

6.4.2.2   Performance analysis.   Inner Product Computation in the FPGA: Analysis indicates that, because of the very sparse input vectors, this simple *inner-product* can easily be computed at a rate that matches the rate at which the weight matrix words are streaming into memory. So the *inner-product* time is determined exclusively by the memory bandwidth. For sequential access the DDR memory interface in the Xilinx Sparten IIE approaches a 1.6GB/sec. bandwidth. However, for non-sequential, random access, the time is approximately about 6x slower or about 60ns to access two 64-bit words.

For the full matrix representation, there are n rows, but only k words are read per row and these are read in a non-sequential, random access mode:

$$\text{Full Inner Product Memory Access Time (sec)} = (nk)\, 3x10^{-8}$$

The number of active nodes, k, is much smaller than the number of rows, n. So on average about k words will be read, since the odds of more than one index falling into the same word is quite low.

k-WTA: For each insertion into the sorted list of k sums, we will assume an average of $k/2$ comparisons are performed. Initially there will be many insertions, however, near the end of the inner product operation there will be few insertions, since the probability of any sum being in the k largest decreases. A pessimistic assumption is that for each item being sorted, the probability of an insertion is $1/2$. Since there are n items examined and if we can do one comparison per 100 MHz clock (10ns), the time for *k-WTA* is:

$$\text{Time for } k\text{-}WTA \text{ (sec)} = (nk/4)\, 10^{-8}$$

Table 2 shows the execution times for the inner product and *k-WTA* operations. We can see that the inner product memory access time dominates. Overlapping the *k-WTA* operation with the inner product saves some time, but not much.

Table 6.2.   Memory Access Time (Inner Product Time) and k-WTA time per Network Size

Vector Size, n	Full Access Time(usec)	k-WTA Time(usec)
1024	307.2	25.6
2048	675.8	56.3
4096	1474.6	122.9
8192	3194.9	266.2
16384	6881.3	573.4

6.5   Performance comparisons

Figure 6.11 shows the node update rate for the P4 and the P3 cluster. The x-axis is the node size. The y-axis is the node update rate (Knodes/sec). The square-dotted curve is the cluster implementation. The diamond-dotted curve is the simulation result from the P4 for full-matrix vector.

Because the computers in the PC cluster are much slower than the P4, and PC133 memory in the PC cluster is much slower than the Rambus memory in P4, the entire PC cluster implementation did not do significantly better than the P4 implementation. This is an interesting result since the cluster machines are only one generation older than the P4 system.

Figure 6.12 shows the expected node update rate comparisons for the P4 and a single FPGA board (one FPGA/DIMM pair). The x-axis is the vector size (number of nodes per vector), n. The y-axis is the node update rate (Knodes/sec). The square-dotted curve is the FPGA implementation. The diamond-dotted curve is the simulation result from the P4 for the full-matrix vector. From Figure 6.12, we can see the FPGA implementation has a performance advantage over the P4 implementation, which increases as the network size increases.

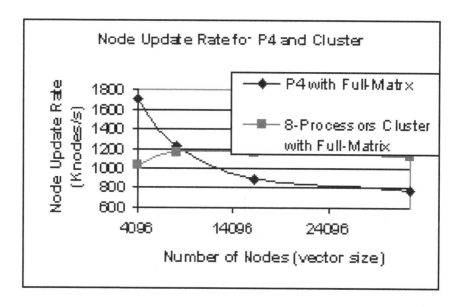

Figure 6.11. shows the node update rate, node outputs computed per second, for the P4 and the P3 cluster vs. the vector dimension.

In Figure 6.13, we show the node update rate for the P4 and the FPGA normalized by each implementation's memory bandwidth. Though it is possible for FPGA based solutions to approximate the memory bandwidth of larger PC workstations, there will be cost as well as power dissipation issues. Consequently, many real applications of computational neurobiological solutions may require less expensive, lower power consuming implementations. We have seen that memory bandwidth is the key factor in such systems, showing the results relative to each system's memory bandwidth gives a sense of the computational power of the system. The assumed bandwidth for each system was 3.2GB/sec for the P4 and 1.6GB/sec for the FPGA respectively.

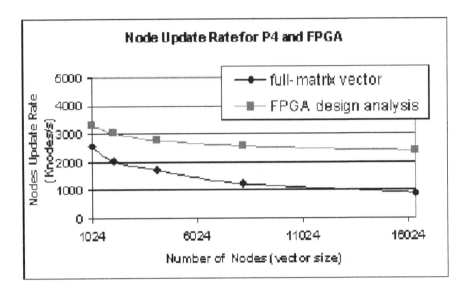

Figure 6.12.   Node update rate, node outputs computed per second, for P4 and FPGA vs. the vector dimension, n

6.6   Summary and conclusions

In this chapter we presented a fairly simple computational model that performs best-match association using distributed representations. We then showed the results of implementing this basic algorithm on three different platforms, a stand-alone high-end PC, an 8-node PC cluster, and an FPGA board that integrates state of the art SDRAM together with state of the art FPGA technology.

Although still preliminary, we believe that the results given here indicate that an FPGA acceleration card based on the tight integration of an FPGA chip and commercial SDRAM creates an effective association network hardware emulation system with a competitive performance-price. Although we have not yet extended our results to include an analysis of multiple FPGA systems, we believe that such systems will have larger incremental performance than PC clusters built from commercial boxes.

Acknowledgements

This work was supported in part by the Research Institute for Advanced Computer Science under Cooperative Agreement NCC 2-1006 between the Universities Space Research Association and NASA Ames Research Center, the Office of Naval Research, Contract N00014-00-1-0257, and by NASA Contracts NCC 2-1253 and NCC-2-1218. And part of this work is sponsored in part by AFRL/SNHC, Hanscom AFB, MA, Contract No. F19628-02-C-0080.

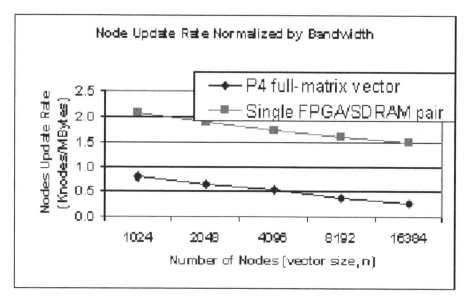

Figure 6.13. Node update rate by memory bandwidth for P4 and FPGA

References

[1] Amari S., 1997. The neural Theory of Association and Concept Formation, *Biol. cyber.* **26**, pp. 175-185

[2] Anderson J. A. and others,1999. Radar Signal Categorization Using a Neural Network, *Proceedings of the IEEE*, 1990, August.

[3] Anderson J. A., 1995. Introduction to Neural Networks, MIT Press, Cambridge, MA.

[4] Batcher K. E., 1977. The Multi-dimensional access in STARAN, *IEEE Transactions on Computers, Special Issue on Parallel Processing*, pp. 174-177

[5] . Braitenberg V. , 2001. Brain Size and Number of Neurons: An Exercise in Synthetic Neuroanatomy, *Journal of Computational Neuroscience*, **10**, pp. 71-77

[6] Braitenberg V. and A. Schuz, 1998. *Cortex: Statistics and Geometry of Neuronal Connectivity*, Springer-Verlag.

[7] Buckingham J. T. and D. J. Willshaw, 1992. Performance characteristics of the associative net, *Network*, **3**, pp. 407-414

[8] Field D. J., 1999. *What is the goal of sensory coding? Unsupervised Learning*, MIT Press, pp. 101-143.

[9] , Gropp W. E. Lusk et al., 1999. Using MPI Portable Parallel Programming with the Message Passing Interface, MIT Press, Cambridge, MA.

[10] Hebb D. O., 1999. *The Organization of Behavior*, Wiley, New York.

[11] Hecht-Nielsen R. Tutorial: Cortronic Neural Networks, 1999, International Joint Conference on Neural Networks, Washington, DC.

[12] Hopfield J., 1982. Neural networks and physical systems with emergent collective computational abilities, *Proc. Natl. Acad. Sci. USA 79*.

[13] Kerr J. R. C. H. Luk et al., 2003. Advanced integrated enhanced vision systems, *SPIE*.

[14] Kohonen T., 1984. *Self-Organization and Associative Memory*, SpringerVerlag, Heidelberg

[15] Lansner A. and A. Holst, 1996. A Higher Order Bayesian Neural Network with Spiking Units, *Int. J. Neural Systems*, **7(2)**, pp. 115-128.

[16] Lansner A. and others, 1997. *Detailed Simulation of Large Scale Neural Networks. Computational Neuroscience: Trends in Research 1997*, Plenum Press, J. M. Bower. Boston, MA, pp. 931-935.

[17] Maass W. and C. M. Bishop, Pulsed Neural Networks, MIT Press, 1999, address=Cambridge MA

[18] Omohundro S. M., 1990. Bumptrees for efficient function, constraint, and classification learning, *Advances in Neural Information Processing Systems*, **3**, Denver, Colorado, Morgan Kauffmann.

[19] O'Reilly R. and Y. Munakata, 2000. *Computational Explorations in Cognitive Neuroscience - Understanding the Mind by Simulating the Brain*, MIT Press., Cambridge, MA

[20] , Palm G., 1980. On Associative Memory, *Biological Cybernetics* , **36**, pp. 19-31

[21] Palm G. F. Schwenker et al., 1997. *Neural Associative Memories. Associative Processing and Processors*, IEEE Computer Society, pp. 284-306, Los Alamitos, CA.

[22] Reschke C. T. Sterling et al. A Design Study of Alternative Network Topologies for the Beowulf Parallel Workstation, *High Performance and Distributed Computing*, 1996

[23] Steinbuch K., 1967. Die Lernmatrix., *Kybernetik*, **1**.

[24] Willshaw D. and B. Graham, 1995. Improving Recall From An Associative Memory, *Biological Cybernetics* , **72**, pp. 337-346.

[25] Zhu S. and D. Hammerstrom, 2002. Simulation of Associative Neural Networks, *ICONIP*, Singapore.

Chapter 7

FPGA IMPLEMENTATIONS OF NEOCOGNITRONS

Alessandro Noriaki Ide
Universidade Federal de São Carlos
Departamento de Computação
Rodovia Washington Luis (SP-310), Km 235, 13565-905, São Carlos, SP, Brasil
noriaki@dc.ufscar.br

José Hiroki Saito
Universidade Federal de São Carlos
Departamento de Computação
Rodovia Washington Luis (SP-310), Km 235, 13565-905, São Carlos, SP, Brasil
saito@dc.ufscar.br

Abstract   In this chapter it is described the implementation of an artificial neural network in a reconfigurable parallel computer architecture using FPGA's, named Reconfigurable Orthogonal Memory Multiprocessor (REOMP), which uses p^2 memory modules connected to p reconfigurable processors, in row access mode, and column access mode. It is described an alternative model of the neural network Neocognitron; the REOMP architecture, and the case study of alternative Neocognitron mapping; the performance analysis considering the computer systems varying the number of processors from 1 to 64; the applications; and the conclusions.

7.1   Introduction

This chapter describes the FPGA (Field Programmable Gate Array) implementations of neural networks, using reconfigurable parallel computer architecture. In spite of FPGA's have been used to rapid prototyping of several kinds of circuits, it is of our special interest the reconfigurable implementation of neural networks. There are several classes of artificial neural networks, which we would like to implement in different time, in analogy to several brain functions which are activated separately, and sequentially. We discuss a reconfig-

A. R. Omondi and J. C. Rajapakse (eds.), FPGA Implementations of Neural Networks, 197–224.

urable architecture using FPGA's which combines the conventional computer with a reconfigurable one, using a software that can recognize and separate the reconfiguration thread to be executed at the reconfigurable computer. This thread activates the reconfiguration of the arithmetic/logic data flow unit, and its control unit.

In the case of neural network implementation, an Orthogonal Memory Multiprocessor (OMP) [15][19], is also of interest. OMP is a multiprocessor where the memory is organized in several modules connected to row access, and column access. Each memory module may be accessed by row and by column, so that when rows are activated, all rows may be accessed, each one by an exclusive processor. Similarly, when columns are activated, all columns may be accessed, each one by an exclusive processor. This kind of memory organization allows a conflict free and efficient memory access in the case of feed-forward neural network implementation.

It is described the reconfiguration of an alternative of the biological vision based neural network, Neocognitron, proposed by Fukushima [6] [7] [9] [8]. Neocognitron is a feed-forward neural network, organized in a sequence of Simple-cell (S-cell) layers, and Complex-cell (C-cell) layers. Each S-cell layer is composed by a number of S-cell planes, each one composed of a matrix of neurons, responsible by the feature extraction using convolution over the preceding C-cell layer. Each C-cell layer is composed by a number of C-cell planes, that is responsible by a shape tolerance using average operation over a region of the preceding S-cell plane. The C-cell layer is also responsible by the reduction of the number of neuron cells at the succeeding layers, until the last layer, which corresponds to the output layer, where each class of recognized object is represented by an unique neuron, which corresponds to the grand mother cell, or "Gnostic" cell, in the human brain model.

In this chapter it is presented an overview of reconfigurable architectures using FPGA's, and the hardware implementation of an alternative to the neural network Neocognitron, using a reconfigurable architecture, named Reconfigurable Orthogonal Multiprocessor (REOMP) [16]. Finally it is presented the results of the reconfiguration, the performance analysis of the resulted parallel computer architecture, its applications, and conclusions.

7.2   Neocognitron

Neocognitron is a massively parallel neural network, composed by several layers of neuron cells, proposed by Fukushima, inspired on the Hubel and Wiesel [14] researches in biological vision. The lowest stage of the network is the input layer U_0. Each of the succeeding i-th stages has a layer U_{S_i} consisting of S-cells, followed by a layer U_{C_i} of C-cells. Each layer of S-cells or C-cells is composed by a number of two-dimensional arrays of cells, called cell-planes. Each S-cell-plane is able to recognize all the same features present at the preceding C-cell-planes, at different positions, because all the S-cells, composing a cell-plane, have the same weight of input connections, limited to a small region corresponding to its receptive field. C-cells are responsible by the distorted features correction, since they process the average of their input connections. Then the C-cell-planes are reduced, since the 8-neighbors of a cell have close values after average. After the succeeding stages the output layer is reduced to a set of planes of only one C-cell. This cell corresponds to the gnostic cell found at the infero-temporal cortex of the brain. Algorithm 1 is used to the computation of the activation of S-cells, and the Algorithm 2 corresponds to the computation of the activation of the C-cells.

Algorithm 1: Computation of the activation of a cell-plane of S-cells

Procedure ComputeLayer U_S *(l)*
begin
   for k=1 to K_l **do**
     begin
       for n=1 to N do
      begin
        for K=1 to K_{l-1} **do**
        begin
          for all v $\in S_v$ **do**
          begin
$$e(n,k) := e(n,k) + a(v,k,K).u_{C_{l-1}}(n+v,k)$$
$$h(n,k) := h(n,k+c(v).(u_{C_{l-1}}(k,n+v))^2$$
          end;
         V(n,k):=sqrt(h(n,k));
$$u_{S_l}(n,k) := (\theta/(1-\theta)).\varphi((1+e(n,k))/$$
$$(1+\theta.b(k).V(n,k)) - 1$$
        end;
       end;
     end;
   end;

Figure 7.1 shows the interconnection environment to the activation of a S-cell. To obtain the S-cell value, $u_{S_l}(n, k)$, it is first computed the weighted sum, $e(n, k)$ and $h(n, k)$, of all inputs coming from all K_{l-1} C-cell-planes of the preceding layer. Here, n is the index of a cell position at the cell-plane, and k is index of the cell-plane at the layer. The value $a(v, k, K)$ and $c(v)$ are the connection weights, respectively, where v corresponds to the connection area, which surrounds the position n, of the preceding layer C-cell, and K corresponds to the index of the preceding layer cell-plane. The $e(n, k)$ is the computation of the excitation input weighted sum, and $h(n, k)$ is the computation of the inhibition input weighted sum. If the $V(n, k) = sqrt(h(n, k))$ is considered as the inhibition cell value, $b(k)$, its input connection weight, and θ the threshold value of the neuron, the $u_{S_l}(n, k)$ value is obtained by the Equation 6.1:

$$u_{S_l}(n, k) = (\theta/(1 - \theta).\varphi((1 + e(n, k))/(1 + \theta.b(k).V(n, k)) - 1 \quad (7.1)$$

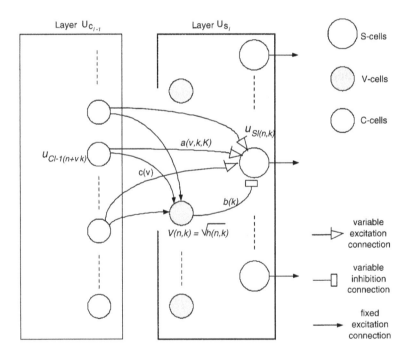

Figure 7.1.   Interconnection environment to the activation of a S-cell

To obtain the C-cell value, $u_{C_l}(n, k)$, it is first computed the weighted sum of all inputs corresponding to the previously obtained $u_{S_l}(n, k)$, in a given connection area S_v. It is used the connection weight $d(v)$. where v corresponds to the connection area of the preceding S-cell layer. Then, a function $\psi(x) = \varphi(x)/(1 + \varphi(x))$, that limits C-cell output to the range $[0,1)$ is computed.

Algorithm 2: Computation of activation of a cell-plane of C-cells

    Procedure ComputeLayer $U_C(l)$

  begin

    for k = 1 to K_l **do**

    begin

      for n = 1 to N **do**

      begin

        for all v $\in S_v$ **do**

        begin

          $u_{C_l}(n, k) := u_{C_l}(n, k) + d(v).u_{S_l}(n + v, k);$

          $u_{C_l}(n, k) := \psi(u_{C_l}(n, k))$

        end;

      end;

    end;

  end;

      Figure 7.2 shows a six-stage Neocognitron structure of S-cells and C-cells. All k cell planes from first stage take the input values from the input layer. The first stage is related to the edge extraction. After the parallel processing of the k cell planes at the first stage, all results are used by all cell planes of the second stage, which is related to the line extraction. The third stage may be processed after the line extraction by the S-cell layer to the bend point extractions. The C-cells of the line extraction stage and the bend point extraction stage are used by the fourth stage to the complex features extraction. The fifth stage joins the complex features of the preceding stage and the sixth stage, which includes the output layer, corresponds to the gnostic cells, found at infero-temporal cortex of the brain.

      In order to improve the Neocognitron learning phase it was implemented an Alternative Neocognitron, described as follows.

7.3   Alternative neocognitron

      The following description is related to an Alternative Neocognitron [2] [1], which differs from the original one on the learning procedure. The Alternative Neocognitron, which is able to manipulate characteristics including surface characteristics, beyond the line segments, is shown at Figure 7.3 . At a first

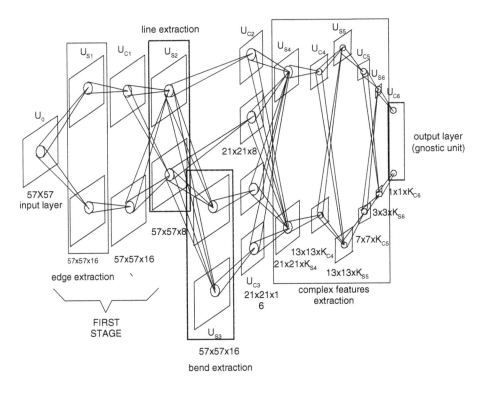

Figure 7.2.   Neocognitron structure with six stages of S-cell layer (U_{S_i}) and C-cell layer (U_{C_i})

glance, the proposed network is very similar to the original Neocognitron, and in fact the network structure is the same except that the input to the fourth stage are the bend points, from the U_{C_3} layer, and the reduced input layer. The difference is present at the network training phase, and at the recognition phase. The network training uses Algorithm 3. Since one S-cell-plane is able to recognize the same feature at all positions, all S-cells of the same cell-plane have the same input connection weights. So, the training of a cell-plane is reduced to the detection of a new feature, and to the extension of the feature to all positions. This is the same as the seed planting over the cell-plane, so it is used a special plane denoted seed-selection-plane, as the last cell-plane of the layer, during the training phase. So, in a moment of the training phase, there are K_l cell-planes at the layer, corresponding to the known features, and a *seed-selection-plane* (SSP). After a new feature detection, by the winner-take-all approach, the input connection weights of the winner, which is the seed cell, is reinforced, and the seed planting occurs to all over the cell-plane. Though, a new cell-plane of a new feature is added to the layer and so the number of cell-planes at the layer increments by one. The training proceeds with a new SSP, until all new features are detected, using different input patterns.

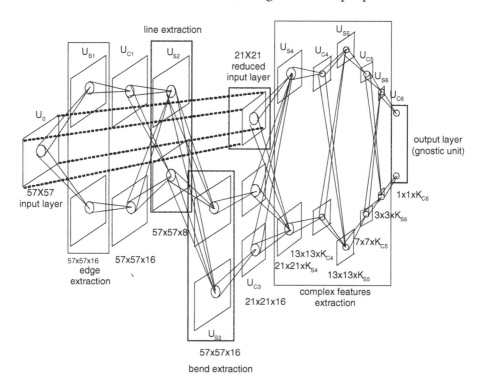

Figure 7.3. Alternative Neocognitron structure: the U_0 layer values are reused, after the bend extraction, composing a reduced input layer

The described algorithm is used at the original Neocognitron, to the training of the layer U_{S_4}, U_{S_5} and U_{S_6}, since the layer U_{S_1}, U_{S_2} and U_{S_3} have fixed input weights. The Alternative Neocognitron uses a little different approach, since it uses again the input layer, as the input to the layer U_{S_4}. This is because it is interested on the features like surface textures present at the input pattern, that is not considered in the original Neocognitron. So the bend points, indicates the position of the input pattern, where the corresponding surface texture is interested to be trained. By this way, a composition of features including textures and lines segments are considered during the training of the following fifth and output stages. Another difference of the Alternative Neocognitron is that although we use the input layer, the first three stages, and the last three stages, during the training phase, at the recognition phase we use only the input layer and the last three stages, so that the network computation would be faster than the original Neocognitron computation. The Alternative Neocognitron is able to be used in patterns, where some line segments are present, and detected by the first three layers, but a complex blending of lines and textures may be present. This is the case of a human face pattern. It may be observed that the threshold associated to the individual neurons during the training phase controls the number of new features extracted. Considering all the modifications in the original Neocognitron structure, the following sections will give a brief description of the main characteristics found in reconfigurable computers and the proposal of a OMP architecture, adequate to the neural networks mapping.

Algorithm 3: Network Training

> *Procedure Training layer_$U_S(l)$*
>> **begin**
>>> **repeat**
>>> ComputeLayer_U_S(l)
>>>> selected := false
>>>>> **repeat**
>>>>>> **if** U_S (next_winner)> 0   **then**
>>>>>> **begin**
>>>>>>> winner : = next_winner;
>>>>>>> selected : = true;
>>>>>>>> **for** k = 1 **to** Kl **do**
>>>>>>>> **begin**
>>>>>>>> **if** U_S (winner, k) > 0 **then**
>>>>>>>> selected := false
>>>>>>>> **end**
>>>>> **until** (selected = true or next_winner = φ) ;
>>>>> **if** selected **then**
>>>>> **begin**
>>>>>> **for** k = 1 **to** K_{l-1} **do**
>>>>>> **begin**
>>>>>>> **for** all v $\in U_S$ **do**
>>>>>>> **begin**
>>>>>>>> $a(v, k, K) := a(v, k, K) + q.c(v).u_{C_{l-1}}(winner + v, k)$;;
>>>>>>>> b(k) := b(k) + q. sqrt(h(winner,k));
>>>>>>>> $K_l := K_l + 1$
>>>>>>> **end;**
>>>>> **until** not(selected)
>>> **end**

7.4   Reconfigurable computer

In this section, it is described an approach to the Alternative Neocognitron hardware implementation, that can be also applied to other neural networks. A reconfigurable computer provides the solution to the hardware implementation of neural networks using the current technology of FPGA's. This approach assumes that not all the neural networks functions are active all the time, so that only the active functions are configured in the computer during a snapshot of operation.

The reconfigurable computer has a reconfigurable unit, and a fixed unit. It uses the components as FPGA's to implement a special function in the reconfigurable unit. A FPGA is an array of processing elements whose function and interconnection can be programmed after fabrication. Most traditional FPGA's use small lookup tables to serve as programmable elements. The lookup tables are wired together with a programmable interconnection, which accounts for most of the area in each FPGA cell. Several commercial devices use four input lookup tables (4-LUT's) for the programmable elements. The commercial architectures have several special purpose features, as carry chains for adders, memory nodes, shared bus lines.

The configurable unit is efficient if it implements the processing elements spatially distributed to exploit the streaming of the datapath [5], as in the neural network. Obviously, some functional systems of the neural networks are always active, that can be implemented at the fixed unit of the computer.

A reconfigurable computer partitions computations between two main groups: (1) reconfigurable units, or fabric; and (2) fixed units. The reconfigurable units exploits the reconfigurable computations, which is efficient when the main computation is executed in a pipelined data path fashion. The fixed units exploit system computations that controls the reconfigurable units. The reconfigurable units are reconfigured to implement a customized circuit for a particular computation, which in a general purpose computer is cost prohibitive, to be implemented in hardware, because of its reduced use. The compiler embeds computations in a single static configuration rather than an instruction sequence, reducing instruction bandwidth and control overhead. Function units are sized properly, and the system can realize all statically detectable parallelism, because the circuit is customized for the computation at hand [16].

A reconfigurable unit can outperform a fixed unit processing in cases that: (a) operate on bit widths different from the processor's basic word size, (b) have data dependencies that enable multiple function units operate in parallel, (c) contain a series of operations that can combine into a specialized operation, (d) enable pipelining, (e) enable constant propagation to reduce operation complexity, or (f) reuse the input values several times [10].

Reconfigurable units give the computational data path more flexibility. However, their utility and applicability depend on the interaction between reconfigurable and fixed computations, the interface between the units, and the way a configuration loads.

It is possible to divide reconfigurable computations into two categories: (1) stream-based functions which corresponds to the processing of large, regular data input streams, producing a large data output stream, and having little control interaction with the rest of the computation; and (2) custom instructions which are characterized with a few inputs, producing a few outputs, executed intermittently, and having tight control interactions with the other processes. Stream-based functions are suitable for a system where the reconfigurable unit is not directly coupled to the processor, whereas custom instructions are usually beneficial only when the reconfigurable unit is closely coupled to the processor [10].

Some examples of reconfigurable architectures implemented using FPGA's are: DISC [22], MATRIX [4] [18], GARP [13] [3], CHIMAERA [12], PIPERENCH [11] [10] and MorphoSys [20] [21] [17]. Next section will describe the proposed reconfigurable computer, REOMP.

7.5   Reconfigurable orthogonal memory multiprocessor

The proposed Reconfigurable Orthogonal Memory Multiprocessor - REOMP, is a computer composed by: (1) a Control Unit that is connected to the Host Processor and the Reconfigurable Processors; (2) Reconfigurable Processors (RP's); and the Memory modules organized in two access modes, row and column.

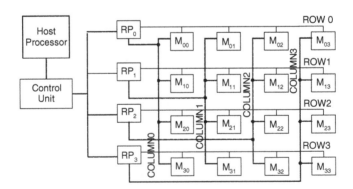

Figure 7.4.   REOMP Architecture

The overall system is an Orthogonal Multi-Processor, OMP, which is characterized by the parallel processing units, each one accessing their memory modules in two ways, column, and row. Each row, and each column, is attributed to one exclusive processor. The row access and column access are performed exclusively, without time-sharing of the buses by the multiple processors, eliminating the memory access conflicts.

The reconfigurable processors (RP's) access the orthogonal memory modules, to provide the input data to the weighted sum operations of the neural network. REOMP presents an interesting solution to the data exchange after one layer processing of a neural network like Neocognitron. REOMP is an orthogonal multiprocessor, with p reconfigurable processors working synchronized, all of them connected to the orthogonal memory. Figure 7.4 shows a REOMP diagram, with four reconfigurable processors, and sixteen memory modules. A control unit is used to interface REOMP with the host processor. If each reconfigurable processor processes stream-based functions data, and each processor must be synchronized with each other, the control unit provides the corresponding control templates to each processor, synchronously. It seems a microprogramming control, except that there is not any conventional level instruction to be executed. Each processing task corresponds to a special optimized microprogram, which is loaded at the Control Unit, and executed to control all the RP's. After the processing of a task, another microprogram is loaded, and the RP's are also reconfigured to the new task.

7.5.1 Reconfigurable Processor

The Reconfigurable Processor structure, Figure 7.5, is composed by (a) fixed units, such as memory address register (MAR), memory buffer register (MBR), set of general registers; and (b) a reconfigurable dataflow unit, implemented in FPGA's, which enables the implementation of special reconfigurations.

In the fixed unit data are loaded/stored from/to memory like in a modern computer. The general registers are efficient when the same data are used several times. So, it is necessary to load data from memory once, and then reuse them from the registers. It becomes the computation faster and more efficient, as a register-register processor.

The reconfigurable unit is composed by an array of arithmetic/logic units (ALU's). The number of ALU's can vary from application to application, using reconfiguration. The reconfigurable processors are fully controlled by control signals, which are responsible by the control of all the arithmetic/logic computations, load and store, and branch conditions (the data passes through an ALU, its computation is done, and a flag signal is sent to the control unit.

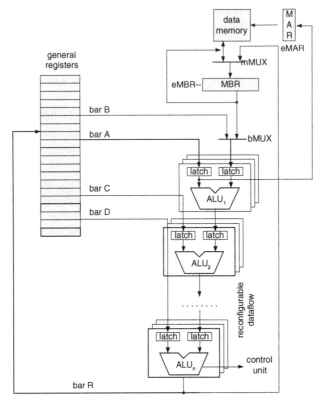

Figure 7.5.   Reconfigurable Processor Structure

7.5.2   Control Unit

The control unit is similar to a microprogramming control, Figure 7.6. It is composed basically by: a template memory, template counter (TC), a template register, and a timing unit. The dataflow is controlled by a set of templates, stored at the template memory, which is modified according to the application. The timing unit synchronizes the template application to the dataflow at RP's. The template counter is responsible to address the next template, at the template memory. However, in some cases the program must branch to another template, and in this case a multiplex (mTC) is selected to branch addressing, which is indicated by the current template. This synchronization enables the architecture to execute several different computations in sequence, parallel and pipeline, outperforming the conventional computer, in special computation applications.

Each template cycle is divided into four sub-cycles. In the first sub-cycle, the control unit loads the template register with the next template; in the second

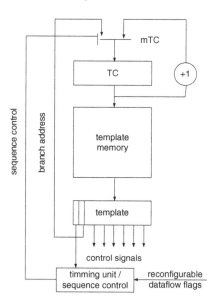

Figure 7.6. Control Unit Structure

one, the data is loaded from the general registers into the ALU latches; in the third one, MAR is loaded, and the ALU's operations are executed; and in the last one, data is loaded from memory to MBR, and are loaded the general registers. It is noticed that the number of sub-cycles does not depend on the number of pipeline stages. So, all operations in the RP's dataflow will spend four sub-cycles, independently of the number of pipeline stages or parallel computations. The increase in the number of ALU's means an increase in the number of template bits.

7.6   Alternative neocognitron hardware implementation

By the analysis of the Alternative Neocognitron described at previous sections it is extracted a set of concurrent computation modules. Five modules E, H, S, C, and ψ, correspond to the Neocognitron computing algorithm.

E-module computes to all N positions of the S-cell-plane, the weighted sum of the input $u_{C_{l-1}}(n + v, K)$, with the weight $a(v, k, K)$, within the connection region S_v, which results in a partial value of $e(n, k)$, corresponding to the preceding plane. Each E-module issue repeats the processing N times to compute all positions of the E-matrix. Its result is added to the E-matrix which will accumulate the weighted sum of all cell-planes, after K issues of the E-module function.

H-module is similar to the E-module but the weight values are $c(v)$ and the input values are squared before weighted sum. It results in the H matrix, which will contain the weighted sum of all preceding layer cell-planes, after K issues of the H-module function.

S-module computes $u_{S_l}(n, k)$, to all N positions of the cell-plane, using the results of the previously described E-module and H-module functions.

C-module corresponds to the computation of the $u_{C_l}(n, k)$, to all N cells. It computes the weighted sum of $u_{S_l}(n + v, k)$, by $d(v)$.

ψ**-module** computes the $\psi(x)$, which limits C-cell output to the range [0,1).

Figure 7.7 shows a two-stage dataflow of the proposed architecture, to the Alternative Neocognitron implementation. This structure may be used to compute the E-module of the Alternative Neocognitron. Figure 7.8 shows the template which controls the E-module. It is a 37-bit template with 12 fields; and two of these twelve fields are responsible by the reconfigurable dataflow control. Tab. 7.1 shows a resumed description of the template fields.

Table 7.1.   Description of the E-module control fields

field	description
cond	branch condition
	00: no branch;
	01: conditional branch;
	10: unconditional branch;
branch	next microinstruction address
mem	memory read/write control
	0: read;
	1: write;
eMAR	load MAR control
	0: load;
	1: not load;
eMBR	load MBR control
	0: load;
	1: not load;
mMUX	multiplex data from memory or ALU
	0: memory;
	1: ALU;
barA, barB	load data into latch A and B
f1/f2	00: A/C
	01: B/D
	10: A+B / C+D;
	11: A-B / C-D;

As follows, it is showed the resumed algorithm to process the E-Module. In this case it was considered a 57x57 input matrix, a 3x3 connection area, and a circular buffer to store the $u_{S_l}(n + v, k)$ values. It is divided into five steps:

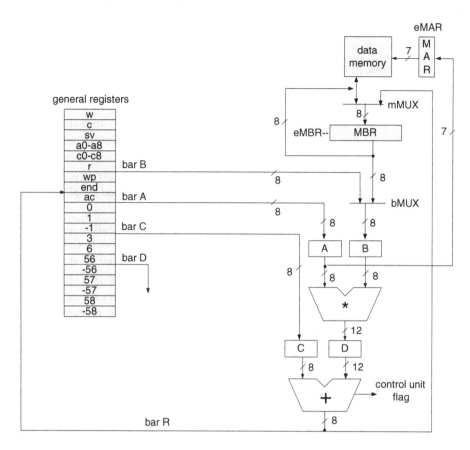

Figure 7.7. Reconfiguration of a RP to the E-module hardware implementation

Figure 7.8. E-module template

- load the weights $a(v, k, K)$ from memory to the general registers;

- load the $u_{C_l}(n + v, k)$ values from memory to the general registers, according to the connection area size; (it fills the full circular buffer)

- load some $u_{C_l}(n+v, k)$ values and replace part of the data in the circular buffer;

- arithmetic processing; and

- loop test.

The first step is related to the weights $a(v, k, K)$ loading into the general registers $(a0\text{-}a8)$. In this case MAR is loaded with the first weight address and the address pointer is incremented; then it is waited for the memory reading; after that, the read value is loaded at the MBR; and finally the first weight is loaded in the general register $a0$. This operation is repeated to all the nine weights $(a0\text{-}a8)$.

After the weights are loaded to the general registers, the same operation is repeated to load the connection area C-cell values from memory to the registers $uc0\text{-}uc8$. But before this, it is necessary to start a counter $(cont)$ and a circular buffer pointer (wp). The counter restarts every time a full column is processed; and wp is restarted when the last position of the circular buffer is occupied.

The circular buffer is implemented via template. For each processed point of the E-matrix, three new values are loaded into the circular buffer. So that, from the nine values, six of them keep unaltered. The substitution control is showed in Figure 7.9, and it proceeds as follows:

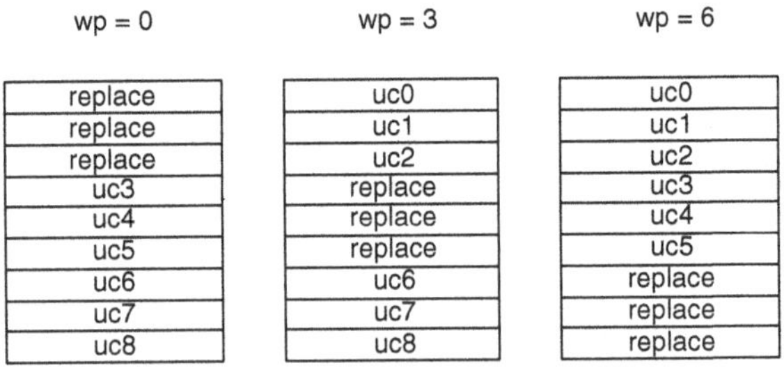

wp = 0	wp = 3	wp = 6
replace	uc0	uc0
replace	uc1	uc1
replace	uc2	uc2
uc3	replace	uc3
uc4	replace	uc4
uc5	replace	uc5
uc6	uc6	replace
uc7	uc7	replace
uc8	uc8	replace

Figure 7.9.   Circular Buffer Control

- if $wp=0$, replace the $uc0\text{-}uc2$ values;

- if $wp=3$, replace the $uc2\text{-}uc5$ values; and

- if $wp=6$, replace the $uc5\text{-}uc8$ values.

Once both the weights and the connection area C-cell values are loaded, the E-matrix processing is started. The data is processed and stored into the memory. In order to finalize the algorithm, there are a set of loop test and updates to control the E-matrix processing.

Figure 7.10, Figure 7.11, and Figure 7.12, show the reconfigurable dataflow for the H-module, S-module, and C-module, respectively. It is noted that the E, H and C-modules are similar; and the S-module is the biggest one, therefore it uses less general registers. Figure 7.13 corresponds to the calculation of the ψ function of the C-module output.

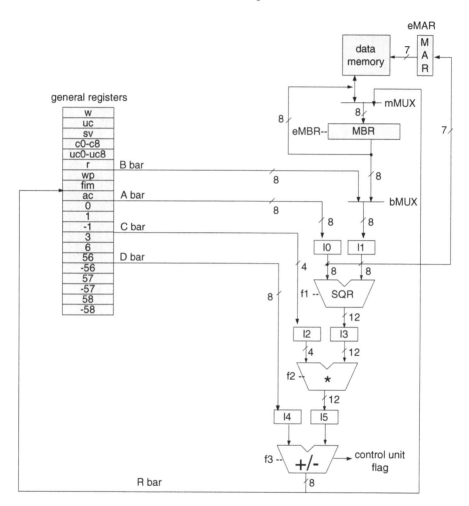

Figure 7.10. H-module reconfiguration

Tab. 7.2 shows the hardware resources used in the implementation of the Alternative Neocognitron using the Altera's Quartus II tool. The first line shows the amount of templates necessary to control each module; the second line shows the number of pipeline stages; the third line shows the template size; the fourth line shows the number of logic elements (term used to define the

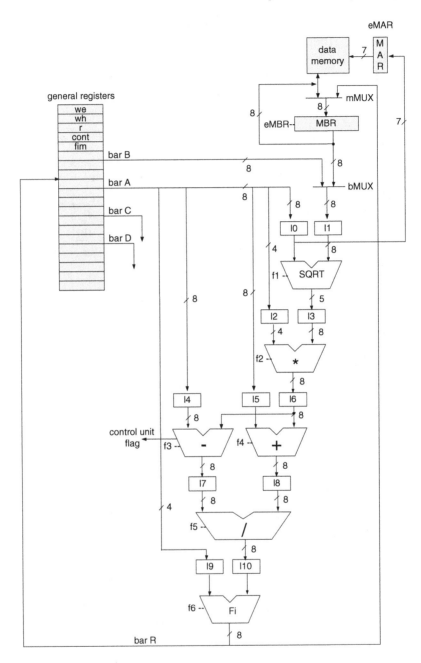

Figure 7.11.   S-module reconfiguration

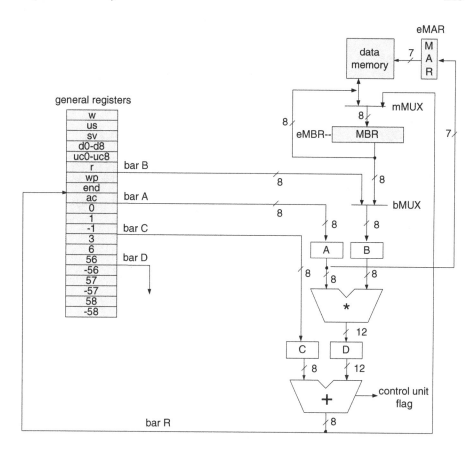

Figure 7.12. C-module reconfiguration

basic logib uilding blocks of Altera's programmable logic devices (PLD)) used to implement each reconfigurable dataflow unit; the fifth row shows the total number of input and output pins; and the last two rows show the configuration time and the processing time, respectively.

7.7   Performance analysis

As an example, the mapping of the Alternative Neocognitron to the proposed architecture may be resumed, as follows. The modules E, H, S, C, and ψ, are processed at the RP's, and the other functions at the host processor.

In a general version of REOMP we propose the use of the orthogonal memory, as follows. When the processors are writing their results in the orthogonal memory, preparing to the next phase of orthogonal access, they write simultaneously, the same data to all accessible memory modules. After that, the

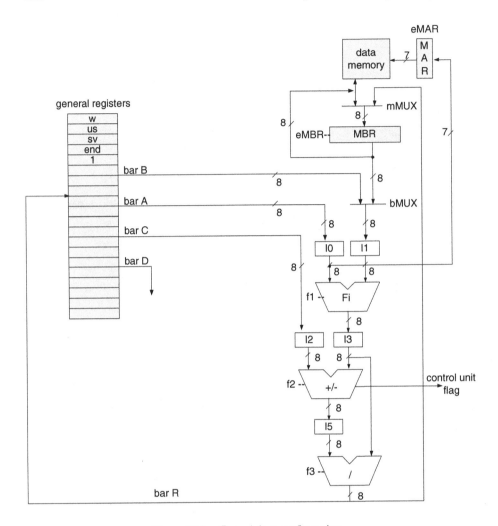

Figure 7.13.   Ψ-module reconfiguration

processors wait for the next orthogonal access, which occurs automatically, after the last processor finishes the processing. At the next orthogonal access,
the access mode is changed, to row or column, depending on the last access.
When the mode is changed, all the processors have the preceding processing
results of all the processors, because of the simultaneous writing of the same
data to all accessible modules, in the preceding access.

Figure 7.14 shows a typical data spreading operation using the REOMP
memory modules. At the left side we can see the diagonal memory modules
which correspond to local memory of the processors, with their respective data.

Table 7.2. Hardware resources used in the reconfigurable datapath implementation

Resources	E	H	S	C
Templates	90	92	16	90/12
Pipeline-stages	2	3	5	2/3
Template Size	37	44	32	37/31
Logic Elements	168	326	426	168/264
I/O Pins	37	43	57	37/37
Configuration Time(μs)	168	326	426	168/264
Processing Time (ms)	2.02	2.08	1.03	2.02/0.78

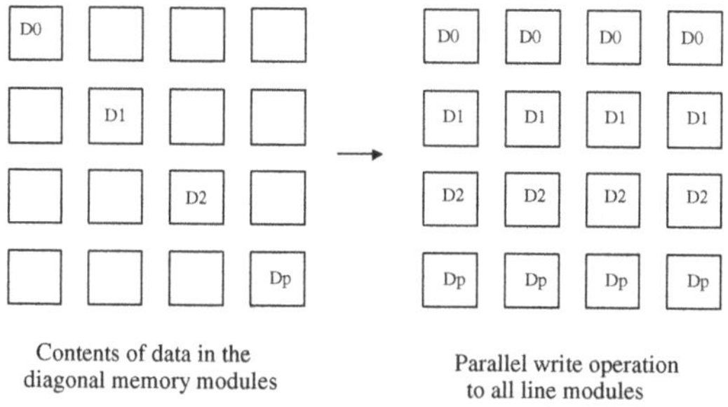

Contents of data in the
diagonal memory modules

Parallel write operation
to all line modules

Figure 7.14. A typical parallel data spreading at the REOMP memory modules

With a row access, the processors connected to the exclusive rows spreads the data to all row modules, simultaneously. At the right side, we see the result of parallel data spreading operation, with all columns containing the same data, distributed in several memory modules. During the following column access, all processors may access their own column memory modules, which contain the spread data.

The computation of all described Alternative Neocognitron modules results in cell plane values. In a layer there are several cell planes, so that each RP is responsible by the computation of a number of them, using the same previously processed, and spread data of the orthogonal memory.

A performance analysis of the REOMP architecture is done considering a typical implementation of the alternative Neocognitron, with an input matrix of size 57x57, a connection area of size 3x3, and a number of 64 cell planes. Tab. 7.3 shows the number of processors (p), varying in a range of 1 to 64 reconfigurable processors; the number of cell planes (NCP) computed by each processor; the reconfiguration time (RT); the orthogonal memory size (OMS); the spread time (ST); the total processing time(Proc.Time) ; and the speed-up.

Table 7.3. Performance Analysis

p	NCP	$RT(\mu s)$	OMS	$ST(ms)$	Proc.Time(ms)	Speed-up
1	64	86.5	220932	0	509	1
2	32	43.2	116964	14.5 (5.4%)	269	1.89
4	16	21.6	64980	21.8 (14.6%)	149	3.41
8	8	10.8	38988	25.4 (28.5%)	89.1	5.71
16	4	5.4	25992	27.2 (46.1%)	59.1	8.61
32	2	2.7	19494	28.2 (63.9%)	44.1	11.5
64	1	1.3	16245	28.6 (78.2%)	36.6	13.9

The distribution of the cell planes is done homogeneously to each reconfigurable processor. When $p=1$ all the 64 cell planes are processed at one RP; otherwise, when the number of processors is 64, each RP processes one cell plane, reducing the reconfiguration time. On the other hand, when the number of processors (p) increases, the number of orthogonal memory modules increases as p^2 and consequently, the spread time also increases. From 16 processors the spread time occupies 46.1% of the total processing time; and with 64 processors, it is 78.2%. These percentages are obtained with the division of the ST by the Proc.Time. It compromises the speed-up which varies from 1 to 13.9. Figure 7.17 shows the speed-up when the number of processors varies from 1 to 64.

7.8   Applications

The Neocognitron model has been originally developed to handwritten characters recognition and classification. It is based on the manipulation of lines, edges and bend points. One of the applications of the Alternative Neocognitron model is on the solution of biometric problems to individuals classification and recognition. Figure 7.18 shows some face images used as input patterns to the Alternative Neocognitron in human face recognition, and the Figure 7.19 shows the results after the edge processing, line segment processing, and the bend point processing. Figure 7.20a shows a fingerprint original image, and Figure 7.20b its enhanced image, used to the minutia detection application. Figure 7.21 shows the 8 minutiae categories been considered by the fingerprint recognition application of the Alternative Neocognitron, and Figure 7.22 illustrates the minutiae position in a fingerprint sample. The Alter-

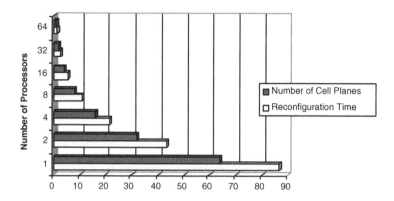

Figure 7.15. Comparison between the number of cell-planes and the reconfiguration time when the number of processors increases

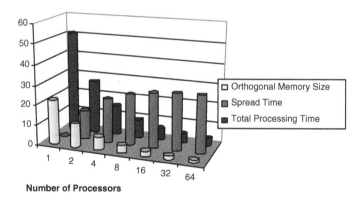

Figure 7.16. Comparison between the orthogonal memory size, spread time and the total processing time when the number of processors increases

native Neocognitron was fully implemented in simulation and it presented an acceptable recognition rate (86%), and rejection rate (14%), using 10 different individuals, in the face recognition, and a number of 10 test patterns [2]. In the fingerprint application, where the Alternative Neocognitron was applied sev-

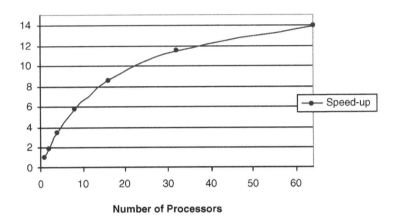

Figure 7.17.   Speed-up x Number of processors

eral times over the full fingerprint image, the correct recognition rate was 96%, the error rate was 1%, and the rejection rate was 3%, using 479 test patterns [1].

Figure 7.18.   Face images used to the Alternative Neocognitron face recognition

Figure 7.19.   Bend points obtained by the first three stages of the Alternative Neocognitron, to the face images of the Figure 7.18, in the respective sequence

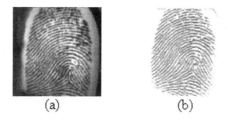

Figure 7.20. (a) Original Fingerprint Image; (b) Enhanced Fingerprint Image

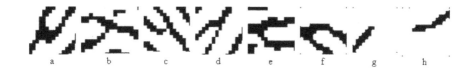

Figure 7.21. Minutia category: (a) class 1; (b) class 2; (c) class 3; (d) class 4; (e) class 5; (f) class 6; (g) class 7; (h) class 8

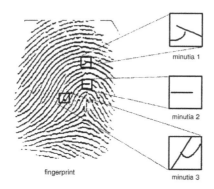

Figure 7.22. An example of the minutiae position in a fingerprint

7.9 Conclusions

This work describes an FPGA implementation of an Alternative Neocognitron, which is a neural network inspired on the biological brain. The proposed neural network was simulated and it was verified that its performance is similar to the original Neocognitron, manipulating other types of input images, than handwritten character recognition. It was also described the proposal of its hardware implementation in a reconfigurable computer. Although the neural networks corresponds to the human brain functions, the proposed architecture is completely different from the brain internal structure based on the neural cells. The proposed RP's are based on FPGA's, which are programmed with configuration data generated by an environment of hardware description language, VHDL, and are reconfigured to execute the algorithms by a microprogramming-like control unit. By the analysis of the orthogonal memory performance, it is possible to conclude about the use of orthogonal memory with reconfigurable components. A REOMP architecture is efficient for a lot of scientific applications but the memory access principle disable the column, and row, data access mix. It reduces the flexibility to those algorithms that do not necessity an orthogonal access. Fortunately, a neural network has a lot of concurrent operations in the neuron processing, which facilitates its mapping in a orthogonal architecture. Another problem in the REOMP architecture laid in the number of memory modules, which grows according to the number of processors (p) in a p^2 rate. This is tolerable until a few number of processors. The results obtained showed that until 8 processors a REOMP architecture is interesting; after that the data spread in the orthogonal memory became to be critic. The algorithms are processed in the RP's which are connected to the orthogonal memory modules. As the brain alternates their functions, it is proposed a neural network hardware implementation which uses a small number of FPGA's simultaneously, each one processing a specific algorithm. If the processing is finished, all the RP's and the Control Unit of the REOMP architecture is reconfigured to another algorithm. As the brain, some algorithms or functions are constantly in operation, and others may be alternated. Several reconfigurable architectures have been proposed, to support demand-driven instruction set modification, most of them using partial reconfiguration, paging instruction modules in and out of an FPGA as demanded by the executing program. In the case, the functions constantly in operation may be at the host processor, and the alternated functions in RP's. The proposed architecture comports hardware implementation of the algorithms, which simulates the brain functions, occupying physically restricted space. Future works are concerned in the REOMP implementation in a System-on-Chip (SoC) approach provided this new technology enables the use of modular design concept.

References

[1] Arantes, M., Ide, A. N., and Saito, J. H.   A System for Fingerprint Minutiae Classification and Recognition. volume 5, pages 2474–2478. *Proceedings of the 9th International Conference on Neural Information Processing - ICONIP'2002.* Singapura, 2002.

[2] Bianchini, A. R. Arquitetura de Redes Neurais para o Reconhecimento Facial Baseado no Neocognitron. Master's thesis, Universidade Federal de São Carlos.

[3] T. J. Callahan, J. R. Hauser, and J. Wawrzyneck. *The Garp Architecture and C Compiler*. volume 33, pages 62–69. IEEE Computer Society, Computer: Innovative Technology for Computer Professionals, April 2002.

[4] A. Dehon. *Reconfigurable Architecture for General Purpose Computing*. PhD thesis, Massachussets Institute of Technology (MIT), 1996.

[5] A. Dehon. The Density Advantage of Configurable Computing. volume 33. *IEEE Computer*, 2000.

[6] K. Fukushima. Neural-Network Model for a Mechanism of Pattern Recognition Unaffected by Shift in Position - Neocognitron. volume 62-A. *Transactions of the IECE*, 1979. Japan.

[7] K. Fukushima and S. Miyake. Neocognitron: A New Algorithm for Pattern Recognition Tolerant of Deformations and Shift in Position. volume 15, pages 455–469. *Pattern Recognition*, 1982.

[8] K. Fukushima and M. Tanigawa. Use of Different Thresholds in Learning and Recognition. volume 11, pages 1–17. *Neurocomputing*, 1996.

[9] K. Fukushima and N. Wake. Improved Neocognitron with Bend-D detecting Cells. *Proceedings of the IEEE - International Joint Conference on Neural Networks*, 1992. Baltimore, Maryland.

[10] S. C. Goldstein, H. Schmit, M. Budiu, S. Cadambi, M. Moe, and R. R. Taylor. Piperench: A Reconfigurable Architecture and Compiler. volume 33, pages 70–77, Issue 4. *IEEE Computer*, April 2000.

[11] S. C. Goldstein, H. Schmit, M. Moe, M. Budiu, S. Cadambi, R. R. Taylor, and R. Laufer. Piperench: A Coprocessor for Streaming Multimedia Acceleration. pages 28–39. *Proceedings of the 26th. Annual International Symposium on Computer Architecture (ISCA'99)*, 1999.

[12] S. Hauck, T. W. Fry, M. M. Hosler, and J. P. Kao. The Chimaera Reconfigurable Functional Unit. pages 87–96. *Proceedings of the IEEE Symposium on Field-Programmable Custom Computing Machines (FCCM'97)*, April 1997.

[13] J. R. Hauser and J. Wawrzynek. Garp: A MIPS Processor with a Reconfigurable Coprocessor. *Proceedings of the IEEE Symposium on Field-Programmable Custom Computing Machines - FCCM '97*, April 1997. USA.

[14] D. H. Hubel and T. N. Wiesel. Receptive Fields and Functional Architecture of Monkey Striate Cortex. volume 165, pages 215–243. *J. Physiology*, 1968.

[15] K. Hwang, P. Tseng, and D. Kim. An Orthogonal Multiprocessor for Parallel Scientific Computations. volume 38, Issue 1, pages 47–61. *IEEE Transactions On Computers*, January, 1989.

[16] A. N. Ide and J. H. Saito. A Reconfigurable Computer REOMP. volume 13, pages 62–69. *Proceedings of the 13th Symposium on Computer Architecture and High Performance Computing - SBAC-PAD'2001*, September 2001. Pirenópolis.

[17] G. Lu. *Modeling, Implementation and Scalability of the MorphoSys Dynamically Reconfigurable Computing Architecture*. PhD thesis, Electrical and Computer Engineering Department, University of California, 2000. Irvine.

[18] E. A. Mirsky. *Coarse-Grain Reconfigurable Computing*. PhD thesis, Massachussetts Institute of Technology, June 1996.

[19] J. H. Saito. A Vector Orthogonal Multiprocessor NEOMP and Its Use in Neural Network Mapping. volume 11. *Proceedings of the SBAC-PAD'99 - 11th Symposium on Computer Architecture and High Performance Computing*, 1999. Natal, RN, Brazil.

[20] H. Singh. *Reconfigurable Architectures for Multimedia and Parallel Application Domains*. PhD thesis, Electrical and Computer Engineering Department, University of California, 2000. Irvine.

[21] H. Singh, M. Lee, G. Lu, F. Kurdahi, N. Bagherzadeh, and E. M. C. Filho. Morphosys: An Integrated Reconfigurable System for Data-Parallel and Computation-Intensive Applications. volume 49, pages 465–481. *IEEE Transactions on Computers*, May 2000.

[22] M. J. Wirthlin and B. I. Hutchings. A Dynamic Instruction Set Computer. *Proceedings of the IEEE Symposium on FPGA's for Custom Computing Machines*, April 1995.

Chapter 8

SELF ORGANIZING FEATURE MAP FOR COLOR QUANTIZATION ON FPGA

Chip-Hong Chang, Menon Shibu and Rui Xiao
Centre for High Performance Embedded Systems,
Nanyang Technological University

Abstract   This chapter presents an efficient architecture of Kohonen Self-Organizing Feature Map (SOFM) based on a new Frequency Adaptive Learning (FAL) algorithm which efficiently replaces the neighborhood adaptation function of the conventional SOFM. For scalability, a broadcast architecture is adopted with homogenous synapses composed of shift register, counter, accumulator and a special SORTING_UNIT. The SORTING_UNIT speeds up the search for neurons with minimal attributes. Dead neurons are reinitialized at preset intervals to improve their adaptation. The proposed SOFM architecture is prototyped on Xilinx Virtex FPGA using the prototyping environment provided by XESS. A robust functional verification environment is developed for rapid prototype development. Rapid prototyping using FPGAs allows us to develop networks of different sizes and compare the performance. Experimental results show that it uses 12k slices and the maximum frequency of operation is 35.8MHz for a 64-neuron network. A 512 X 512 pixel color image can be quantized in about 1.003s at 35MHz clock rate without the use of subsampling. The Peak Signal to Noise Ratio (PSNR) of the quantized images is used as a measure of the quality of the algorithm and the hardware implementation.

Keywords: Artificial Neural Networks, FPGA, Self Organizing Feature Maps, Color Quantization

8.1   Introduction

The color of a given pixel on the video display is determined by the amount of color provided by each of the respective R, G and B electron guns. Each of the three basic colors of a 24-bit RGB formatted digitized color image can contribute to one of the 256 shades of that color. This relatively large number of possible color combinations produces a true-to-life color image on

A. R. Omondi and J. C. Rajapakse (eds.), FPGA Implementations of Neural Networks, 225–245.

the video display. However, problems arise when superior image quality is of a secondary concern than the scarce valuable memory in applications like transparency displays, animation and displaying continuous tone color images on monitors that lack full color frame buffers. The image size becomes more important a concern for portable applications, where the available memory for storage of the image is less. This calls for a compression scheme which achieves a good compression performance, while maintaining the best possible image quality.

One of the most common compression methods for color images involves their coding using Vector Quantization (VQ). Color Quantization (CQ) [1–7] is a subset of VQ that provides color reduction by selecting a small number of code words (colors) from a universal set of available colors to represent a high color resolution image with minimum perceptual distortion. Reduction of the image colors is an important factor for compression of a color image. A global compression scheme utilizes the DCT transforms and a low pass filter to preprocess the image before being quantized [8]. Since human vision has considerably less spatial acuity for chrominance than luminance [9], in addition to the standard lossy compression aiming at removing the unperceivable high frequency components, further compression is achieved with Color Quantization (CQ) [5, 7, 9].

CQ can be considered as a two-step process involving "color palette design" and "quantizer mapping" [5]. The color palette design is to find a reduced set of pixel values called the codebook, which is most representative of the original image. Let the set of all pixels of the image be I. The color value of I is represented by $I = (r, g, b)$, where r, g and b are the associated pixel color values. Thus, the color palette design is to partition the input colors into M disjoint sets, $C_s, (0 \leq s \leq M)$, where M is the number of colors in the quantized image, which is usually limited by the hardware. For every color set C_s, there is a representative color q_s, which constitutes the color palette. Thus, the main issue in colormap design is to select the best possible set of representative colors q_s for the images, and partition the colors into M color sets, C_s. Quantizer mapping, on the other hand, involves the mapping of each pixel of the image to an entry in the codebook to produce the highest possible quality image. Quantizer mapping, thus involves the association of each pixel of the image with a color from q_s to yield the highest quality image. By optimizing the color palette design to choose a codebook using a finer scale on vectors occurring more frequently, the loss of precision due to the quantization scheme is minimized.

A number of approaches have been suggested for the design of CQ including the LBG algorithm [10], popularity algorithm [11], variance-minimization CQ algorithm [12] and the Octree algorithm [13]. However, the approach that has the most potential in terms of the compression ratio, computational complexity and perceptual distortion is found to be the one using Neural Network (NN) algorithms [14, 29, 16, 17]. The frequently used clustering techniques for color quantization are the Kohonen Self-Organizing Feature Map (SOFM) [18], Fuzzy C-means [19], C-means [20] and K-means [21]. The use of NN for CQ has a number of significant advantages. First, algorithms based on NN lend themselves to highly parallel digital implementation and offer the potential for real time processing. Second, the large body of training techniques for NN can be adapted to yield new and better algorithm for codebook design. Finally most NN training algorithms are adaptive, thus NN based CQ design algorithms can be used to build adaptive vector quantizers. One class of NN structures, called self-organizing or competitive learning networks [29, 22, 23], appears to be particularly suited for image clustering, and a number of relevant studies have been reported [24], [25, 26]. In comparison with other neural networks, and especially with supervised learning, it is known that competitive learning is highly suited to discovering a feature set and to create new classes automatically [27]. Basically, there are two different models of self-organizing neural networks originally proposed by Willshaw and Von Der Malsburg [28] and Kohonen [29], respectively. The model developed by Willshaw et al. is specialized to the types of mappings where the input dimension is the same as the output dimension. However, Kohonen SOFM is more general in the sense that it is capable of generating mappings from high dimensional signal spaces to lower dimensional topological structure.

Kohonen Self Organizing Feature Map (SOFM) [30–36], wherein a sequence of inputs is presented to a network of neurons that self-organize to quantize the input space as optimally as possible has been reported to have achieved phenomenal success in image color quantization [3–5, 8, 31, 35, 15]. When used in color quantization, its advantages are manifested in the inherent massively parallel computing structure and the ability to adapt to the input data through unsupervised learning.

The Kohonen SOFM adopts a massively parallel computing structure. This inherent parallelism arises out of the independent processing ability of individual neurons and is of immense benefit when a VLSI implementation of the network is considered. This structure is also ideally suited for building a scalable architecture. Another feature of the Kohonen SOFM that makes it attractive is that it is essentially an on-line algorithm where codebook updating is done while the training vectors are being processed, thus needing very minimal memory to store the data. These features along with another important characteristic of Kohonen SOFM, namely graceful degradation makes it the algorithm of choice for most hardware implementation of feature mapped networks.

The traditional Von Neumann based architecture of the general-purpose computers and high speed digital signal processors have been optimized to process arbitrary code and in such application, the performance will be short of the required speed by direct mapping of the algorithm of interest onto programmable codes running on a generic computing architecture. Software implementations of neural network algorithms suffer from the inherent weakness of being inefficient and hence time-consuming. Kohonen SOFM especially is a highly parallel algorithm which needs large CPU times for training large data sets to the network [37]. Thus a need for specialized hardware systems dedicated to SOFM computations and which maximize the utilization of resources for exploiting the inherent parallelism in the algorithm is called for. Moreover, in portable applications, where the time taken for compression is critical, a specialized hardware for speeding up the process of quantization is justified.

Re-configurable devices such as FPGAs offer a low cost middle ground between the efficiency of VLSI ASIC implementations and the flexibility of a software solution. An FPGA approach allows fast time-to-market, customizability and just-in-time production. Moreover, since different compression rates can be obtained with different network sizes, FPGAs are a suitable way to study the comparative features of the different networks. An implementation on an FPGA also gives us the option of run time programmability in portable applications to obtain different compression ratios.

In this chapter we propose a novel hardware implementation of the Kohonen SOFM using Frequency Adaptive Learning (FAL) algorithm with frequency reinitialization at set intervals. In such a scheme, the learning rate is localized to each individual neuron and is dependent on the number of times the neuron has won instead of the length of training. This change in the fundamental learning equation improves the utilization of trained neurons in the mapping process and eliminates the dead neuron problems due to poor initialization. As a result, neighborhood updating can be ignored without severely affecting the adaptation rate.

8.2   Algorithmic adjustment

In VLSI implementation of the Kohonen SOFM, each neuron is typically implemented as a processing element. The more complex the neuron structure is, the lesser is the number of neurons that can be accommodated within reasonable area constraints. Given the need for a decent number of neurons in any application, a direct implementation of the Kohonen algorithm is implausible as it requires an impractical number of multipliers and I/O pins per neuron [31, 32, 34]. This section discusses the modified Kohonen algorithm and the novelties that we have implemented.

8.2.1 Conventional algorithmic simplification

Most present digital implementations of the Kohonen network rely on modifications to the original algorithm to simplify the hardware architecture [31, 32]. Past researches have focused on how to optimally implement the Kohonen SOFM on a digital platform without overtly affecting the effectiveness of the algorithm [3]. The general trend has been towards simplifying those stages of the Kohonen algorithm that involve digitally impractical calculations. The most common modifications involve the use of Manhattan distance for Euclidean distance [31, 34] and the use of simplified learning rate functions, both of which cut down on the usage of digital multipliers. Typical algorithm tailored for hardware implementation is presented here:

1. The neurons are initialized with random weights.

2. The topological distance between the neuron weight, $W = \{W_r, W_g, W_b\}$ and the input pixel, $X = \{X_r, X_g, X_b\}$ is calculated. Manhattan distance is used instead of the Euclidean distance to avoid the computationally intensive square root and squaring functions.

3. The neuron with the least topological distance is deemed the winning neuron. The weights of the winning neuron are updated based on some learning rule.

$$W_w(t+1) = W_w(t) + \alpha(t)(X(t) - W_w(t)) \qquad (8.1)$$

where $\alpha(t)$ is the learning rate function, $W_w(t)$ and $X(t)$ are respectively, the weight of the winning neuron and the input pixel presented at time t. For hardware simplicity, unit learning rate [31] or learning rate decreasing as integer multiples of $\frac{1}{2}$ [32] have been adopted.

4. The neighborhood neurons of the winning neuron are determined and their weights are updated according to a similar learning rule:

$$W_n(t+1) = W_n(t) + \beta(t)(X(t) - W_n(t)) \qquad (8.2)$$

where $\beta(t)$ is the learning rate function of the neighborhood, which reduces with their topological distances from the winning neuron. As $\beta(t) < \alpha(t) \forall t$, a good neighborhood learning rate function is difficult to accommodate in hardware without significantly degrading its area and time performances. On the other hand, ignoring neighborhood learning leads to slow rate of adaptation.

8.2.2 Frequency Adaptive Learning Algorithm

In the conventional SOFM algorithm, the learning rate $\alpha(t)$ is set high initially. As more input pixels are presented to the network, the learning rates $\alpha(t)$

and $\beta(t)$ are gradually reduced according to a linear or exponential function. This approach can however lead to the problem of over utilization of some and under utilization of other neurons in the pixel mapping process owing to the localization of responses. The resulting visual artifacts are more severe in hardware than software due to the limited number of neurons imposed by the size constraints. The situation is alleviated by randomizing the input presentation [5]. Again, the hardware cost required for the input randomization has often been overlooked.

We propose an implementation based on a novel Frequency Adaptive Learning algorithm (FAL) [30]. This method requires each neuron to track the number of times it has been updated with a winning frequency counter. The learning rate of the winner is then a direct function of the frequency count. At predefined intervals in the training process, the "dead" neurons (as denoted by neurons with zero frequency count) are reinitialized with the weights of the "saturated" neuron (with maximum frequency count). To implement FAL, a localized learning rate function $\alpha(F)$ is proposed as follows:

$$\alpha(F) = K^{10 \cdot log(F)} \tag{8.3}$$

where K is a constant value ranging from 0 to 1, and F is the local frequency count normalized according to the network and image sizes. Figure 8.1 compares the proposed learning rate function with some popular learning functions [5, 35]. It is noticed that the proposed learning rates strive to drag the winning neuron towards the input cluster as quickly as possible within its first 10 to 50 updates, thereafter the neuron is allowed to fine tune gradually in its later wins. For other learning rate functions, the neurons are almost frozen after 50 times of winning.

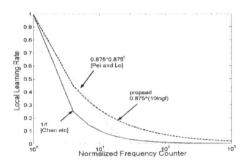

Figure 8.1.   Comparisons of Different Learning Rates

For ease of hardware implementation, learning rates as a function of the sum and difference of power-of-two terms were chosen. Instead of using the original continuous, complex learning rate function of (3), several discrete rates are selected for different ranges of normalized winning frequencies, F, as shown in Table 8.1. The lower limits for the normalized frequency ranges are chosen such that various learning rates can be multiplexed with simple selector logics formed by the bits of the frequency counter. When the normalized frequency is greater than 2047, the neuron is inhibited from further updating to avoid unnecessary oscillations around the cluster centroid. The advantages that this method

Table 8.1. Comparisons of Different Learning Rates

F	$0 - 7$	$8 - 63$	$64 - 1023$	$1024 - 2047$	> 2048
α	0.75	0.375	0.09	0.0156	0

offers are ample. Firstly, this implementation achieves a truly utilization dependent learning as opposed to unity learning rate [31] or a broadcast common learning rate [34]. Consequently, neurons that have been utilized quite heavily undergo smaller change of magnitudes. This is equivalent to not disturbing those neurons that are already representative of some majority colors. Secondly, training can be divided into stages, so that dead neurons can be detected at each stage and reinitialized towards the active neurons. Software simulation shows that adaptation can be sped up by eight times and utilization rate at the final pixel mapping process improved significantly [30]. It has also been shown that one pass of all pixels of an image is sufficient to achieve a good reconstruction quality. This deterministic convergence has eliminated substantial amount of hardware cost and computation time to evaluate termination criteria in conventional SOFM. Lastly, by decoupling learning rate function from the order of presentation, input randomization can be foregone. Most VLSI implementations ignore neighborhood function without compensating for their severe degradation on the adaptation results. With the rate of adaptation accelerated by FAL, neighborhood updating can be ignored in our proposed implementation. Re-initialization of dead or undertrained neurons achieves a similar effect as neighborhood updating to the clustering of neurons.

8.3 Architecture

This section deals with the architectural features of the implemented SOFM. The most critical aspect of any hardware design is the selection and design of the architecture that provides the most efficient and effective implementation of the algorithm available. It is a general requirement of neural nets that the neurons be richly interconnected for the purpose of inter-neuron communication. In an ideal situation, any neuron must be able to communicate with any other neuron. A direct digital implementation of this rich interconnection can only be achieved at the cost of an impractical number of I/O and routing resources. An FPGA has only limited routing resources. Moreover, when the routing gets congested, long routing paths may be necessary to complete a signal path. This could slow down the design and cause routing problems for the signal that must travel across or around the congested areas.

An architecture that suits the need for interconnected networks admirably without taxing hardware needs very much is the broadcast architecture [31, 34]. A broadcast architecture fully exploits the property of computational independence of neurons. A distinct advantage of such a network is the leeway provided for scalability with respect to the network size. The network structure is illustrated in Figure 8.2.

The Control Unit is the backbone of the network and it controls the operation of the network by broadcasting the control and data values on the control and data buses, respectively. The individual neurons are also capable of taking control of the bi-directional data bus to broadcast their data values.

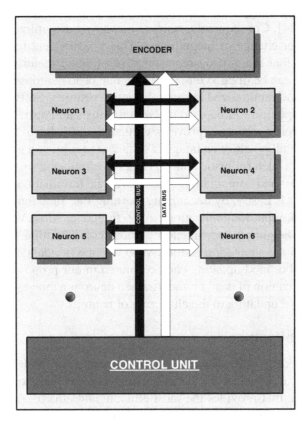

Figure 8.2.   Broadcast Architecture

Each neuron in Figure 8.2 has a structure shown in Figure 8.3. The internal arithmetic circuits are designed to allow each processing element to autonomously compute its distance to the input and update its weight. Another salient feature of this neuron architecture is that all the computations can be carried out with just a single adder module. The SORTING_UNIT is used to allow the controller to read any bit of the internal register through the single bit feedback line by placing the bit mask on the data-in lines. This unit is very useful for finding the neuron with the minimum Manhattan distance or minimum frequency count. Using a down-counting frequency counter offers two distinct advantages: The SORTING_UNIT can be used to determine the neuron with the maximum frequency and saturated neurons can be easily determined as those with a frequency count of 0, thus cutting down on unnecessary comparators. With the SORTING_UNIT, the number of cycles taken to find the neuron with minimum attributes is almost half that of the binary search method. The tradeoff is in terms of an increase in hardware resources used.

During the initialization phase, the Control Unit sends an n-bit address on the data bus (for a network of $N = 2^n$ neurons) and asserts the add_cyc signal. The addressed neuron acknowledged by setting its en_neuron flag. In the next cycle, the Control Unit broadcasts the 8-bit data on the data bus and deasserts the add_cyc signal to initialize the addressed neuron with $W_r = W_g = W_b = 255 \cdot A(N - 1)$ where A is the address of the neuron. This process is repeated with consecutive addresses until all neurons have been initialized. The frequency counters of all neurons can be initialized in a similar manner by asserting the $init\_freq$ signal on the control bus and putting the initial frequency value on the data bus. The frequency counters are initialized to f_{max} instead of 0 so that the winning frequency, $F = f_{max} - f$, where f is the counter value. This arrangement allows the SORTING_UNIT to search for the most active neuron during the reinitialization. In addition, saturated neurons can be identified by zero frequency count so that they can be inhibited from further updating.

After initialization, the system performs the following operations as the input pixels are presented:

- **Step 1**: The Control Unit broadcasts the input pixel values: $r(t)$, $g(t)$ and $b(t)$ on the data bus. The neurons calculate the metric $|X_i - W_i|$ for $i = r, g, b$ and store the three difference values in the D registers, D_R, D_G and D_B.

- **Step 2**: The Manhattan distance $|(X_r - W_r) + (X_g - W_g) + (X_b - W_b)|$ is calculated and stored in the internal register of the SORTING_UNIT. Through the SORTING_UNIT, the neuron with the minimum Manhattan distance is found and its wnr flag is set.

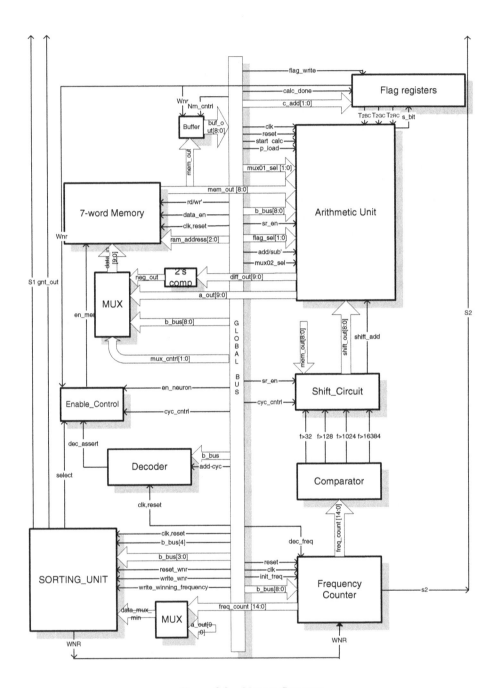

Figure 8.3.   Neuron Structure

- **Step 3**: The Control Unit asserts the dec_freq signal to decrement the frequency counter of the winning neuron, which is identified by the local wnr flag.

- **Step 4**: The neuron with wnr asserted will be updated, i.e., $W_c(t+1) = W_c(t) + \gamma_i((X(t) - W_c(t+1))$ is executed with the shift-add operations through the shift register and the Arithmetic unit. The amount of shifts is determined by the range of frequency counter value, f.

- **Step 5**: At this stage, the Control Unit checks if the predefined cycle time for re-initialization has reached. If the re-initialization cycle is not due, it proceeds directly to Step 8. During re-initialization, the Control Unit asserts the $load\_frequency$ signal on the control bus to load the frequency counter value to the internal register of the SORTING_UNIT. By accessing the bits of this register, the neuron with the minimum frequency count (denoting maximum updates) is found and its wnr flag is asserted.

- **Step 6**: The neuron whose wnr flag is set takes control of the data bus and places its W_r, W_g and W_b values on the bus.

- **Step 7**: The neurons with $f = f_{max}$ have their no_update flag set. The no_update flag enables the neurons to reinitialize their weights to the weights on the data bus.

- **Step 8**: Steps 1 to 8 are repeated for the next input pixel.

After all the pixels of an image have been exhausted, the weights stored in the neurons form the entries of the LUT indexed by the addresses of the neurons. The quantizer mapping is done in a manner similar to the color palette design.

8.4   Implementation

The SOFM architecture discussed in earlier sections was implemented on a Xilinx FPGA (XCV1000-4bg560). The design and verification was accomplished using Verilog HDL (Hardware Description Language). Verilog HDL is an all purpose language that affords the user the ability to design hardware elements in abstract terms as well as the ability to design the verification environment. Compilation and Simulation were achieved using MODELSIM. Synthesis, which is the stage where the abstract Verilog elements is translated into a gate-level net-list, was accomplished using Xilinx ISE and FPGA EXPRESS.

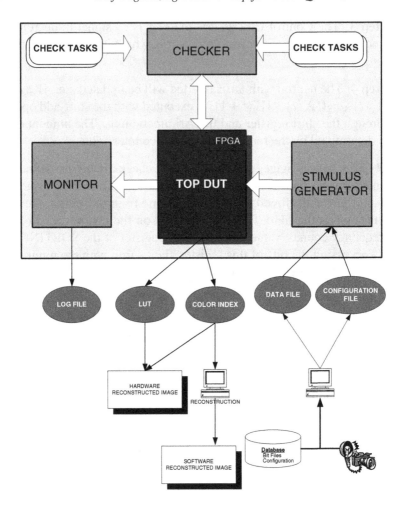

Figure 8.4.   Complete Design and Verification Environment

8.4.1   Development Environment

Figure 8.4 shows the complete design and verification environment. The TOP DUT (Device Under Test) is the synthesizable HDL description of the hardware. This module is functionally verified using a combination of a stimulus generator, which generates test stimuli and a checker module that compares the obtained response with the expected response. Pre-written check tasks provide the checker module with the expected responses.

Functional simulation with actual images involves, firstly the software deconstruction of an image into a configuration file and a data file. The configuration file provides the stimulus generator with details such as the number of neurons, the number of pixels in the image, the initialization values, etc. The data file, on the other hand contains the R, G and B values of all pixels of the image. The stimulus generator generates stimuli based on these two files and the monitor generates the log file, which carries information regarding the quantization process. Information printed includes the weights of all neurons after every learning cycle. The output of the DUT, meanwhile is stored separately as the LUT values in one file and the color indices (weights) in the other file.

To compare the hardware and software encoding qualities, the color indices are used to encode the data using a software implementation and this is used to obtain a software reconstruction of the image. Similarly, the LUT generated for the hardware implementation is used to reconstruct the image. The image quality is quantified using the PSNR.

8.4.2   Control Unit Description

The Control unit is described as a big Finite State Machine (FSM) that has an almost one-to-one correspondence with the algorithm implemented. Figure 8.5 shows the state diagram simplified for presentation. Each of the states shown in this figure can be seen as a super state that has child states. As can be seen from this figure, the state machine closely follows the algorithmic steps.

All registers and sequential elements of the design are reset initially. This is represented by the RESET state. This state is followed by the LOAD CONFIGURATION state where the configuration data is loaded. Configuration data includes such essential items as the number of neurons, number of pixels and initial frequency. Once these data have been loaded as seen by the setting of the flag *load_complete*, the control unit goes about initializing frequency counters of the neuron. This is followed by the INITIALIZE WEIGHT state where the neuron weights are initialized. When all the neurons have been initialized, which is reflected in *initialize_count* being equal to *neuron_count*, the CALCULATE MANHATTAN DISTANCE state is arrived at. This state is responsible for broadcasting the data, generating the control signals for the calculation of the Manhattan distance, and storing the computed distance in each neuron. At the end of this state, as witnessed by the setting of the *calculation_done* flag, the FIND WINNER state is entered. As the name suggests, this state involves the generation of control and data signals to expedite the sorting step for finding the neuron with the minimum Manhattan distance. When this step has been accomplished, the *wnr* signal that is feedback from the neurons to the control unit is asserted and the state machine moves on to the UPDATE WINNER state. In this state, the winning neuron is updated based on preset learning rates.

Following this comes the SELECT PATH state. The sole responsibility of this state is to determine the path the state machine should take from thereon.

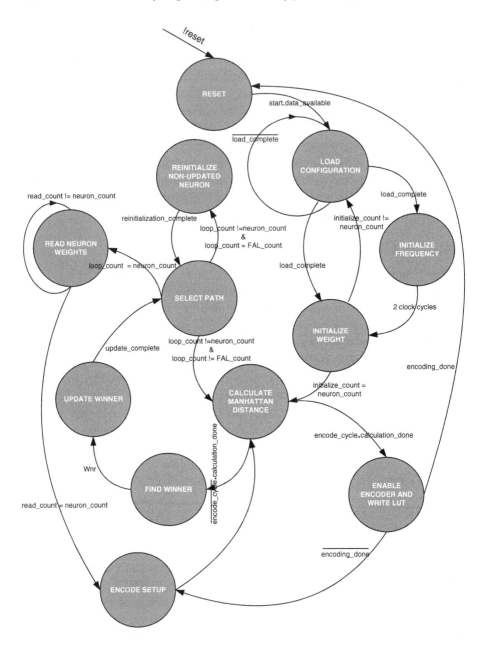

Figure 8.5.   Simplified State Diagram for the Controller

If the number of pixels that have already been trained (represented in Figure 8.5 by $loop\_count$) is not equal to the total number of pixels and also the preset interval for frequency based re-initialization (represented by FAL_count) has not been reached, then the next pixel is presented for training by repeating the CALCULATE MANHATTAN DISTANCE state. However, if the preset interval for frequency based re-initialization has been reached, then the next state is the REINITIALIZE NON-UPDATED NEURON state. On the other hand if the number of pixels that have already been trained is equal to the total number of pixels, then the training process is deemed to have been completed and the state machine moves on to the READ NEURON WEIGHTS state. In this state, the weights of the neuron are read and written as the "color index" file shown in Figure 6.4. When all the neuron weights have been read the *read_count* becomes equal to the *neuron_count*. In such a case, the next state involves setting up the encoding process, by reloading the pixel values. This state is called the ENCODE SETUP state. Encoding of pixel values also involves the Manhattan distance calculation which is reflected in the state diagram. This is followed by the ENABLE ENCODER AND WRITE LUT state. When all the pixels in the data file have been encoded, the flag *encoding_done* is set. After encoding, the state machine goes back to the reset state.

8.5 Experimental results

Table 8.2 shows the computational complexity of the network in terms of the processing time spent on a P pixel image by a network with N neurons with M cycles of frequency re-initialization.

Table 8.2. Time Complexity for Color Quantization for P pixels and N neurons

Operation	Number of Clock Cycles
Weights Initialization	$3N$
Frequency Initialization	2
Distance Calculation	$17P$
Winner Search	$(22+N)P$
Frequency Update	$1P$
Winner Update	$30P$
Dead neuron reinitialization	$58M$
Reading LUT	$9N$
Quantizer Mapping	$17P$

The time complexity of the training and quantizer mapping processes are expressible as $(3 + P) \cdot N + 70 \cdot P + 58 \cdot M + 2$ and $9 \cdot N + 17 \cdot P$, respectively. The weights initialization depends on the number of neurons in the network owing to the fact that different neurons are initialized with different values through the data bus from the controller. The initial frequency on the other hand is common for all neurons and thus takes only 2 steps regardless of the neural network size. The Manhattan distance calculation for a pixel takes 17 cycles owing to the need for distance calculation for the three color components and the addition of these three individual components. The search for the winning neuron for each pixel needs 22 cycles for winner determination and N cycles for arbitration to resolve tie. For a pixel, the Winner Updating process takes a good amount of time because of its frequency dependence and the need for multiple shift and add operations. Dead neuron re-initialization is an involved process where each reinitialization takes up 58 clock cycles. This arises due to the need to have to do a minimum frequency search (using the same resources as those used for Winner search) followed by a transmission of the weights from a minimum frequency neuron to dead neurons. The saving grace however is that this step has to be done only a few times in the course of the learning process. Reading LUT values involves reading the weights of the neurons and mapping them into a tabular format with the neuron addresses representing the index values.

The synthesis results of the entire SOFM architecture along with the breakup into individual components from XILINX ISE using Synopsys FPGA Express is given in Table 8.3. The TOP DUT in Table 8.3 consists of the neural network, the synchronous encoder and the control unit. The design has been mapped on to the Xilinx Virtex series of target device XCV1000-4bg560. The datapath is dominated by the processing element and the critical path delay of the top level architecture is 27.866ns. This gives a maximum operating frequency of 35.8MHz irrespective of the network size. It can be seen from this table that most of the area is taken up by the neural network.

Table 8.3.   Synthesis Results

	16N			32N			64N		
	% Slice	No. of FF	Eqvt. Gate Count	% Slice	No. of FF	Eqvt. Gate Count	% Slice	No. of FF	Eqvt. Gate Count
Top DUT	37%	2734	57,424	68%	5281	107,480	99%	10369	226,703
Synchronous Encoder	1%	5	172	1%	6	342	1%	752	7
Control unit	2%	125	4210	2%	125	4210	2%	125	4210

Most literatures on hardware implementations of SOFM [32, 3, 34] report only their area-time complexity. The hardware simplifications and truncation of several fractional decimal digit precision to integer representation has significant impact on the quality of the output due to cumulative errors. Unfortunately, this critical sacrifice of the output quality has often been omitted in hardware implementation and no objective quality assessment for image reconstructed from hardware based SOFM has been reported. The images are trained by our hardware SOFM and its reconstructed PSNR in dB is presented in Table 8.4. The second column for each network size specifies the performance for a process where the color palette design is done in hardware and the Quantizer Mapping is done in software

Table 8.4. Quality of reconstructed image in PSNR

	16N		32N		64N	
	H/W	S/W mapped	H/W	S/W mapped	H/W	S/W mapped
Lena (512 X 512)	19.87	21.97	22.48	26.04	23.84	27.50
Pepper (512 X 512)	17.64	18.53	21.10	23.21	22.32	25.30
Baboon (512 X 512)	16.58	18.51	18.19	21.72	18.73	23.25

Figure 6.6 shows a few samples of images that have been quantized and mapped using the hardware architecture developed. The top row shows the original images Lena, Pepper and Baboon respectively. The bottom row shows the corresponding images reconstructed from the quantized data. The quantized images are for 64 neuron implementation.

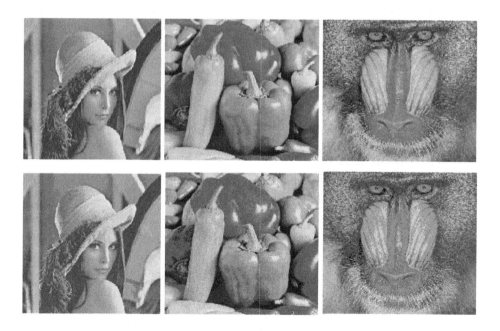

Figure 8.6. Comparison between Original (top) and Reconstructed (bottom) Images

8.6   Conclusions

Unlike digital video, successive digital still images like those taken from the digital camera, may vary drastically in their color distribution. Therefore, it is imperative to have an adaptive colormap to minimize the distortion for a given bit budget. VLSI chip for color quantization in such portable application need not be real time but should be power efficient to extend the battery life. Several important considerations that have significant impact on the trade-off between hardware complexity and rate-distortion performance have been omitted in conventional approaches to hardware implementation of adaptive colormap design. We observe from software simulations that the variations of neuron initialization, input pixel randomization and the training time have non-negligible effect on the neuron utilization and adaptation rates. These factual results prompt us to adopt a drastically different philosophy to revising the SOFM algorithm by endowing the low complexity architecture with an effective learning mechanism.

In this chapter, an efficient architecture of SOFM dedicated for color quantization has been proposed. The frequency adaptive learning (FAL) function with dead neuron re-initialization is employed to enhance the learning efficiency. A frequency down counter is used in each processing element for finding the minimum Manhattan distance and the maximum frequency count to save the number of comparators required by the FAL algorithm. The architecture has been successfully implemented on the Xilinx Virtex XCV1000Ú4bg560 FPGA and tested for network size of up to 64 neurons.

References

[1] J. P. Braquelaire and L. Brun, "Comparison and optimization of methods of color image quantization," *IEEE Trans. Image Processing*,vol. 6, no. 7, pp. 1048-1052, 1997.

[2] M. Orchard and C. Bouman, "Color quantization of image," *IEEE Trans. Signal Processing*,vol. 39, no.12, pp. 2667-2690, 1991.

[3] N. Papamarkos, "Color reduction using local features and a SOFM neural network," *Int. J. Imaging Systems and Technology*, vol. 10, no.5, pp. 404-409, 1999.

[4] N. Papamarkos, A. E. Atsalakis and C. P. Strouthopoulos, "Adaptive color reduction," *IEEE Trans. Syst., Man, and Cybern.*, Part B, vol. 32, no. 1, pp. 44-56, Feb. 2002.

[5] S. C. Pei and Y.S. Lo, "Color image compression and limited display using self-organization Kohonen map," *IEEE Trans. Circuits Syst. Video Technol.*, vol. 8, no. 2, pp. 191-205, 1998.

[6] G. Sharma and H. J. Trusell, "Digital color imaging," *IEEE Trans. Image Process.*, vol. 6, no. 7, pp. 901-932, 1997.

[7] C.H. Chang, R. Xiao and T. Srikanthan, "A MSB-biased selforganizing feature map for still color image compression," *in Proc. IEEE Asia-Pacific Conf. on Circuits and Syst.*, Bali, Indonesia, vol. 2, pp. 85-88, 2002.

[8] C. Amerijckx, M. Verleysen, P. Thissen and J. Legat, "Image compression by Self-Organized Kohonen map," *IEEE Trans. Neural Networks*, vol. 9, no. 3, pp. 503-507, 1998.

[9] C.H. Chang, R. Xiao and T. Srikanthan, "An adaptive initialization technique for color quantization by self organizing feature map," *in Proc. IEEE Int. Conf. on Acoustics, Speech, and Signal Processing (ICASSPŠ03)*, Hong Kong, Apr. 2003.

[10] Y. Linde, A. Buzo, and R. M. Gray, "An algorithm for vector quantizer design," *IEEE Trans. Commun.*, vol. COMM-28, no. 1, pp. 84-95, Jan. 1980.

[11] P. Heckbert, "Color image quantization for frame buffer display," *ACM Comput. Graphics*, vol. 16, no. 3, pp. 297-307, Jul. 1982.

[12] S. J. Wan, P. Prusinkiewicz and S. K. M. Wong, "Variance-based colour image quantization for frame buffer display," *Colour Research and Application*, vol. 15, no. 1, pp. 52-58, 1990.

[13] M. Gervautz and W. Purgathofer, "A simple method for color quantization: octree quantization," in *Graphics Gems*. A. Glassner, Ed., pp. 287-293, New York: Academic Press, 1990.

[14] E. S. Grossberg, *Neural Networks and Natural Intelligence*. Cambridge, MA: M. I. T. Press, 1988.

[15] T. Kohonen, *Self-Organization and Associative Memory*. New York: Springer-Verlag, 1989

[16] R. P. Lippmann, "An introduction to computing with neural nets," *IEEE Acoustics, Speech, and Signal Processing Mag.*, pp. 4-22, Apr. 1987.

[17] D. E. Rumelhart, J. L. McClell and PDP Research Group, *Parallel Distributed Processing*. Cambridge, MA: M. I. T. Press, 1986.

[18] A. H. Dekker, "Kohonen neural networks for optimal color quantization," *Network: Computtat. Neural Syst.*, vol. 5, pp. 351-367, 1994.

[19] Y. W. Lim and S. U. Lee, "On the color image segmentation algorithm based on the thresholding and the fuzzy C-means techniques," *Pattern Recognit.*, vol. 23, no. 9, pp. 935-952, 1990.

[20] S. A. Shafer and T. Kanade, "Color vision," in*Encyclopedia of Artificial Intelligence*, S. C. Shapiro and D. Eckroth, Eds., pp. 124-131, New York: Wiley, 1978.

[21] O. Vereka, "The local K-means algorithm for color image quantization," M.Sc. dissertation, Univ. Alberta, Edmonton, AB, Canada, 1995.

[22] N. R. Pal, J. C. Bezdek and E.K. Tsao, "Generalized clustering networks and Kohonen's self organizing scheme." *IEEE Trans. Neural Networks*, vol. 4, no. 4, pp. 549-557, July 1993.

[23] E. Yair, K. Zeger and A. Gersho, "Competitive learning and soft competition for vector quantizer design," *IEEE Trans. Signal Processing*, vol. 40, no. 2, pp. 294-309, Feb. 1992.

[24] A. K. Krishnamurthy, S. C.Ahalt, D. E. Melton and P. Chen, "Neural networks for vector quantization of speech and images," *IEEE J. Select. Areas Commun.*, vol. 8, no. 8, pp. 1449Ű1457, Oct. 1990.

[25] T. C. Lee and A. M. Peterson, "Adaptive vector quantization using a self-development neural network," *IEEE J. Select. Areas Commun.*, vol. 8, no. 8, pp. 1458-1471, Oct. 1990.

[26] T. D. Chiueh, T. T Tang and L. G. Chen, "Vector quantization using tree-structured self-organizing feature maps," *IEEE J. Select. Areas Commun.*, vol. 12, no. 9, pp. 1594-1599, Dec. 1994.

[27] Haykin, *Neural Networks: A comprehensive foundation*. New York: MacMillan College Publishing Company, 1994.

[28] D. J. Willshaw and C. V. D. Malsburg, "How patterned neural connections can be set up by self-organization," in *Proc. Roy. Soc. London B*, vol. 194, pp. 431-445, 1976.

[29] T. Kohonen, "Self-organized formation of topologically correct feature maps," *Biological Cybernetics*, vol. 43, pp. 59-69, 1982.

[30] R. Xiao, C.H. Chang and T. Srikanthan, "An efficient learning rate updating scheme for the self-organizing feature maps," in *Proc. 2nd Int. Conf. on Visualization, Imaging and Image Processing*, Malaga, Spain, pp. 261-264, Sep, 2002.

[31] M.S. Melton, T. Phan, D.S. Reeves and D.E.V.D. Bout, "The TinMANN VLSI chip," *IEEE Trans. on N. N.*, vol. 3, no. 3, pp. 375-383, 1992.

[32] S. Rueping, K. Goser and U. Rueckert, "A chip for selforganizing feature maps," in *Proc. IEEE Conf. On Microelectronics for Neural Network and Fuzzy Syst.*, vol. 15, no. 3, pp. 26-33, Jun. 1995.

[33] P. Ienne, P. Thiran and N. Vassilas, "Modified self-organizing feature map algorithms for efficient digital hardware implementation," *IEEE Trans. on Neural Network*, vol. 8, no. 2, pp. 315-330, 1997.

[34] E.T. Carlen and H.S. Abdel-Aty-Zohdy, "VLSI Implementation of a feature mapping neural network," in *Proc. 36th Midwest Symp. on Circuits and Syst.*, vol. 2, pp. 958-962, 1993

[35] O.T.C Chen, B.J.Sheu and W.C. Fang, "Image compression using self-organization networks," *IEEE Trans. Circuits Syst. Video Techno.*, vol. 4. no. 5, pp. 480 Ű 489, 1994.

[36] T. Kohonen, *Self-Organizing Maps*. New York: Springer, 1995.

[37] X. Fang, P. Thole, J. Goppert and W. Rosenstiel, "A hardware supported system for a special online application of self-organizing map," in *Proc. IEEE Int. Conf. on Neural Networks*, vol. 2, pp. 956-961, Jun. 1996

Chapter 9

IMPLEMENTATION OF SELF-ORGANIZING FEATURE MAPS IN RECONFIGURABLE HARDWARE

Mario Porrmann, Ulf Witkowski, and Ulrich Rückert

Heinz Nixdorf Institute, University of Paderborn, Germany
System and Circuit Technology

porrmann@hni.upb.de

Abstract   In this chapter we discuss an implementation of self-organizing feature maps in reconfigurable hardware. Based on the universal rapid prototyping system RAPTOR2000 a hardware accelerator for self-organizing feature maps has been developed. Using state of the art Xilinx FPGAs, RAPTOR2000 is capable of emulating hardware implementations with a complexity of more than 15 million system gates. RAPTOR2000 is linked to its host – a standard personal computer or workstation – via the PCI bus. For the simulation of self-organizing feature maps a module has been designed for the RAPTOR2000 system, that embodies an FPGA of the Xilinx Virtex (-E) series and optionally up to 128 MBytes of SDRAM. A speed-up of up to 190 is achieved with five FPGA modules on the RAPTOR2000 system compared to a software implementation on a state of the art personal computer for typical applications of self-organizing feature maps.

Keywords: SOM, Self-Organizing Feature Maps, Reconfigurable Hardware, RAPTOR2000

9.1   Introduction

Self-organizing feature maps (SOMs) [4] are successfully used for a wide range of technical applications, in particular, dealing with noisy or incomplete data. Examples of use are explorative data analysis, pattern matching, and controlling tasks. In cooperation with an industrial partner we are using SOMs for the analysis of IC (Integrated Circuits) fabrication processes. The large amount of data, that is captured during fabrication, has to be analyzed in order to optimize the process and to avoid a decrease of yield [5, 6]. Currently, we are analyzing data sets with more than 15,000 input vectors and a vector

A. R. Omondi and J. C. Rajapakse (eds.), FPGA Implementations of Neural Networks, 247–269.
© 2006 *Springer. Printed in the Netherlands.*

dimension of about 100. Software simulations of medium sized maps (i.e. about 100×100 neurons) on state of the art workstations require calculation times of several hours for these data sets. The simulation of larger maps (more than one million neurons with vector dimensions in the order of hundreds) seems promising but is not feasible with state of the art PCs or workstations. From the various possibilities to speed up neural computations we have chosen the design of a hardware accelerator for neural networks. Our goal is to integrate the system into state of the art workstations if very high performance is required and to enable access to the accelerator via the internet if the accelerator is only sporadically used.

In recent years various hardware implementations for different neural network architectures have been presented [1–3]. The main benefit of special purpose hardware is the higher resource efficiency compared to software implementations. On the one hand, special purpose hardware is well suited for low-power implementations, on the other hand, much higher performance can be achieved, compared to software implementations on a sequential processor. However, many of the proposed architectures are dedicated to single neural network algorithms or groups of similar algorithms. The aim of our project is to deliver a system that is capable of accelerating a wide range of different neural algorithms. Additionally, in most applications, different methods of information processing are combined. For example, artificial neural networks are combined with fuzzy logic or with techniques for knowledge-based information processing. In contrast to implementing different components for data pre- and postprocessing and for neural networks we use a dynamically (i.e. during runtime) configurable hardware accelerator to implement all of the algorithms that are required for a special application. The system can be reconfigured for the different tasks in one application (e.g. different configurations for pre- and postprocessing may be selected). Because of the reconfigurability the hardware can be adapted to the changing requirements of the application, thus allowing an easy expansion by new algorithms to improve flexibility and performance.

9.2   Using reconfigurable hardware for neural networks

Dynamic reconfiguration (or runtime reconfiguration) offers an opportunity for application specific implementations on a dedicated hardware environment. Different levels of dynamic reconfiguration can be distinguished. For example in [7] three categories are presented: algorithmic reconfiguration, architectural reconfiguration and functional reconfiguration.

The goal in algorithmic reconfiguration is to reconfigure the system with a different computational algorithm that implements the same functionality, but with different performance, accuracy, power, or resource requirements [7].

In the field of neural network hardware, algorithmic reconfiguration can be used e.g. to implement algorithms with variable precision. For self-organizing feature maps a low precision of e.g. 8 bit is sufficient for a rough ordering of the map in the beginning of the learning process. For fine tuning of the map, the precision of the hardware is increased (e.g. to 16 bit). Using a lower precision allows us to set up an optimized architecture that can be faster, smaller or more energy efficient than a high precision architecture.

In architectural reconfiguration the hardware topology and the computation topology are modified by reallocating resources to computations. The need for this type of reconfiguration arises e.g. if resources become unavailable or if additional resources become available. Designing massively parallel architectures for neural networks with a large number of processing elements (PEs), an interesting approach for architectural reconfiguration is to check the functionality of the processing elements during start up. Processing elements that are not working correctly can be disabled. This makes it possible to enhance the yield, because problems in single processing elements can be tolerated in this way.

Functional reconfiguration is used to execute different algorithms on the same resources. Thus, limited hardware resources can be used to implement a wide range of different algorithms. In neural network simulation we are often interested in providing as much computing power as possible to the simulation of the algorithm. But pre- and postprocessing of the input and output data often also requires quite a lot of calculations. In this case dynamic reconfiguration offers us the opportunity to implement special preprocessing algorithms in the beginning, switch to the neural network simulation and in the end reconfigure the system for postprocessing. Thus, we do not require the system resources that would be necessary to calculate all algorithms in parallel. An example for functional reconfiguration in this sense is the use of self-organizing feature maps as a preprocessing stage for radial basis function networks, as proposed e.g. by [8].

Our implementation allows us to add additional functionality for data postprocessing like U-matrix calculation. The integration of postprocessing algorithms on the one hand speeds up computation time, on the other hand it often drastically reduces communication time because the amount of data, that has to be transferred from the hardware accelerator to the host computer is reduced by these algorithms. In the case of data visualization we are able to offer a speed up compared to a postprocessing algorithm on the host computer, that enables us to display U-matrices online, during learning with nearly no loss in performance.

9.3   The dynamically reconfigurable rapid prototyping system RAPTOR2000

The hardware accelerator for self-organizing feature maps, that is presented here is based on the modular rapid prototyping system RAPTOR2000. The system consists of a motherboard and up to six application specific modules (ASMs). Basically, the motherboard provides the communication infrastructure between the ASMs and links the RAPTOR2000 system via the PCI bus to a host computer. Additionally, management functions like bus arbitration, memory management and error detection services are integrated in two *Complex Programmable Logic Devices* (CPLD).

The various communication schemes that can be used between different ASMs and between the ASMs and the host computer are depicted in the block diagram in figure 9.1. Every ASM slot is connected to the *Local Bus* for internal communication with other devices or ASMs and for external communication with the host processor or with other PCI bus devices. An additional *Broadcast Bus* can be used for simultaneous communication between the ASMs. Additionally, a dual port SRAM can be accessed by all ASMs via the Broadcast Bus (e.g. utilized as a buffer for fast direct memory accesses to

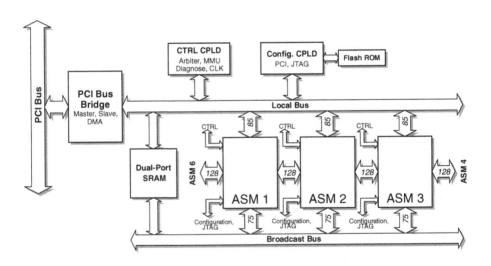

Figure 9.1.   Architecture of the RAPTOR2000 system

A crucial aspect concerning FPGA designs is the configuration of the devices. Each ASM that carries an FPGA has to be configured by an application specific data stream that determines the function of the device. In order to uti-

Figure 9.2. Photo of the RAPTOR2000 system with two ASMs of type DB-VS

lize dynamic reconfiguration (i.e. during runtime) it is necessary to minimize this reconfiguration time, therefor the configuration algorithms have been implemented in hardware. Reconfiguration of an ASM can be started by the host computer, another PCI bus device or by another ASM. Thus, it is possible that an FPGA autonomously reconfigures itself by configuration data that is located anywhere in the system. Due to the hardware implementation of the reconfiguration algorithm, a Xilinx Virtex 1000 FPGA can be completely reconfigured within less than 20 ms. The configuration algorithm implemented into the hardware also supports the partial reconfiguration of the system [9].

For the simulation of self-organizing feature maps the module DB-VS has been designed for the RAPTOR2000 system (figure 9.3). This ASM embodies an FPGA of the Xilinx Virtex (-E) series and optionally up to 128 MBytes of SDRAM. The ASM can be equipped with various chips, that emulate circuits with a complexity of 400,000 to 2.5 million system gates. The SDRAM controller is integrated into the FPGA logic. A photo of the RAPTOR2000 system with two DB-VS modules is shown in figure 9.2. In the context of this article we focus on the implementation of self-organizing feature maps on RAPTOR2000. Because of the flexibility of the system many other neural and conventional algorithms may be mapped to the system. As another example for neural networks we have analyzed the implementation of neural associative memories and of radial basis function networks on the RAPTOR2000 system [10]. Another case study focuses on the implementation of octree based 3D graphics [11].

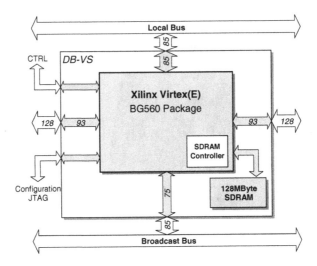

Figure 9.3.   Block diagram of the application specific module DB-VS

9.4   Implementing self-organizing feature maps on RAPTOR2000

Self-organizing feature maps as proposed by Kohonen [4] use an unsupervised learning algorithm to form a nonlinear mapping from a given high dimensional input space to a lower dimensional (in most cases two-dimensional) map of neurons (figure 9.4). A very important feature of this mapping is its topology-preserving quality, i.e. two vectors that are neighbors in the input space will be represented close to each other on the map.

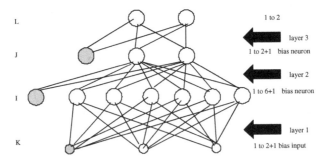

Figure 9.4.   Arrangement of neurons in a two-dimensional SOM

An n-dimensional vector, called weight vector, is assigned to every neuron. The simulation of self-organizing feature maps can be divided into two phases. In the initial learning phase, the map is trained with a set of input vectors. After learning, the weights of the map remain unchanged and the map is used in recall mode. The learning phase starts with an appropriate initialization of the weight vectors $(\vec{m}_i)$. If no information about the input data is given, then, e.g., a random initialization of the map may be used. Subsequently, the input vectors $(\vec{x})$ are presented to the map in multiple iterations. For each input vector the distance to all weight vectors is calculated and the neuron with the smallest distance is determined. This neuron is called the best matching element $(\vec{m}_{bm})$. In most applications, when dealing with two-dimensional grids of neurons, the Euclidean distance is used. After the best match has been determined, this neuron and the vectors in its neighborhood are adapted to the input vector by

$$\vec{m}_i(t+1) = \vec{m}_i(t) + h_{ci}(t)[\vec{x}(t) - \vec{m}_i(t)] \tag{9.1}$$

with the neighborhood function $h_{ci}(t)$, which is a function of the distance between the best matching element $(\vec{m}_{bm})$ and the indexed weight vector $(\vec{m}_i)$. The neighborhood function can be calculated e.g. by the term

$$h_{ci} = \alpha(t) \cdot \exp\left(-\frac{\|\vec{r}_{bm} - \vec{r}_i\|^2}{2\sigma^2(t)}\right). \tag{9.2}$$

The expression $\|\vec{r}_{bm} - \vec{r}_i\|^2$ determines the distance between the actual weight vector and the best matching element on the two-dimensional grid of neurons. By varying $\sigma(t)$, the width of the neighborhood function can be changed. The learning rate factor $\alpha(t)$ defines the strength of the adaptation and has to be chosen application-specific. A large value at the beginning of the learning process leads to a fast organization of the map while smaller values are used for its fine-tuning.

9.4.1 Modifications of the Original SOM Algorithm

Our goal is to find an efficient implementation on state of the art FPGAs that, on the one hand delivers the required high performance and, on the other hand, fits into the limited resources of current FPGAs. Because of their internal structure FPGAs are well suited for the implementation of neural networks. Previous work of the authors concerning highly optimized ASIC implementations of self-organizing feature maps has emphasized, that avoiding memory bottlenecks by using on-chip memory is a must in order to achieve optimal performance [12]. We use Xilinx Virtex FPGAs for our implementations, because these devices come with large internal SRAM blocks that can be used for internal weight storage.

In order to facilitate an efficient implementation in hardware, the original SOM-algorithm has been modified. In particular, the Manhattan distance, cf. (9.3), is used for calculating the distance between the input vector and the model vectors to avoid multiplications as required for the Euclidean distance (which is typically used in SOM implementations on PCs or workstations).

$$d = \|\vec{x} - \vec{m}\|_1 = \sum_{j=1}^{l} |x_j - m_j| \tag{9.3}$$

The internal precision is set to 16 bit and the precision of the input vector components and of the model vectors is set to eight bit. Restricting the values of the neighborhood function to negative powers of two, cf. (9.4), gives us the opportunity to replace the multiplications that are required for adaptation by shift operations.

$$h_{ci} \in \left\{ 1, \frac{1}{2}, \frac{1}{4}, \frac{1}{8}, \cdots \right\} \tag{9.4}$$

The impact of the modifications on the quality of the results and on the convergence speed has been analyzed in [13]. It has been shown that the simplified algorithm is well suited for a lot of applications. Furthermore the actual generation of Xilinx FPGAs (Virtex II) comes with integrated multipliers. These devices are integrated in the latest RAPTOR2000 ASMs and our implementations on these chips will thus be able to use Euclidean distance instead of Manhattan distance with no loss in performance.

9.4.2 FPGA Implementation

Our SOM-implementation consists of processing elements (PE) that are working in SIMD-manner. The elements are controlled by an additional controller as depicted in figure 9.5 for a matrix of $k \cdot l$ PEs. Multiple FPGAs can work in parallel in order to increase the size of the simulated maps and the performance of the system. In this case, each FPGA contains its own internal controller together with a matrix of processing elements. Nearly all calculations are performed in parallel on all processing elements. A bidirectional bus is used for data transfers to dedicated elements and for broadcasting data to groups of processor elements (or to all processor elements). Single elements and groups of elements are addressed by row and column lines that are connected to the two-dimensional matrix.

Every processing element has the capability to do all calculations required for learning and recall. In order to minimize the resource requirements for our FPGA implementation, the chip-internal controller performs those tasks that are identical for all PEs, such as decoding the instructions and calculating the memory addresses. The whole system is synchronously clocked with a single clock signal. The externally generated instructions are transferred to the processor elements via a dedicated control bus. Two additional signals are used for status messages from and to the controller. The twelve instructions that can be executed by the processing elements are explained in table 9.1.

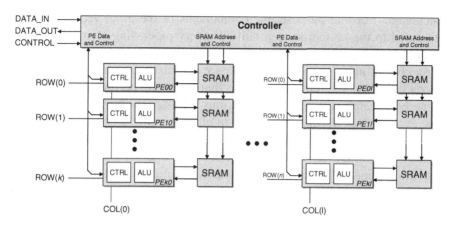

Figure 9.5. Architecture of the hardware accelerator – the processing elements are addressed by row and column lines

The different processing steps that are required for recall and for learning of self-organizing feature maps when using our FPGA based neuroprocessor are illustrated in figure 9.6.

Table 9.1. Instruction set of the implemented processing elements

Instruction	Description
RESET	Reset all internal registers of the processing elements
WLOAD	Load the initial weight vectors
CDIST	Calculate the distance between the input vector and the weight vectors
FMINI	Search for the best matching element, i.e. the element with the minimum distance between input vector and weight vector
ALOAD	Load the adaptation factors
ADAPT	Adapt the weights
SELUM	Select the element that distributes its weights over the data bus during the U-Matrix calculation
CALUM	Calculate the U-Matrix
WAIT	Insert a wait cycle
BMSEA	Find the best matching element
WDIST	Write the distance between a selected element and the input vector to the external controller
WRITE	Write the weight vectors to the external controller

The computation of the recall phase is divided into three phases. First, the distance between the actually presented input vector and the weight vectors is calculated. For that purpose, the components of the input vector have to be presented to the system sequentially. If every processing element represents one neuron all processing elements calculate their local distance in parallel. For a given dimension l of the input space, the distance calculation is finished after l clock cycles. The second step that has to be performed during recall is the best match search. For all neurons a bit-serial comparison of the distance values is performed in parallel in all processing elements. Using eight bit precision for the weights and for the input vectors, our implementation requires $\lceil \mathrm{ld}(l \cdot 255) \rceil$ steps (where ld means $\log_2$) of comparison to find the best matching element. The bit-serial comparison requires a global AND operation between all processing elements and a feedback to the PEs. In order to achieve a high clock frequency for large numbers of processing elements a two stage pipeline has been implemented in this path, while the rest of the design is not pipelined in order to minimize the area requirements. Thus, the best match search takes $2 \cdot \lceil \mathrm{ld}(l \cdot 255) \rceil$ clock cycles. In the last step, called "Output of BM Position" in figure 9.6, the best match position is sent to the external controller in two clock cycles. Thus, the number of clock cycles c_{recall} that are required for the recall of one input vector is given by

$$c_{recall} = l + 2 \cdot \lceil \mathrm{ld}(l \cdot 255) \rceil + 2. \qquad (9.5)$$

For the adaptation phase a maximum of eight clock cycles is required to distribute the adaptation factors (depending on the number of different adaptation factors). Figure 9.7 illustrates this process. First, the processing elements

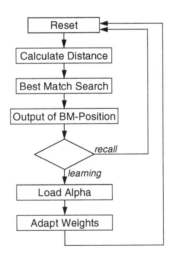

Figure 9.6.   Flow of operations that are required for the recall and the learning phase of self-organizing feature maps when using the FPGA based hardware accelerator

are initialized with the smallest adaptation factor (in figure 9.7 the PEs are initialized with $1/8$). Subsequently, the next adaptation factor (in our example $1/4$) is presented to the processor array. The PEs that are addressed via the row and column lines overwrite their previous value of the adaptation factor. During the next steps the adaptation factor is increased and the number of selected processing elements is decreased according to the selected neighborhood   ads
the

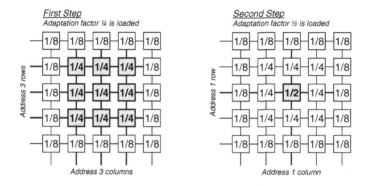

Figure 9.7.   Exemplary distribution of the adaptation factor

After all processing elements have received their adaptation factors, the input vector is presented to the map again. One clock cycle is required for each input vector component. Hence, the number of clock cycles c_{adapt} for the adaptation of the whole map to one input vector is given by

$$c_{adapt} = l + 8. \tag{9.6}$$

The number of clock cycles that are required for one learning step (c_{learn}) includes the time for the recall of this input vector (which is required prior to any adaptation) and the time for the adaptation itself. As can be seen from the given equations, the times that are required for classification and for training depend linearly on the vector dimension and are independent of the matrix size. The architecture is optimized for two-dimensional rectangular maps but any other kind of network topology can be implemented (e.g. one-dimensional or toroidal maps). Additionally, the values for the adaptation factors are provided by the external controller, thus any adaptation function and neighborhood function may be realized with the proposed hardware without any changes in the FPGA configuration.

9.4.3   Simulation of Virtual Maps

An essential limitation of many hardware implementations of neural networks is their restriction to a relatively small number of neurons because each processing element represents one (or a fixed small number) of neurons. In the case of our implementation this would mean that a large amount of weight memory remains unused when simulating maps with smaller vector dimensions than provided by the actual hardware implementation. In order to achieve an optimal utilization of the available hardware resources (i.e. computing power and internal memory) our FPGA based hardware implementation supports the simulation of virtual maps, i.e. it is possible to simulate maps that are larger than the array of processor elements that is implemented. Thus, the size of the maps that can be simulated with our FPGA implementation is not limited by the number of processing elements but only by the amount of on-chip memory.

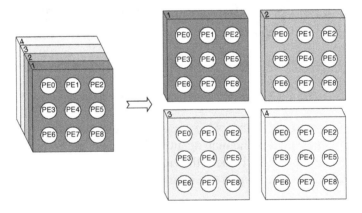

Figure 9.8.   For the simulation of virtual maps the internal weight memory of the processing elements is divided into sections, each representing a part of the virtual map

Using virtual maps, a chip that integrates $N_N = 10 \times 10$ processing elements with a maximum vector dimension of $l_{max} = 2000$ can be used to simulate a map of the same size but can also be used to simulate a virtual map with 100×100 elements and $l = 20$ vector components. For the simulation of n_v neurons with one processing element, the SRAM memory of all processing elements is divided into sections of equal size. Now each memory section of a PE represents a weight vector of the self-organizing feature map.

Figure 9.8 illustrates this process for a self-organizing feature map with 6×6 neurons, that is mapped onto an array of 3×3 processing elements. Thus, the maximum number of neurons $n_{v,max}$, that can be mapped onto one PE is

$$n_{v,max} = \left\lfloor \frac{l_{max}}{l} \right\rfloor . \tag{9.7}$$

Where l is the vector dimension of the virtual map and l_{max} is the maximum vector dimension that is supported by the processing elements. Because the memory sections are processed sequentially, the maximum degree of parallelism that can be achieved is still equal to the number of implemented processing elements. During recall the best match search is performed separately for all memory sections. The global best match is calculated in the external controller of the map, two additional clock cycles are required for this operation, cf. (9.5). After n_v local best match searches, the global best match is detected. The numbers of clock cycles $c_{recall,v}$ that are required for recall using a virtual map is given by

$$c_{recall,v} = n_v \cdot (l + 2 \cdot \lceil \mathrm{ld}(l \cdot 255) \rceil + 4); \quad \text{with} \quad n_v \leq \left\lfloor \frac{l_{max}}{l} \right\rfloor . \quad (9.8)$$

In the training phase the adaptation of the neurons is executed separately for every level. Hence, the adaptation of one input vector requires

$$c_{adapt,v} = n_v \cdot (l + 8); \quad \text{with} \quad n_v \leq \left\lfloor \frac{l_{max}}{l} \right\rfloor \quad (9.9)$$

clock cycles. With a decreasing width of the neighborhood function it becomes more and more probable that neurons have to be updated only in some of the memory sections. If – in the end of training – only the best matching element is updated, only one section has to be updated (in contrast to n_v sections as supposed in 9.9). Therefore, equation 9.9 gives the maximum number of clock cycles for adaptation.

9.4.4 Hardware Acceleration of Data Postprocessing

Data pre- and postprocessing is a crucial and often ignored aspect in neurocomputer design. The use of reconfigurable hardware enables us to implement optimally fitting hardware implementations - not only for neural networks but also for pre- and postprocessing. As an example, we have integrated the main visualization techniques for self-organizing feature maps, that had to be performed in software so far, into hardware.

As mentioned earlier, the result of the unsupervised SOM learning algorithm is an array of weight vectors which shows a representation of the input data. Although the input space has an arbitrary dimension, the result is a two-dimensional array (assuming a two-dimensional grid of neurons). Several visualization techniques for self-organizing feature maps are known, which help to analyze the learning results and therefore the structure of the investigated data. A *component map* is a picture that visualizes the values of a single component of the weight vectors. Component maps can be used to analyze the value of a component – usually representing a physical value, e.g. voltage – in respect to other components.

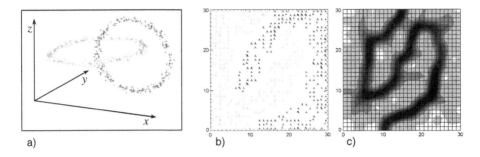

Figure 9.9. Visualization of a SOM that has been trained with a simple three-dimensional data set (a): pattern position map (b) and U-Matrix (c)

The *U-Matrix* [16] is a representation of two-dimensional SOMs which takes into account the distance of a weight vector to the weights in the neighboring neurons. For each neuron the mean of the distance between its weight vector and the weight vector of the neighboring neurons is calculated. The matrix of these values can be visualized e.g. as a grayscale picture whose bright areas depict small distances and whose dark areas represent large distances between the neurons.

Another representation of the SOM learning result is the *pattern position map* which takes into account previous knowledge about the input data. If information about the input data is available, attributes for the map areas can be defined corresponding to this knowledge. After learning, the whole set of input data is presented to the map again. For each input pattern the best matching element is calculated and labeled with an attribute corresponding to the type of the input vector.

Figure 9.9 illustrates the visualization of the learning results for a simple example. The input data set consists of 500 vectors that represent two rings in the three-dimensional input space (figure 9.9a). After learning, the SOM (consisting of 30×30 neurons) has mapped these rings onto two separate regions on the map. In the pattern position map (figure 9.9b) the best matching neurons are labeled with *A* or *B* according to the ring, they represent. The separation of the two regions is illustrated by a U-Matrix in figure 9.9c), where the dark areas correspond to a large difference between the weights of neighboring neurons.

Implementing the post-processing algorithms in hardware significantly reduces I/O bandwidth requirements and thus enables a more efficient utilization of the hardware accelerator. The data that is required for the presentation of component maps can be received by downloading the appropriate weight vector components from the internal SRAM to the external controller unit. Pattern position maps are generated by doing a recall on the whole input data set. Both representations can be generated very quickly because the calculations are part of the instruction set that is typically required for the implementation of SOMs. For an efficient implementation of the U-Matrix visualization, additional logic and two special instructions have been implemented. As a fully parallel implementation would have required an immoderate increase in the complexity of the processing elements, we have implemented a mechanism that fits best into the given architecture.

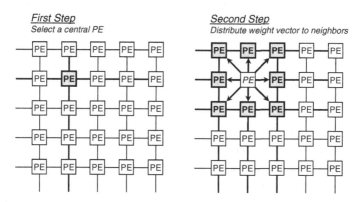

Figure 9.10. The U-Matrix calculation is performed in two steps: First a PE is selected, then its distances to the neighboring PEs are calculated

The main bottleneck for the U-Matrix calculation is the 8-bit broadcast bus that is connected to all processing elements. In order to avoid major changes to the architecture this bus is used to transmit the weight vectors between the processing elements. In the first step an element of the matrix is selected via the row and column lines as depicted in figure 9.10. In the second step this element distributes its weight vectors over the data bus and the neighboring processing elements calculate their distance to this element. If necessary, the distance values are accumulated automatically. These calculations are performed for all processing elements. Following these calculations, the results are transmitted to the external control unit, which does the necessary normalization (i.e. division by 8, 5 or 3, depending on the position of the processing element). The U-Matrix calculation requires $2N_N \cdot (l + 1)$ clock cycles for the calculations (with N_N being the number of neurons) and $2N_N$ clock cycles for the output of the data. Thus, the total number of clock cycles for U-Matrix calculation c_{UM} is given by

$$c_{UM} = 2N_N \cdot (l + 2). \tag{9.10}$$

The main drawback of this implementation is the dependence of the calculation time on the number of neurons. But, as has been shown in [14] and [15], this technique offers sufficient performance up to the order of 10,000 elements. For very large maps it is possible to accelerate the U-Matrix calculation even more by means of a hierarchical internal bus system. In [15] an optimal division into bus-segments depending on the number of simulated neurons is derived. The speed up that can be achieved with this hardware accelerated U-Matrix calculation is, on the one hand, based on the parallel distance calculation. On the other hand, only $2N_N$ 8-bit data words have to be communicated to the host computer while the transmission of all weight vectors (as required if the U-Matrix is calculated on the host computer) results in $l \cdot N_N$ 8-bit data words.

9.4.5   System Integration and Performance

In order to implement the proposed architecture on the RAPTOR2000 rapid prototyping system, five DB-VS modules are applied. Figure 9.11 shows the architecture of the hardware accelerator. Four ASMs are used to implement a matrix of processing elements while the fifth is used to implement the matrix controller, an I/O controller for the connection to the local bus of the RAPTOR2000 system and a dual port SRAM that is used to buffer input- and output-data. An integrated SDRAM interface controls the external 128 MBytes of SDRAM. The dual port SRAM is used to store single input vectors, commands from the host computer and the results (e.g. best match positions or postprocessed visualizations of the map).

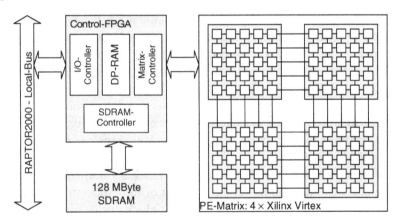

Figure 9.11.   Architecture of the hardware accelerator for self-organizing feature maps on the basis of the RAPTOR2000-system

The large SDRAM is used to store one or more input data sets. During learning of a self-organizing feature map, the whole input data set has to be presented to the map for several times. Thus, it is recommendable to transfer the whole data set in one fast block transfer via the PCI bus to the hardware accelerator in order to minimize the load of the PCI bus.

The design has been described in VHDL and synthesized using the Synopsys FPGA compiler and the Xilinx Alliance tools for place and route. An important design goal was the optimal utilization of the embedded SRAM blocks in order to be able to simulate maps as large as possible. The (meanwhile old-fashioned) Xilinx Virtex XCV1000 offers only 32 internal SRAM Blocks, each of the size 512 Byte. Hence, a straight forward implementation would integrate 32 processing elements. But synthesis has shown that 32 processing elements result in a very small utilization of the FPGA's slices. Thus we decided to implement two processing elements per SRAM block, using the SRAM blocks in the dual-port mode. This implementation results in an FPGA slice-utilization of 48% (i.e. 48% of the FPGA slices have been used, while 100% of the integrated block RAM is used). Theoretically, it is possible to further increase the device utilization by implementing three or even four processing elements per SRAM block but this would require a more complex memory controller. We skipped this, because all newer FPGAs offer a larger number of SRAM blocks per slices and are thus better suited for our implementation.

The design was synthesized to several FPGAs of the Xilinx Virtex-E and Virtex-II series. Because of the large number of integrated block RAMs each processing element has been equipped with an integer number of block RAMs. By various synthesis flows, the number of block RAMs per processing element has been chosen as small as possible in order to be able to implement as many PEs as possible. When increasing the number of processing elements, the utilization of the block RAMs was always fixed to 100% while the utilization of the slices had to be chosen small enough in order to fit into the design. Table 9.2 shows the number of processing elements and the maximum vector dimension that can be achieved with some typical devices of the Xilinx Virtex family (using only one Xilinx device). In all examined cases the resulting utilization of logic blocks was small enough to enable synthesis to the clock frequency that is given in the table. The implementation on Xilinx Virtex-II FPGAs is not yet realized (the presented data is based on synthesis and place&route results) and does not take into account the special features of these FPGAs (like integrated multipliers). The final performance, that can be achieved with these devices will thus be even larger than the values presented here.

For the evaluation of the performance that a system achieves for neural network simulation, the performance is determined separately for the recall phase and for the learning phase. We are using the well established performance metrics MCPS (Million Connections per Second) for the recall phase and MCUPS (Million Connection Updates per Second) for the learning phase. The complexity of the calculations differs strongly for different models of neural networks. Hence, a direct comparison of the performance-values of different neural networks is not possible, but the values are well suited to compare different implementations of self-organizing feature maps. The performance during

Table 9.2. Number of processing elements, maximum vector dimension and performance of the proposed implementation of self-organizing feature maps on various FPGAs of the Xilinx Virtex series

Device	N_{PE}	l_{max}	f_{FPGA} [MHz]	$P_{C,r}$ [MCPS]	$P_{C,l}$ [MCUPS]	Utilization of Slices
XCV1000-6	64	256	50	2825	325	48
XCV812E-8	70	2048	65	4463	495	68
XCV2000E-8	160	512	65	9717	1097	77
XCV3200E-8	208	512	65	12632	1426	59
XC2V6000-6	288	1024	105	29158	3253	78
XC2V10000-6	384	1024	105	38877	4338	58

recall ($P_{C,r}$) and the performance during learning ($P_{C,l}$) can be calculated as follows:

$$P_{C,r} := \frac{l \cdot N_N}{T_r} \qquad P_{C,l} := \frac{l \cdot N_{NA}}{T_l} \tag{9.11}$$

Here, l is he input vector dimension, N_N is the number of neurons and N_{NA} is the number of neurons that are updated during learning. T_r is the time that is required to find the best matching element for a given input vector and T_l is the time for one learning step (including recall and adaptation of an input vector). It has to be mentioned, that, for the calculation of the performance $P_{C,l}$ during learning, only those neurons are taken into account, that are updated during learning (in agreement with many other authors, e.g. G. Myklebust [17]). Nevertheless, some authors take into account all neurons during learning [18]. Of cause, this leads to larger performance numbers. In [15] it has been shown, that the mean of the number of neurons that are updated during learning is approximately 22% for typical learning parameters. This allows an easy transformation of the different performance values into one another.

With 9.5 and 9.6 the performance of a Matrix of N_N processing elements, each simulating one neuron with a clock frequency of f, is given by

$$P_{C,r} = \frac{l \cdot N_N \cdot f}{c_{recall}} = N_N \cdot f \cdot \frac{l}{l + 2 \cdot \lceil \mathrm{ld}(l \cdot 255) \rceil + 2}, \tag{9.12}$$

$$P_{C,l} = \frac{l \cdot N_{NA} \cdot f}{c_{learn}} = N_{NA} \cdot f \cdot \frac{l}{2l + 2 \cdot \lceil \mathrm{ld}(l \cdot 255) \rceil + 10}. \tag{9.13}$$

Using Virtex XCV812E-8 devices, $N_{PE} = 70$ processing elements can be implemented, each equipped with 2 kBytes of internal SRAM. The utilization of the FPGAs is about 68% and a clock frequency of 65 MHz has been achieved. The maximum performance is achieved, if every processing element represents one neuron ($n_v = 1$) and if the maximum available vector dimension is used. For our system implementation we use the RAPTOR2000 configuration that has been described earlier, consisting of one DB-VS module that acts as a system controller and four DB-VS modules that embed the processing elements. Thus, the number of processing elements (and the achieved performance) is four times higher than the values that are given in table 9.2. With the proposed architecture and Xilinx Virtex XCV812E-8 devices, about $P_{C,r} = 17900$ MCPS can be achieved during recall and $P_{C,l} = 2000$ MCUPS during learning. With an optimized software implementation on a state of the art personal computer (Pentium IV, 2.4 GHz) a performance of 205 MCPS and 51 MCUPS has been achieved for this application. Thus, a maximum speedup of 87 during recall and of 39 during learning can be achieved. Of cause, the speed up can be increased by using larger FPGAs. Using four Virtex XCV2000E-8 devices, e.g., a maximum speedup of 190 during recall and of 86 during learning can be achieved (corresponding to 38900 MCPS and 4400 MCUPS).

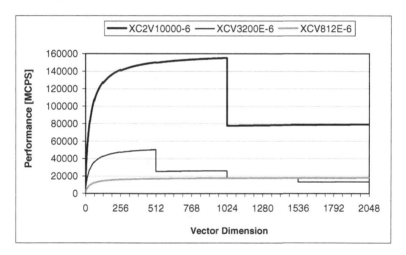

Figure 9.12. Performance that can be achieved with the proposed system, composed of five DB-VS modules, for different Xilinx devices

When simulating virtual maps, i.e. emulating more than one neuron with each processing element, the performance of our implementation decreases due to additional sequential processing. In order to prove our calculated performance data, various benchmark data sets have been applied. Using a benchmark data set with a vector dimension of $l = 9$ [19], maps with up to 250×250 neurons can be simulated on RAPTOR2000 equipped with four Xilinx Virtex XCV812E FPGAs for the implementation of the processing elements and an additional system controller ASM. The performance that can be achieved with this environment is $P_{C,r} = 4400$ MCPS and $P_{C,l} = 665$ MCUPS, respectively.

The performance data that is given in table 9.2 is based on the assumption that the size of the simulated neural network is identical to the number of neurons and the vector dimension is maximal. Figure 9.12 shows the performance that can be achieved with the proposed system, composed of five DB-VS modules, during recall for different Xilinx Virtex devices. For the performance estimation it has been assumed, that for a given vector dimension the size of the simulated map is set to the maximum possible size. Performance increases until the maximum vector dimension is reached for the particular FPGA, that is given in table 9.2. Larger vector dimensions are implemented by reconfiguring the FPGA for large vector dimensions with less processing elements and thus with smaller performance.

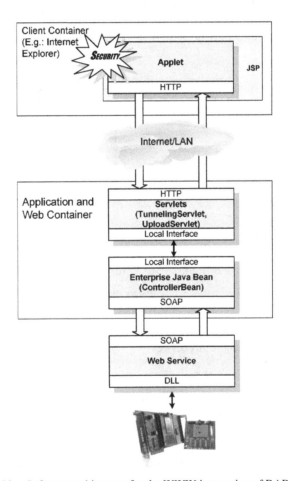

Figure 9.13.   Software architecture for the WWW integration of RAPTOR2000

In addition to the possibility of using the hardware accelerator from any workstation in the local Intranet, we provide an environment that enables access to the RAPTOR2000 system via the World Wide Web (*http:\\www.raptor2000.de*). The RAPTOR2000 web interface offers access to the functionality of the RAPTOR2000 system including, e.g. setting the clock frequency of the local bus and configuring any available FPGA in the system. Additionally, read and write access to the on-board dual-port SRAM and to the ASMs is provided. Figure 9.13 shows the architecture of the software that integrates RAPTOR2000 into the WWW. The software system consists of two services. The low-level service is the C++ based SOAP-(Simple Object Access Protocol)-server which manages the hardware requests to the RAPTOR2000 system.

The communication between the SOAP-server and the higher level *Application and Web Container* is based on messages, that are exchanged between the two software modules. The *Application and Web Container* handles communication between the client and the RAPTOR2000 system. It provides multi user handling, session management and basic security features concerning the RAPTOR2000 system. Communication to a client java applet is facilitated via the http-protocol. This applet offers a graphical user interface that allows the user to configure the RAPTOR2000 system, limited by the security restrictions only.

9.5   Conclusions

A dynamically reconfigurable hardware accelerator for the simulation of self-organizing feature maps has been presented. Equipped with five FPGA modules, the system achieves a maximum performance of about 39 GCPS (Giga Connections per Second) during recall and more than 4 GCUPS (Giga Connection Updates per Second) during learning. Even higher performance numbers can be achieved by using new FPGA architectures like the Xilinx Virtex-II series. Apart from the high performance the system is capable of doing pre- and postprocessing tasks – either by use of the implemented visualization features or by dynamically reconfiguring the devices during runtime. The latter is supported by the RAPTOR2000 rapid prototyping system by means of the ability to reconfigure the FPGAs very fast via the PCI bus.

References

[1] Glesner, M., Pochmüller, W.: "Neurocomputers: An Overview of Neural Networks in VLSI", Chapman Hall, 1994.

[2] Rückert, U.: "ULSI Implementations for Artificial Neural Networks", 9th Euromicro Workshop on Parallel and Distr. Processing 2001, Feb. 7-9, 2001, Mantova, Italien, pp. 436–442.

[3] Ienne, P.: "Digital Connectionist Hardware: Current Problems and Future Challenges", Biological and Artificial Computation: From Neuroscience to Technology, Vol. 1240 of Lecture Notes in Computer Science, pp. 688–713, 1997, Springer, Berlin.

[4] Kohonen, T.: Self-Organizing Maps. Springer-Verlag, Berlin, (1995).

[5] Marks, K.M., Goser, K. "Analysis of VLSI Process Data Based on Self-Organizing Feature Maps", Proc. of First Int. Workshop on Neural Networks and their Applications, Neuro-Nimes, pp. 337 - 348, France, Nov. 1988.

[6] Rüping, S., Müller, J.: "Analysis of IC Fabrication Processing using Self-Organizing Maps", Proc. of ICANNt'99, Edinburgh, 7.-10. Sept. 1999, pp. 631–636.

[7] Neema, S., Bapty, T., Scott, J.: "Adaptive Computing and Runtime Reconfiguration", 2nd Military and Aerospace Applications of Programmable Devices and Technologies Conference, MAPLD99, Laurel, Maryland, USA, September 1999.

[8] Tinós, R., Terra, M. H.: "Fault detection and isolation in robotic manipulators and the radial basis function network trained by Kohonen's SOM". In Proc. of the 5th Brazilian Symposium on Neural Networks (SBRN98), Belo Horizonte, Brazil, pp. 85–90, 1998.

[9] Porrmann, M., Kalte, H., Witkowski, U., Niemann, J.-C., Rückert, U.: "A Dynamically Reconfigurable Hardware Accelerator for Self-Organizing Feature Maps", Proc. of SCI 2001 Orlando, Florida USA, 22.-25. Juli, 2001, pp. 242–247.

[10] Porrmann, M., Witkowski, U., Kalte, H., Rückert, U.: "Implementation of Artificial Neural Networks on a Reconfigurable Hardware Accelerator", 10th Euromicro Workshop on Parallel, Distributed and Network-based Processing (PDP 2002), 9.-11. Januar 2002, Gran Canaria Island, Spain.

[11] Kalte, H., Porrmann, M., Rückert, U.: "Using a Dynamically Reconfigurable System to Accelerate Octree Based 3D Graphics", PDPTA'2000, June 26-29, 2000 Monte Carlo Resort, Las Vegas, Nevada, USA, pp. 2819–2824.

[12] Porrmann, M., Rüping, S., Rückert, U.: "The Impact of Communication on Hardware Accelerators for Neural Networks", Proc. of SCI 2001 Orlando, Floriada USA, 22.-25. Juli 2001, pp. 248–253.

[13] Rüping, S., Porrmann, M., Rückert, U., "SOM Accelerator System", Neurocomputing 21, pp. 31–50, 1998.

[14] Porrmann, M., Rüping, S., Rückert, U., "SOM Hardware with Acceleration Module for Graphical Representation of the Learning Process", Proc.

of the 7th Int. Conference on Microelectronics for Neural, Fuzzy and Bio-Inspired Systems, pp. 380–386, Granada, Spain, 1999.

[15] Porrmann, M.: "Leistungsbewertung eingebetteter Neurocomputersysteme", Phd thesis, University of Paderborn, System and Circuit Technology, 2001.

[16] Ultsch, A.: "Knowledge Extraction from Self-organizing Neural Networks", in Opitz, O., Lausen, B. and Klar, R. (editors), Information and Classification, pp. 301-306, London, UK, 1993.

[17] Myklebust, G.: "Implementations of an unsupervised neural network model on an experimental multiprocessor system", Phd thesis, Norwegian Institute of Technology, University of Trondheim, Trondheim, Norway, 1996.

[18] Hämäläinen, T., Klapuri, H., Saarinen, J., and Kaski, K.: "Mapping of SOM and LVQ algorithms on a tree shape parallel computer system", Parallel Computing, 23(3), pp. 271-289, 1997.

[19] Mangasarian, O. L., Setiono, R., and Wolberg, W.H.: "Pattern recognition via linear programming: Theory and application to medical diagnosis". In Coleman, Thomas F. and Yuying Li: Large-scale numerical optimization, pp. 22-30, SIAM Publications, Philadelphia, 1990.

Chapter 10

FPGA IMPLEMENTATION OF A FULLY AND PARTIALLY CONNECTED MLP

Application to Automatic Speech Recognition

Antonio Canas[1], Eva M. Ortigosa[1], Eduardo Ros[1] and Pilar M. Ortigosa[2]

[1]*Dept. of Computer Architecture and Technology, University of Granada, Spain*
[2]*Dept. of Computer Architecture and Electronics, University of Almeria, Spain*

Abstract   In this work, we present several hardware implementations of a standard Multi-Layer Perceptron (MLP) and a modified version called eXtended Multi-Layer Perceptron (XMLP). This extended version is an MLP-like feed-forward network with two-dimensional layers and configurable connection pathways. The interlayer connectivity can be restricted according to well-defined patterns. This aspect produces a faster and smaller system with similar classification capabilities. The presented hardware implementations of this network model take full advantage of this optimization feature. Furthermore the software version of the XMLP allows configurable activation functions and batched backpropagation with different smoothing-momentum alternatives. The hardware implementations have been developed and tested on an FPGA prototyping board. The designs have been defined using two different abstraction levels: register transfer level (VHDL) and a higher algorithmic-like level (Handel-C). We compare the two description strategies. Furthermore we study different implementation versions with diverse degrees of parallelism. The test bed application addressed is speech recognition. The implementations described here could be used for low-cost portable systems. We include a short study of the implementation costs (silicon area), speed and required computational resources.

Keywords: FPGA, Multi-Layer Perceptron (MLP), Artificial Neural Network (ANN), VHDL, Handel-C, activation function, discretization.

10.1   Introduction

An Artificial Neural Network (ANN) is an information processing paradigm inspired by the way biological nervous systems process information. An ANN

A. R. Omondi and J. C. Rajapakse (eds.), FPGA Implementations of Neural Networks, 271–296.
© 2006 *Springer. Printed in the Netherlands.*

is configured for a specific application, such as pattern recognition or data classification, through a learning process. As in biological systems, learning involves adjustments of the synaptic connections that exist between the neurons.

An interesting feature of the ANN models is their intrinsic parallel processing strategies. However, in most cases, the ANN is implemented using sequential algorithms, that run on single processor architectures, and do not take advantage of this inherent parallelism.

Software implementations of ANNs are appearing in an ever increasing number of real-world applications [17, 2, 7]: OCR (Optical Character Recognition), data mining, image compression, medical diagnosis, ASR (Automatic Speech Recognition), etc. Currently, ANN hardware implementations and bio-inspired circuits are used in a few niche areas [9, 7]: in application fields with very high performance requirements (e.g. high energy physics), in embedded applications of simple hard-wired networks (e.g. speech recognition chips), and in neuromorphic approaches that directly implement a desired function (e.g. artificial cochleas and silicon retinas).

The work presented here studies the implementation viability and efficiency of ANNs into reconfigurable hardware (FPGA) for embedded systems, such as portable real-time ASR devices for consumer applications, vehicle equipment (GPS navigator interface), toys, aids for disabled persons, etc.

Among the different ANN models available used for ASR, we have focused on the Multi-Layer Perceptron (MLP) [16].

A recent trend in neural network design for large-scale problems is to split a task into simpler subtasks, each one handled by a separate module. The modules are then combined to obtain the final solution. A number of these modular neural networks (MNNs) have been proposed and successfully incorporated in different systems [3]. Some of the advantages of the MNNs include reduction in the number of parameters (i.e., weights), faster training, improved generalization, suitability for parallel implementation, etc. In this way, we propose a modified version of the MLP called eXtended Multi-Layer Perceptron (XMLP). This new architecture considers image and speech recognition characteristics that usually make use of two-dimensional processing schemes [14], and its hardware implementability (silicon area and processing speed). Talking about two dimensions only makes sense if partial connectivity is assumed. Therefore, our model allows defining neighborhood-restricted connection patterns, which can be seen as a sort of modularity.

In this work, we introduce two implementation versions, a parallel and a sequential design of MLP and XMLP. Both parallel and sequential versions are described using two different abstraction levels: register transfer level (RTL) and a higher algorithmic-like level. These implementations have been defined using two hardware description languages, VHDL and Handel-C respectively.

Final computation results will show that RTL description generates more optimized systems. However, one of the main advantages of the high level description is the short design time.

Let us focus on a voice-controlled phone dial system. This can be of interest for drivers that should pay attention to driving, for example. This application defines the MLP/XMLP parameters to be implemented and the word set required (numbers from 0 to 9). All classification results provided in the next sections have been obtained in a speaker-independent scheme using these words, extracted from a multi-speaker database [13].

The chapter is organized as follows. Section 10.2 briefly describes the MLP and XMLP models. Section 10.3 addresses different aspects about activation functions and discretization, which need to be studied before designing the system. Then we described the detailed hardware implementation strategies of the MLP (Section 10.4) and the XMLP (Section 10.5). Finally, Section 10.6 summarizes the conclusions.

10.2   MLP/XMLP and speech recognition

Automatic speech recognition is the process by which a computer maps an acoustic signal to text [14]. Typically, speech recognition starts with the digital sampling of the voice signal. The raw waveform sampled is not suitable for direct input for a recognition system. The commonly adopted approach is to convert the sampled waveform into a sequence of feature vectors using techniques such as filter bank analysis and linear prediction analysis. The next stage is the recognition of phonemes, groups of phonemes or words. This last stage is achieved in this work by ANNs (MLP and XMLP) [13], although other techniques can be used, such as HMMs (Hidden Markov Models) [11], DTW (Dynamic Time Warping), expert systems or a combination of these.

10.2.1   Multi-Layer Perceptron

The MLP is an ANN with processing elements or neurons organized in a regular structure with several layers (Figure 10.1): an input layer (that is simply an input vector), some hidden layers and an output layer. For classification problems, only one winning node of the output layer is active for each input pattern.

Each layer is fully connected with its adjacent layers. There are no connections between non-adjacent layers and there are no recurrent connections. Each of these connections is defined by an associated weight. Each neuron calculates the weighted sum of its inputs and applies an activation function that forces the neuron output to be high or low, as shown in Eqn. (10.1).

$$Z_{li} = f(S_{li}); S_{li} = sum_j w_{lij} Z_{(l-1)j} \tag{10.1}$$

In this equation, z_{li} is the output of the neuron i in layer l, s_{li} is the weighted sum in that neuron, f is the activation function and w_{lij} is the weight of the connection coming from neuron j in the previous layer ($l-1$).

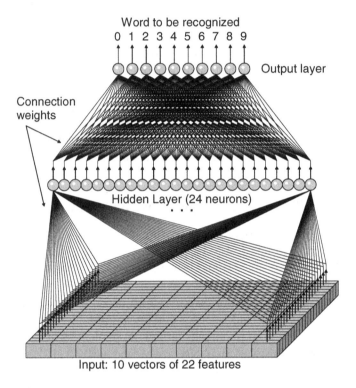

Figure 10.1.   Example of the MLP for isolated word recognition

In this way, propagating the output of each layer, the MLP generates an output vector from each input pattern. The synaptic weights are adjusted through a supervised training algorithm called backpropagation [16].

Different activation functions have been proposed to transform the activity level (weighted sum of the node inputs) into an output signal. The most frequently used is the sigmoid, although there are other choices such as a ramp function, a hyperbolic tangent, etc. All of these are continuous functions, with a smooth S-like waveform, that transform an arbitrary large real value to another value in a much restricted range. More details about activation functions are discussed in Section 10.3.1.

10.2.2 Extended Multi-Layer Perceptron

The XMLP is a feed-forward neural network with an input layer (without neurons), a number of hidden layers selectable from zero to two, and an output layer. In addition to the usual MLP connectivity, any layer can be two-dimensional and partially connected to adjacent layers. As illustrated in Figure 10.2, connections come out from each layer in overlapped rectangular groups. The size of a layer l and its partial connectivity pattern are defined by six parameters in the following form: $x(g_x, s_x) \times y(g_y, s_y)$, where x and y are the sizes of the axes, and g and s specify the size of a group of neurons and the step between two consecutive groups, both in abscissas (g_x, s_x) and ordinates (g_y, s_y). A neuron i in the X-axis at layer $l+1$ (the upper one in Figure 10.2) is fed from all the neurons belonging to the i-the group in the Xaxis at layer l (the lower one). The same connectivity definition is used in the Y-axis. When g and s are not specified for a particular dimension, the connectivity assumed for that dimension is $g_x = x$ and $s_x = 0$, or $g_y = y$ and $s_y = 0$. Thus, MLP is a particular case of XMLP where $g_x = x$, $s_x = 0$, $g_y = y$ and $s_y = 0$ for all layers.

The second dimension in each layer can be considered as a real spatial dimension for image processing applications or as the temporal dimension for time related problems. Two particularizations of the XMLP in time-related problems are the *Scaly Multi-Layer Perceptron* (SMLP) used in isolated word recognition [8], and the *Time Delay Neural Network* (TDNN), used in phoneme recognition [10].

10.2.3 Configurations Used for Speech Recognition

To illustrate the hardware implementation of the MLP/XMLP system we have chosen a specific speaker-independent isolated word recognition application. Nevertheless, many other applications require embedded systems in portable devices (low cost, low power and reduced physical size).

For our test bed application, we need an MLP/XMLP with 220 scalar data in the input layer and 10 output nodes in the output layer. The network input consists of 10 vectors of 22 components (10 cepstrum, 10 Δcepstrum, energy, Δenergy) obtained after preprocessing the speech signal. The output nodes correspond to 10 recognizable words extracted from a multi-speaker database [15]. After testing different architectures [4], the best classification results (96.83% of correct classification rate in a speaker-independent scheme) have been obtained using 24 nodes in a single hidden layer, with the connectivity of the XMLP defined by $10(4,2) \times 22$ in the input layer and 4×6 in the hidden layer.

10.3   Activation functions and discretization problem

For hardware implementations, we have chosen a two's complement representation and different bit depths for the stored data (inputs, weights, outputs, etc). In order to accomplish the hardware implementation, it is also necessary to discretize the activation function. Next, we present different activation functions used in the MLP and some details about their discretization procedure.

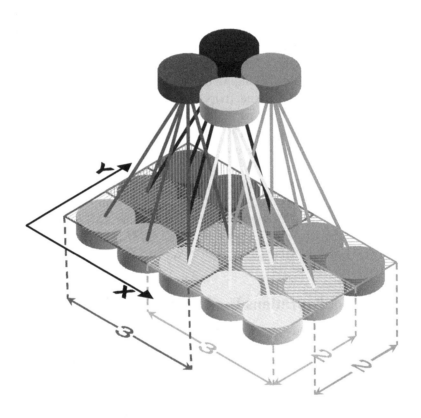

Figure 10.2.   Structure of an example XMLP interlayer connectivity pattern defined by the expression $5(3, 2) \times 3(2, 1)$

10.3.1   Activation Functions

One of the difficult problems encountered when implementing ANNs in hardware is the nonlinear activation function used after the weighted summation of the inputs in each node (Eqn. 10.1). There are three main nonlinear activation functions: threshold (hard limited), ramp and various sigmoid curves. We have studied different options: classical sigmoid, hyperbolic tangent, arc

tangent and ramp activation functions. In the hardware implementation, we
have focused on the sigmoid activation function.

In order to achieve generality, an activation function $f(x)$ is defined depending on three parameters: $f_0\prime$ (slope at $x = 0$), f_{max} (maximum value of $f(x)$)
and f_{min} (minimum value). The generic expressions for the four functions
considered and their respective derivatives, needed in the backpropagation algorithm, are given in Table 10.1. An example for three particular values of the
parameters is plotted in Figure 10.3. For simplicity, f_R is defined as $f_{max}-$
f_{min}.

Table 10.1. Activation functions and their derivatives

sigmoid	$f(x) = \dfrac{f_R}{1+e^{-\frac{4f_0\prime}{f_R}x}} + f_{min}$	$f_0\prime = \frac{1}{4}, f_{max} = 1, f_{min} = 0$
		$(f_R = 1)$
		define the standard sigmoid function
	$g(x) = \dfrac{1}{1+e^{-\frac{4f_0\prime}{f_R}x}}$, we can obtain a simple expression of f(x) as a function	
	of g(x): $f\prime(x) = f_R g\prime(x) = 4f_0\prime g(x)(1 - g(x))$	
tanh	$f(x) = \frac{f_R}{2} \tanh \frac{2f_0\prime}{f_R}x + (f_{max} + f_{min})/2$	This function coincides with
		the sigmoid defined above.
	$f\prime(x) = f_0\prime \left(1 - \tanh^2 \frac{2f_0\prime}{f_R}x\right)$	
arctan	$f(x) = \frac{f_R}{\pi} \arctan \left(\frac{f_0\prime \pi}{f_R}x\right) + (f_{max} + f_{min})/2$	
	$f\prime(x) = \dfrac{f_0\prime}{1+\left(\frac{f_0\prime \pi}{f_R}x\right)^2}$	
ramp	$f(x) = \begin{cases} f_{min} & \text{if } x \leq -f_R/2f_0\prime \\ f_0\prime x + (f_{max} + f_{min})/2 & \text{if } -f_R/2f_0\prime < x < f_R/2f_0\prime \\ f_{max} & \text{if } x \geq f_R/2f_0\prime \end{cases}$	
	$f\prime(x) = \begin{cases} f_0\prime & \text{if } -f_R/2f_0\prime < x < f_R/2f_0\prime \\ 0 & \text{otherwise.} \end{cases}$	

10.3.2 Discretization

One particular difficulty regarding the migration of ANNs towards hardware
is that the software simulations use floating point arithmetic and either double
or simple precision weights, inputs and outputs. Any hardware implementation would become unreasonably expensive if incorporating floating point
operations and therefore needing to store too many bits for each weight. Fixed
point arithmetic is better suited for hardware ANNs because a limited precision requires fewer bits for storing the weights and also simpler calculations.
This causes a reduction in the size of the required silicon area and a considerable speed-up. A complete revision of different aspects concerning limited
precision ANNs can be found in [6].

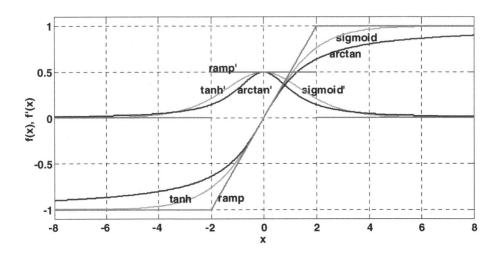

Figure 10.3. Activation functions and their derivatives with $f'_0 = 1/2, f_{max} = 1, f_{min} = -1(f_R = 2)$

In this contribution, learning is carried out offline using floating points, while hardware implementations use discrete variables and computations. However, classification results in hardware are very similar to the ones obtained with the software approach. For instance, in phoneme recognition application with the MLP, we obtained 69.33% correct classification with the continuous model and 69% when using the discretized model.

In order to use limited precision values, it is necessary to discretize three different sets of variables: network inputs and neuron outputs (both with the same number of bits), weights and the activation function input. Each of these sets has a different range, specified by the number of bits (n) of the discrete variable and the maximum absolute value (M) of the corresponding continuous variable. The expressions for the conversion between continuous (c) and discrete (d) values are:

$$d = round\left(c\frac{2^n - 2}{2M}\right) ; c = d\frac{2M}{2^n - 2} \quad (10.2)$$

10.3.2.1   Issues related to the discretization of the activation function.

In digital implementations only the threshold nonlinearity can be matched exactly in hardware; the ramp and sigmoidal nonlinearities have to be approximated. As summarized in [1, 7], classical digital solutions fall into two main trends: Look-Up Tables (LUTs) and expansion into Taylor series. We have adopted the first strategy because specific memory resources on FPGAs have become common in this kind of devices.

Next, we describe how to compute the values to be stored in the LUT and how to obtain the output of a neuron using this LUT.

Notation. M_z: maximum absolute value of the continuous MLP/XMLP inputs and the continuous neuron output

M_w: maximum absolute value of the continuous weights

n_z: number of bits of the discrete MLP/XMLP inputs and the discrete neuron output

n_w: number of bits of the discrete weights

#w: maximum number of incoming connections to a neuron (fan-in)

$n_{\#w}$: number of bits to specify #w (Eqn. 10.3)

$$n_{\#w} = \lceil \log_2 \#w \rceil \tag{10.3}$$

M_s: maximum absolute value of a continuous weighted sum (Eqn. 10.4)

$$M_s = \#w M_z M_w \tag{10.4}$$

N_s: number of bits of the discrete weighted sum (Eqn. 10.5)

$$N_s = \left\lceil \log_2 \left(\#w \left(2^{n_z-1} - 1 \right) \left(2^{n_w-1} - 1 \right) \right) \right\rceil + 1 \tag{10.5}$$

f_{min_d}: minimum value of the discrete activation function
f_{max_d}: maximum value of the discrete activation function

LUT indexed with n_s bits.   We split the N_s-bits weighted sum into two parts: the n_s most significant bits (used as LUT input) and the m_s least significant bits (Eqn. 10.6).

$$N_s = n_s + m_s; 1 \leq n_s \leq N_s \tag{10.6}$$

The greater n_s, the more accurate activation function. We can choose a trade-off value for n_s labeled $n_{s_{opt}}$, for optimal. This is the minimum n_s that satisfies the condition that all the possible output values of the discrete activation function will be present in the LUT. We have obtained a good approximation of $n_{s_{opt}}$ with the following expression (10.7)

$$\lceil n_{s_{opt}} = \left\lceil \log_2 \left(f_0' \#w M_w \frac{2^{N_s} \left(2^{n_z} - 2 \right)}{2^{N_s} - 2} \right) \right\rceil \tag{10.7}$$

Let us look at the following example: $M_z = 1.0$, $M_w = 5.0$, $n_z = 4$, $n_w = 5$, $n_{\#w} = 4$. The maximum absolute value of the discrete weighted sum will be $4 \cdot 7 \cdot 15 = 420$. Hence, the minimum number of bits necessary to represent all the values (N_s) is 10 (Eqn. 10.5). Instead of using all the bits of the weighted sum for the index of the LUT that stores the discrete activation function, we use only the most significant ones (n_s). In this example, we choose $n_s = 7$. With 7 bits for the activation function input and 4 bits for its output, the LUT should have 128 entries, each one storing a value between -7 and 7. Figure 10.4 illustrates the calculation of some entries around zero for this example. The upper plot in Figure 10.5 shows the whole discrete activation function for the nine possible values of n_s (1 . . . 10).

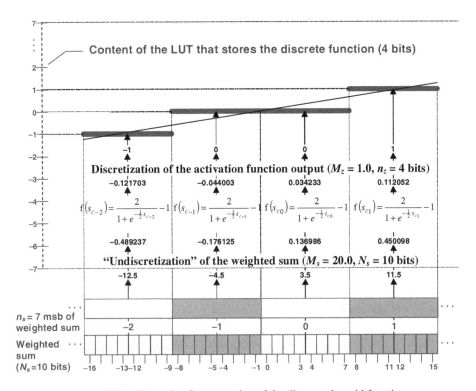

Figure 10.4.   Example of computation of the discrete sigmoid function

For greater values of N_s, most of the values of the discrete activation function are highly replicated. In our implementation, we have chosen $n_s = 15$ (the most significant bits of a weighted sum of $N_s = 23$ bits). If we calculate the discrete activation function in this case, most of the values are equal to f_{min_d} (left side of plots in Figure 10.5) and f_{max_d} (right side of plots in Figure 10.5).

Only a small central interval (0.5% of outputs) is different to f_{min_d} and f_{max_d}. Due to this, we can use a smaller LUT that only stores the central interval $[i_{min}, i_{max}]$ of the n_s-bits input where the discrete activation function is not constant.

We now calculate general expressions for the limits i_{min} and i_{max} of that central interval. These limits depend on the activation function, as expressed in Eqn. 10.8-10.13.

- sigmoid / tanh activation function:

$$i_{min} = \left\lceil \frac{-\frac{f_R(2^{N_s}-2)}{4f_0' M_s} \ln\left(\frac{f_R}{\left(\left\lfloor f_{min} \frac{2^{n_z}-2}{M_z} \right\rfloor + \left\lfloor f_{min} \frac{2^{n_z}-2}{M_z} \right\rfloor \bmod 2 + 1 \right) \frac{M_z}{2^{n_z}-2} - f_{min}} - 1 \right) + 1}{2^{m_s}+1} + \frac{1}{2} \right\rceil$$

$$(10.8)$$

$$i_{max} = \left\lceil \frac{-\frac{f_R(2^{N_s}-2)}{4f_0' M_s} \ln\left(\frac{f_R}{\left(\left\lfloor f_{max} \frac{2^{n_z}-2}{M_z} \right\rfloor + \left\lfloor f_{max} \frac{2^{n_z}-2}{M_z} \right\rfloor \bmod 2 - 1 \right) \frac{M_z}{2^{n_z}-2} - f_{min}} - 1 \right) + 1}{2^{m_s}+1} - \frac{3}{2} \right\rceil$$

$$(10.9)$$

-arctan activation function:

$$i_{min} = \left\lceil \frac{\frac{f_R(2^{N_s}-2)}{f_0' \pi M_s} \tan\left(\frac{\pi}{f_R} \left(\left(\left\lfloor f_{min} \frac{2^{n_z}-2}{M_z} \right\rfloor + \left\lfloor f_{min} \frac{2^{n_z}-2}{M_z} \right\rfloor \bmod 2 + 1 \right) \frac{M_z}{2^{n_z}-2} - \frac{f_{max}+f_{min}}{2} \right) \right) + 1}{2^{m_s}+1} + \frac{1}{2} \right\rceil$$

$$(10.10)$$

$$i_{max} = \left\lceil \frac{\frac{f_R(2^{N_s}-2)}{f_0' \pi M_s} \tan\left(\frac{\pi}{f_R} \left(\left(\left\lfloor f_{max} \frac{2^{n_z}-2}{M_z} \right\rfloor + \left\lfloor f_{max} \frac{2^{n_z}-2}{M_z} \right\rfloor \bmod 2 - 1 \right) \frac{M_z}{2^{n_z}-2} - \frac{f_{max}+f_{min}}{2} \right) \right) + 1}{2^{m_s}+1} - \frac{3}{2} \right\rceil$$

$$(10.11)$$

- ramp activation function:

$$i_{min} = \left\lceil \frac{\frac{2^{N_s}-2}{f_0' M_s} \left(\left(\left\lfloor f_{min} \frac{2^{n_z}-2}{M_z} \right\rfloor + \left\lfloor f_{min} \frac{2^{n_z}-2}{M_z} \right\rfloor \bmod 2 + 1 \right) \frac{M_z}{2^{n_z}-2} - \frac{f_{max}+f_{min}}{2} \right) + 1}{2^{m_s}+1} + \frac{1}{2} \right\rceil$$

$$(10.12)$$

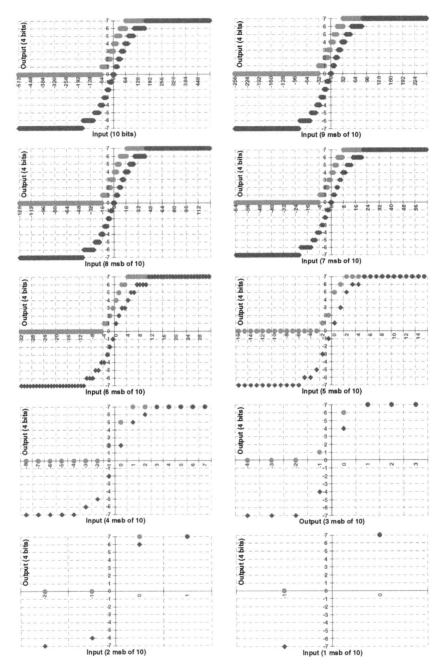

Figure 10.5.   Discretization of f(x) = 1 / (1+e^{-x} (lighter plots) and f(x) = 2/(1+$e^{-x/2}$) - 1 (darker plots), with n_s = 1 . . . 10, in the case: M_z = 1.0, M_w = 5.0, nz = 4, nw = 5, #w = 4, $n_{\#w}$ = 2, M_s = 20.0, N_s = 10)

$$i_{max} = \left\lceil \frac{\frac{2^{N_s}-2}{f_0' M_s}\left(\left(\left\lfloor f_{max}\frac{2^{n_z}-2}{M_z}\right\rceil + \left\lfloor f_{max}\frac{2^{n_z}-2}{M_z}\right\rfloor \bmod 2 - 1\right)\frac{M_z}{2^{n_z}-2} - \frac{f_{max}+f_{min}}{2}\right) + 1}{2^{m_s}+1} - \frac{3}{2}\right\rceil$$

$$(10.13)$$

The number of bits (n_{LUT}) needed to address that central interval $[i_{min}, i_{max}]$ is given in Eqn. 10.14.

$$n_{LUT} = \lceil \log_2(i_{max} - i_{min} + 1) \rceil \qquad (10.14)$$

The algorithm to calculate a LUT that only includes the central non-constant interval of the activation function is shown below.

for $i = i_{min}$ to i_{max}:

1 Calculate the N_s-bits value s_{d_i} corresponding to the center of an interval of width 2^{m_s} (where $m_s = N_s - n_s$ and n_s can be $n_{s_{opt}}$ (7) or an arbitrary value in the range $[1, N_s]$):

$$s_{d_i} = \frac{2^{m_s} - 1}{2} + i2^{m_s} \qquad (10.15)$$

2 Calculate the continuous value S_{c_i} corresponding to S_{d_i}:

$$s_{c_i} = s_{d_i}\frac{2M_s}{2^{N_s} - 2} \qquad (10.16)$$

3 Apply to S_{c_i} the continuous activation function to obtain $f(S_{c_i})$.

4 Discretize $f(S_{c_i})$ and store it in the LUT:

$$LUT(i - i_{min}) = f(s_{d_i}) = round\left(f(s_{c_i})\frac{2^{n_z} - 2}{2M_z}\right) \qquad (10.17)$$

The algorithm to calculate the discrete output of a neuron from an N_s-bits weighted sum is the following:

1 Take the n_s most significant bits of the discrete N_s-bits weighted sum $s_d[N_s - 1 : 0]:s_d[N_s - 1 : m_s]$

2a If $s_d[N_s - 1 : m_s] < i_{min}$ then $f(s_d) = f_{min_d}$

2b If $s_d[N_s - 1 : m_s] > i_{max}$ then $f(s_d) = f_{max_d}$

2c If $i_{min} \leq s_d[N_s - 1 : m_s] \leq i_{max}$ then:

2c.1 Take the n_{LUT} least significant bits of

$$s_d[N_s - 1 : m_s]:s_d[m_s + n_{LUT} - 1 : m_s]$$

2c.2 $f(s_d) = \mathrm{LUT}(s_d[m_s + n_{LUT} - 1 : m_s] - i_{min})$

10.4   Hardware implementations of MLP

In this section, a sequential and a parallel version of the MLP architecture (Figure 10.1) are described using two different abstraction levels: an algorithmic-like level (Handel-C) and a register transfer level (VHDL).

10.4.1   High Level Description (Handel-C)

The high level design of MLP has been defined using Handel-C [5] as the hardware description language. Handel-C is a language for rapidly implementing algorithms in hardware straight from a standard C-based representation. Based on ANSI-C, Handel-C has added a simple set of extensions required for hardware development. These include flexible data widths, parallel processing and communication between parallel threads. The language also utilizes a simple timing model that gives designers control over pipelining without needing to add specific hardware definition to the models. Handel-C also eliminates the need to exhaustively code finite state machines by proving the ability to describe serial and parallel execution flows. Therefore, Handel-C provides a familiar language with formal semantics for describing system functionality and complex algorithms that result in substantially shorter and more readable code than RTL-based representations.

Considering these Handel-C properties, both serial and parallel architectures of the MLP have been implemented in an algorithmic description level by defining their functionality as shown in the next subsections.

Although a detailed study of the possible architecture for both implementations has not been carried out in this work, it should be noted that a previous analysis on how the information will be coded and stored is required. Consequently, the designer must specify aspects such as the variable widths (number of bits), the required bits for each calculation, the variables that should be stored in RAM or ROM modules, the kind of memory storage (distributed in FPGA general proposed LUT resources or using specific Embedded Memory Blocks (EMBs)), etc. Therefore, although Handel-C dramatically reduces the design time, to obtain an efficient system it is required the analysis of the possible options in detail. In other words, a priori small modification of the

Handel-C code can result in a large change in the final architecture, and hence either the area resources or the maximum clock rate can vary considerably.

The whole design process was done using the DK1 Design Suite tool from Celoxica [5]. Multiple higher level languages are supported for simulation (C, C++ and Handel-C). Sequential and parallel designs have been finally compiled using the synthesis tool Xilinx Foundation 3.5i [18].

10.4.1.1   Sequential Version.   To describe the sequential MLP, only its functionality is required. Thus, the MLP computes the synaptic signals for each neuron in the hidden layer by processing the inputs sequentially and later on, the obtained outputs are similarly processed by the final neurons.

From an algorithmic point of view, this functional description implies the use of two loops, one for calculating the results of the neurons in the hidden layer, and other one for neurons in the output layer. As an example of the high level description reached when designing with Handel-C, the following shows the programmed code of the first loop, whose functionality corresponds to the weighted sum of the inputs of each node (Eqn. (10.1)).

```
for (i=0; i<NumHidden; i++)
{
    Sum = 0;
    for (j=0; j<NumInput; j++)
        Sum = Sum+(W[i][j]*In[j]);
}
```

NumHidden is the number of neurons in the hidden layer (24 in our case). *NumInput* is the number of inputs (220 in our case). *W* is the array containing the weight values of the incoming connections to the hidden layer (24×220). *In* is the input array. *Sum* is a variable that stores the accumulated weighted sum of the inputs. If no parallel directive is used in Handel-C, the produced system calculates all the operations sequentially.

10.4.1.2   Parallel Version.   The functionality of the parallel version is such that all neurons belonging to the same layer calculate their results simultaneously, in parallel, except for accessing the activation function that is carried out in a serial way.

The parallel designs have been defined using the Handel-C *"par"* directive that requires the implementation of dedicated circuits for a certain part of an algorithm to be computed in parallel. To optimally use this *"par"* directive, it is necessary to analyze the set of calculations and information transfers that can be done in parallel. For example, the designer must know that RAM modules only allow one location to be read in a single clock cycle.

As a result, the serial loop shown in the previous subsection can be parallelized as:

```
par (i=0; i<NumHidden; i++)
    Sum[i] = 0;
par (i=0; i<NumHidden; i++)
{
    for (j=0; j<NumInput; j++)
        Sum[i] = Sum[i]+(W[i][j])*In[j]);
}
```

10.4.2   Register transfer level description (VHDL)

The Register Transfer Level design of the MLP has been carried out using standard VHDL as the hardware description language. This language allows three different levels of description. We have chosen RTL to implement both sequential and parallel architectures.

The entire design process was done using the FPGA Advantage 5.3 tool, from Mentor Graphics [12].

Figure 10.6 describes the basic structure of the functional unit (or processing element) that implements the calculations associated with Eqn. 10.1. It mainly consists of an 8-bit multiplier and a 23-bit accumulative adder. In order to connect the 16-bit multiplier output with the 23-bit adder input, a sign extension unit is introduced. Note that we need 23 bits to accumulate the multiplications of the 220 inputs by the corresponding weights.

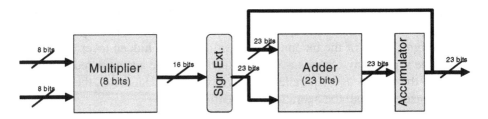

Figure 10.6.   Basic structure of the functional unit

10.4.2.1   Sequential Version.   The sequential architecture has been designed to estimate the minimum area needed to implement the MLP, even though it implies a long execution time. Therefore, this serial version of the MLP consists on a single functional unit that carries out all calculations for all neurons, as is shown in Figure 10.7.

The inputs of the functional unit are both the synaptic signals and their associated weights, which have been stored in different RAM modules. In particular, there is a single RAM module to store all the weights, and two different RAM modules to store the synaptic signals, one for the values of the input layer and the other one for the values of the hidden neurons. Separating these RAM modules allows the MLP to read from the module associated with the input layer and to store the output of the hidden neurons in the same clock cycle.

The output of the functional unit is connected to the activation function module. The activation function output is stored either in the hidden or in the final neuron RAMs, depending on the layer that is being computed.

The different RAMs are addressed with an 8-bit and a 5-bit counter. The 8-bit counter addresses the synaptic signal RAM modules when reading them, and the 5-bit counter addresses them when writing to memory.

The 13-bits address of the weights RAM module is calculated by merging the addresses of both 5-bit and 8-bit counters, in such a way that the most significant bits of the merged address correspond to the 5-bit counter (see Figure 10.7).

The control unit associated with this data path has been designed as a finite state machine, and hence, the detailed description of all necessary states, their outputs and transitions have been implemented.

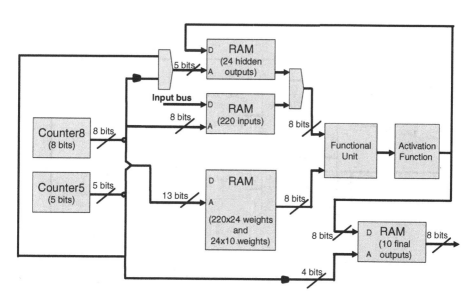

Figure 10.7. Structure of the serial MLP

10.4.2.2 Parallel Version. The proposed parallel architecture describes a kind of node parallelism, in the sense that requires one functional unit per neuron when working at a specific layer. With this strategy, all neurons of a layer work in parallel and therefore produce their outputs simultaneously. This is not a fully parallel strategy because the outputs are obtained in serial from the "activation function" block (Figure 10.4).

For our particular MLP with 24 neurons in the hidden layer and 10 nodes in the output layer, 24 functional units are required. All of them will work in parallel when computing the weighted sums of the hidden layer and only 10 of them will work for the output layer. Figure 10.8 shows the basic structure of this parallel version.

As all functional units work in parallel for each synaptic signal, they need to access the associated weights simultaneously, and hence, the weight RAM should be private to each one.

Every functional unit also needs some local storage for output data because the data between layers are transmitted in serial. We have decided to use 24 parallel registers instead of the RAM module of the serial version, because this reduces the writing time. The data in these parallel registers are introduced to a single activation function unit by a 24:1 multiplexer. The select signals of the multiplexer are the 5-bit counter outputs. The output of the activation unit is either used as a synaptic signal for the 10 neurons at the output layer or stored in the output RAM module. A new control unit has been designed for this parallel data path.

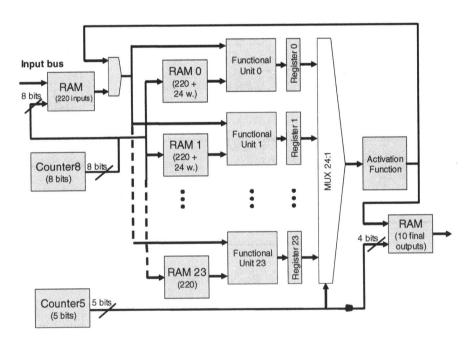

Figure 10.8. Structure of the parallel MLP

10.4.3 Implementation characteristics of MLP with different design strategies

To extract the EDIF files, the systems have been designed using the development environments *FPGA advantage*, for VHDL, and *DK1.1*, for Handel-C. All designs have been finally placed and routed onto a Virtex-E 2000 FPGA, using the synthesis tool Xilinx Foundation 3.5i [18].

The basic building block of the Virtex-E CLB (Configurable Logic Block) is the logic cell (LC). Each CLB contains four LCs organized in two similar slices. An LC includes a 4-input function generator that is implemented as 4-input LUT. A LUT can provide a 16×1 synchronous RAM. Virtex-E also incorporates large Embedded Memory Blocks (EMBs) (4096 bits each) that complement the distributed RAM memory available in the CLBs.

Table 2 presents the results obtained after synthesizing the sequential and parallel versions of the MLP using Handel-C. These results are characterized by the following parameters: number and percentage of slices, number and percentage of EMBs RAM, minimum clock period, the number of clock cycles and the total time required for each input vector evaluation. The percentage values apply to the Virtex-E 2000 device.

As mentioned in Section 10.4.1, obtaining an efficient system requires a detailed analysis of the possible choices. When programming the MLP, there are different options for data storage in the RAM. In order to analyze the effects of the several techniques for distributing and storing data in RAM, we have studied three different choices labeled (a), (b) and (c) in Table 10.2. In (a), only distributed RAM for the whole designs has been used; in (b), the weights associated with synaptic signals (large array) are stored in EMBs RAM modules, while the remaining data is stored in a distributed mode; and finally, (c) only uses memory grouped in EMBs RAM modules.

Table 10.2. Implementation characteristics of the MLP designs described with Handel-C. (a) Only distributed RAM. (b) Both EMBs and distributed RAM. (c) Only EMBs RAM

MLP Design		♯ Slices	% Slices	♯EMBs RAM	%EMBs RAM	Clock (ns)	♯ Cycles	EvaluationTime (ms)
(a)	Serial	2582	13	0	0	50.620	5588	282.864
	Parallel	6321	32	0	0	58.162	282	16.402
(b)	Serial	710	3	24	15	62.148	5588	347.283
	Parallel	4411	22	24	15	59.774	282	16.856
(c)	Serial	547	2	36	22	65.456	5588	365.768
	Parallel	4270	22	36	22	64.838	282	18.284

The results in Table 10.2 show that regardless of the memory option, the parallel version requires more area resources than its associated serial version. Taking into account that the code for implementing the functionality of a single

Table 10.3.   Implementation characteristics of the MLP designs described with VHDL

MLP Design	♯ Slices	% Slices	♯EMBs RAM	%EMBs RAM	Clock (ns)	♯ Cycles	EvaluationTime (ms)
Serial	379	1.5	19	11	49.024	5630	276.005
Parallel	1614	8.5	26	16	53.142	258	13.710

functional unit uses about 70 slices after synthesis, the parallel version requires at least 1680 slices for the 24 functional units. Actually, the increment in the amount of resources is larger due to the replication of other registers and variables. The computation time for each input vector is much shorter (about 20 times) in the parallel version.

Comparing the different options (a), (b) and (c) in Table 10.2, some general characteristics of our synthesized MLP architecture can be noted. When employing only distributed memory (a), the clock cycle and consequently the evaluation time are shortest. In contrast, when only EMBs RAM are used (c), the evaluation time is longest but the number of consumed slices is lower. Therefore, it can be concluded that it is useful to employ EMBs RAM for storing arrays whose size are similar to the size of an EMB RAM and use distributed memory for smaller amounts of data, as in (b) (Table 10.2). Due to the almost shortest evaluation time and the percentage of area resources and of memory blocks, this option is more efficient than the other options. Eventually, the choice of a specific option will depend on the area and time constraints. Nevertheless, defining the design with Handel-C makes it easy to change in order to meet the area and computation time requirement for a specific application.

Table 10.3 shows the implementation characteristics obtained after synthesizing both sequential and parallel versions of the MLP using VHDL. This design corresponds to option (b) described in Table 10.2.

As was expected, results indicate that the sequential version with only a functional unit requires less area resources than the parallel one. Furthermore, it is interesting to note that this increment in the amount of resources for the parallel version is proportional to the number of functional units. Therefore, if a single functional unit requires 53 slices, the parallel version needs at least 24 times more slices (1272). With regard to the computation time for each input vector (last column), it can be seen that it is 20 times shorter in the parallel version. In this way, we are taking advantage of the inherent parallelism of the ANN computation scheme.

When comparing Tables 10.2 and 10.3, it can be seen that the RTL implementation results in better performance and a more optimized approach. However, one of the main advantages of the high level description is the time saved in designing a system and in making future modifications. Therefore, the design time for the serial case when using high level description is about 10 times shorter than the RTL one. In the parallel case, the design time for the high level was relatively short (just dealing with "par" directives), while in the RTL description a new architecture and control unit was designed which resulted in a greater difference in the system redefinition.

10.5 Hardware implementations of XMLP

This section introduces a sequential and a parallel version of the XMLP. For reasons of clarity, only the hardware implementation for the parallel version has been described in detail. However, the implementation characteristics of both the sequential and parallel designs are presented.

In this particular XMLP implementation, the connectivity is defined by $10(4, 2) \times 22$ in the input layer and 4×6 in the hidden layer.

10.5.1 High Level Description (Handel-C)

From an algorithmic point of view, two-dimensional layers require the use of nested loops to index both dimensions. The following example shows the programmed code equivalent to that of the parallel MLP implementation described in Section 10.4.1.2. Its functionality corresponds to the weighted sum of the inputs of each node in the hidden layer. Note that the external loops have been parallelized using the *par* directive.

```
par (X=0; X<NumHiddenX; X++)
{
    FirstX[X] = X*InStepX;
    par (Y=0; Y<NumHiddenX; Y++)
    {
        FirstY[Y] = Y*InStepY;
        Sum[X][Y] = 0;
    }
}
par (X=0; X<NumHiddenX; X++)
{
    par (Y=0; Y<NumHiddenY; Y++)
    {
        for (XGrp=0; XGrp<InGrpX; XGrp++)
        {
            Mul[X][Y] = W[X][Y][XGrp][YGrp]*
            In[FirstX[X]+XGrp][FirstY[Y]];
            for (YGrp=1; YGrp<InGrpY; YGrp++)
                par{
                    Sum[X][Y] = Sum[X][Y]+Mul[X][Y];
                    Mul[X][Y] = W[X][Y][XGrp][YGrp]*
                        In[FirstX[X]+XGrp][FirstY[Y]+YGrp];}
        }
    }
```

```
            Sum[X][Y] = Sum[X][Y]+Mul[X][Y];
       }
   }
```

NumHiddenX and *NumHiddenY* are the sizes of the axes in the hidden layer. *InGrp* and *InStep* specify the size of a group of inputs and the step between two consecutive groups (in the input layer), both in abscissas (*InGrpX*, *InStepX*) and ordinates (*InGrpY*, *InStepY*). *W* is the array containing the weight values. *In* is the input array. Finally, *Sum* is a variable that stores the accumulated weighted sum of the inputs.

In order to compare the XMLP to the MLP, similar design alternatives (a), (b) and (c) to the ones considered in the MLP (Table 10.2) have been chosen.

Table 10.4. Implementation characteristics for the XMLP designs described with Handel-C. (a) Only distributed RAM. (b) Both EMBs and distributed RAM. (c) Only EMBs RAM

MLP Design		♯ Slices	% Slices	♯EMBs RAM	%EMBs RAM	Clock (ns)	♯ Cycles	EvaluationTime (ms)
(a)	Serial	2389	12	0	0	44.851	2566	115.087
	Parallel	5754	29	0	0	47.964	143	6.858
(b)	Serial	1700	8	96	60	71.568	2566	183.643
	Parallel	5032	26	96	60	64.270	143	9.190
(c)	Serial	1608	8	140	91	77.220	2566	198.146
	Parallel	4923	25	147	91	64.830	143	9.271

Similar considerations to the MLP model can be made after taking into account the implementation characteristics included in Table 10.4.

As in the MLP (Section 10.4.3), we observe that the occupation rate of the parallel version increases. Also the computation time of the parallel version is much shorter, on average 18 times.

As expected, the XMLP approaches result in systems twice as fast compared to the fully connected MLP version. This gain in speed depends on the connectivity pattern defined for the XMLP model. In the case studied, the XMLP requires only 2352 multiplications compared to the 5520 needed by the MLP.

10.5.2   Register Transfer Level Description (VHDL)

The parallel architecture of the XMLP and the MLP are similar. Keeping in mind that the connectivity pattern of the XMLP is different, only modifications related to this feature need to be made, as in Figure 10.9.

Since each hidden neuron is only connected to 88 input values (4×22), the global input RAM module with 220 MLP inputs has been replaced by 24 local RAM modules. These local modules store the 88 necessary input values for each functional unit. As each functional unit only computes 88 input values, local weight RAMs can be reduced to 112-word RAM modules for the first ten units (that also compute the output layer), and 88-word RAM modules for the rest.

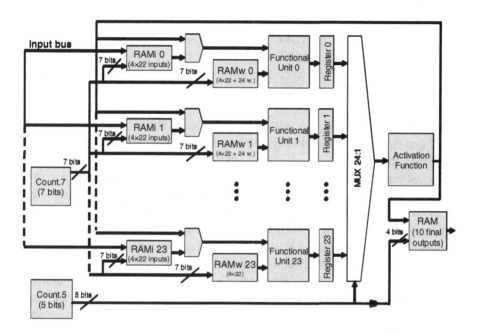

Figure 10.9.   Structure of the parallel XMLP

Table 10.5 shows the implementation characteristics obtained after synthesizing both sequential and parallel versions of the XMLP using VHDL. This design corresponds to the option (b) described in Table 10.4.

Table 10.5.   Implementation characteristics of the XMLP designs described with VHDL

MLP Design	♯ Slices	% Slices	♯EMBs RAM	%EMBs RAM	Clock (ns)	♯ Cycles	EvaluationTime (ms)
Serial	267	2	11	6	37.780	2478	93.618
Parallel	1747	9	49	30	53.752	152	8.170

The performance improvement of the XMLP compared to the MLP is similar to that described for the Handel-C approaches.

10.6   Conclusions

We have presented an FPGA implementation of fully and partially connected MLP-like networks for a speech recognition application. Both sequential and parallel versions of the MLP/XMLP models have been described using two different abstraction levels: register transfer level (VHDL) and a higher algorithmic-like level (Handel-C). Results show that RTL implementation produces more optimized systems. However, one of the main advantages of the high level description is the reduction of design time. The Handel-C design is completely defined with less than 100 code lines.

In both (VHDL and Handel-C) described systems, the parallel versions lead to approaches 20 times faster on average for the MLP, and around 18 times faster for the XMLP. This speed-up corresponds to the degree of parallelism (24 functional units). Therefore, it depends on the number of hidden neurons that are computed in parallel.

Finally on comparing the XMLP approaches (Tables 10.4 and 10.5) to the MLP ones (Tables 10.2 and 10.3), we see that the XMLP computes faster than the MLP. In the best case, it reduces the computation time from 13.7 to 6.9 microseconds for the parallel version. The advantages of XMLP are due to the partial connectivity patterns, which reduce the number of multiplications from 5520, with a fully connected configuration (MLP), to 2352 with the XMLP configured as described in Section 10.2.3. It can also be observed that XMLP connectivity reduces the RAM storage requirements, once more because it requires less connection weights to be stored.

For the speech recognition application we obtain a speaker-independent correct classification rate of 96.83% with a computation time of around 14-16 microseconds per sample. This amply fulfills the time restrictions imposed by the application. Therefore, the implementation can be seen as a low-cost design where the whole system, even the parallel version, would fit into low-cost FPGA device. The system could be embedded in a portable speech recognition platform for voice-controlled systems.

A pipeline processing scheme taking one neural layer in each stage would lead to a faster approach. The processing bottleneck is imposed by the maximum neural fan-in, 220 in a hidden node, because of the need for 220 multiplications. With a pipeline structure, we could overlap the computation time of the hidden layer with the computation time of the output layer (24 multiplications per node). This speeds up the data path by a maximum of 10%. Here we did not study the pipeline choice because our design fulfills the application requirements (portability, low-cost and computation time).

Acknowledgments

This work is supported by the SpikeFORCE (IST-2001-35271) and CICYT TIC2002-00228 projects.

References

[1] Beiu V. Peperstraete J. A.Vandewalle J. and Lauwereins R. Close Approximations of Sigmoid Functions by Sum of Steps for VLSI Implementation of Neural Networks, *Scientific Annals, Section on Informatics*, 1994, pp. 31-50.

[2] Bishop C. M. *Neural Networks for Pattern Recognition.*, 1995.

[3] Caelli T. Guan L. and Wen W.. Modularity in Neural Computing. *Proceedings of the IEEE*, **87**, no. 9, 1999.

[4] Canas A. Ortigosa E. M. Diaz A. F and Ortega J.. XMLP: a Feed-Forward Neural Network with Two-Dimensional Layers and Partial Connectivity. Lecture Notes in Computer Science, LNCS, **2687**, 2003, pp. 89-96

[5] Celoxica, http://www.celoxica.com/

[6] Draghici S. On the capabilities of neural networks using limited precision weights, *Neural Networks*, **15**, 2002, no. 3, pp. 395-414.

[7] Fiesler E. and Beale R. *Handbook of Neural Computation,*, IOP Publishing Ltd and Oxford University Press, 1997.

[8] Huang X. D. Ariki Y. and Jack M. A. *Hidden Markov Models for Speech Recognition*, Edinburgh University Press, 1990.

[9] Ienne P. Cornu T. and Gary K. Special-Purpose Digital Hardware for Neural Networks: An Architectural Survey. *Journal of VLSI Signal Processing*, **13**, 1996, pp. 5-25.

[10] Krause A. and Hackbarth H. Scaly Artificial Neural Networks for Speaker-Independent Recognition of Isolated Words, In: *Proc. of the IEEE Int. Conf*, On Acoustics, Speech and Signal Processing, ICASSP '89, 1989, pp. 21-24.

[11] Lippmann Richard P. Review of neural networks for speech recognition, *Neural Computation*, **1**, 1989, pp. 1-38.

[12] Mentor Graphics, http://www.mentorg.com/

[13] Peinado A. M. Lopez J. M. Sanchez V. E. Segura J. C. and Rubio A. J. Improvements in HMM-based isolated word recognition system, Communications, Speech and Vision, *IEE Proceedings I* , **138**, 1991 pp. 201 -206

[14] Rabiner L. and Juang B. H. *Fundamentals of Speech Recognition*, Prentice-Hall, 1993.

[15] Waibel A. Hanazawa T. Hinton G. Shikano K. and Lang K. Phoneme Recognition Using Time-Delay Neural Networks. *IEEE Transactions on Acoustics, Speech, and Signal Processing*, **37**, 1989.

[16] Widrow B. and Lehr M. 30 years of adaptive neural networks: Perceptron, Madaline and Backpropagation. *Proceedings of the IEEE*, **78**, 1990, pp. 1415-1442.

[17] Widrow B. Rumenlhart D. and Lehr M. Neural networks: Applications in industry, business and science. *Communications of the ACM, 37(3)*, 1994.

[18] Xilinx, http://www.xilinx.com/

Chapter 11

FPGA IMPLEMENTATION OF NON-LINEAR PREDICTORS

Application in Video Compression

Rafael Gadea-Girones

Laboratory of Design of Digital Systems, Universidad Politcnica de Valencia

rgadea@eln.upv.es

Agustn Ramrez-Agundis

Instituto Tecnoløgico de Celaya

agraag@itc.mx

Abstract   The paper describes the implementation of a systolic array for a non-linear predictor for image and video compression. We use a multilayer perceptron with
a hardware-friendly learning algorithm. Until now, mask ASICs (full and semi-
custom) offered the preferred method for obtaining large, fast, and complete
neural networks for designers who implement neural networks. Now, we can
implement very large interconnection layers by using large Xilinx and Altera
devices with embedded memories and multipliers alongside the projection used
in the systolic architecture. These physical and architectural features – together
with the combination of FPGA reconfiguration properties and a design flow
based on generic VHDL – create a reusable, flexible, and fast method of designing a complete ANN on FPGAs . Our predictors with training on the fly, are
completely achievable on a single FPGA. This implementation works, both in recall and learning modes, with a throughput of 50 MHz in XC2V6000-BF957-6
of XILINX, reaching the necessary speed for real-time training in video applications and enabling more typical applications to be added to the image compression processing

Keywords: Non-Linear Prediction, Pipeline Backpropagation, Embedded FPGA Resources.

297

A. R. Omondi and J. C. Rajapakse (eds.), FPGA Implementations of Neural Networks, 297–323.

11.1   Introduction

In recent years, it has been shown that neural networks can provide solutions to many problems in the areas of pattern recognition, signal processing, and time series analysis, etc. Software simulations are useful for investigating the capabilities of neural network models and creating new algorithms; but hardware implementations remain essential for taking full advantage of the inherent parallelism of neural networks.

Traditionally, ANNs have been implemented directly on special purpose digital and analogue hardware. More recently, ANNs have been implemented with re-configurable FPGAs. Although FPGAs do not achieve the power, clock rate, or gate density, of custom chips; they are much faster than software simulations [1]. Until now, a principal restriction to this approach has been the limited logic density of FPGAs.

Although some current commercial FPGAs maintain very complex array logic blocks, the processing element (PE) of an artificial neural network is unlikely to be mapped onto a single logic block. Often, a single PE could be mapped onto an entire FPGA device, and if a larger FPGA is chosen, it would be possible to implement some PEs – perhaps a small layer of neurons – but, never a complete neural network. In this way, we can understand the implementations in [2] and [3] – in which simple multilayer perceptrons (MLPs) are mapped using arrays of almost 30 Xilinx XC3000 family devices. These ANNs perform the training phase off-chip and so save considerable space.

A second solution for overcoming the problem of limited FPGA density is the use of pulse-stream arithmetic. With this technique, the signals are stochastically coded in pulse sequences and therefore can be summed and multiplied using simple logic gates. This type of arithmetic can be observed with fine-grained FPGAs, such as the ATMEL AT6005 [4]; or with coarse-grained FPGAs, such as the Xilinx XC4005 [5] and XC4003 [6]. These implementations use an off-chip training phase, however, a simple ANN can be mapped onto a single device. In the same way, [7] presents an FPGA prototyping implementation of an on-chip backpropagation algorithm that uses parallel stochastic bit-streams.

A third solution is to implement separate parts of the same system by time-multiplexing a single FPGA chip through run-time reconfiguration. This technique has been used mainly in standard backpropagation algorithms; dividing the algorithm into three sequential stages: forward, backward, and update. When the stage computations are completed, the FPGA is reconfigured for the following stage. We can observe this solution by using the Xilinx XC3090 [8], or the Altera Flex10K [9]. Evidently, the efficiency of this method depends on the reconfiguration time when compared to computational time.

Finally, another typical solution is to use time-division multiplexing and a single shared multiplier per neuron [10, 11]. This solution enables mapping an MLP for the XOR problem (3-5-2) onto a single Xilinx XC4020 with the training phase off-chip.

This paper offers advances in two basic respects to previously reported neural implementations on FPGAs. The first is the use of an aspect of back-propagation and stems from the fact that forward and backward passes of different training patterns can be processed in parallel[12]. This possibility was noted, but unimplemented, in a work by Rosemberg and Belloch [13] with the Connection Machine. Later, A. Petrowski et al.[14], describe a theoretical analysis and experimental results with transputers. However, only the batch-line version of the backpropagation algorithm was shown. The possibility of an on-line version was noted by the authors in general terms, but it was not implemented with systematic experiments and theoretical investigations.

The second point we contribute is to produce a completed ANN with on-chip training, and good throughput for the recall phase – on a single FPGA. This is necessary, for example, in industrial machine vision [2, 3], and for the training phase, with continual online training (COT) [15].

In Section 2, a pipelined on-line backpropagation is presented and proposed. Section 3 describes an alternating orthogonal systolic array, the design of synapses, and the synthesis of non-linear functions of neurons. Finally, Section 4 reviews some of the physical design issues that arose when mapping an adaptive non linear predictor onto FPGA devices, and appraises the performance of the network.

11.2   Pipeline and back-propagation algorithm

11.2.1   Initial point

The starting point of this study is the on-line version of the backpropagation algorithm. We assume we have a multilayer perceptron with three layers: two hidden layers and the output layer (2-6-2-2 of Fig. 1)

The phases involved in backpropagation – taking one pattern m at a time and updating the weights after each pattern (on-line version) – are as follows:

1 Forward phase. Apply the pattern a_i^K to the input layer and propagate the signal forward through the network until the final outputs a_i^L have been calculated for each i (index of neuron) and l(index of layer)

$$a_i^l = f(u_i^l) \quad u_i^l = \sum_{j=0}^{N_{l-1}} w_{ij}^l a_j^{l-1} \quad 1 \leq i \leq N_l, \ 1 \leq l \leq L \quad (11.1)$$

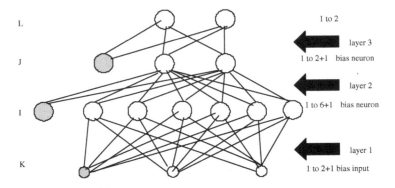

Figure 11.1.   Structure of a multiplayer perceptron

where a is the activation, w the weights and f the non-linear function

2 Error calculation step. Compute the δ's for the output layer L and compute the δ's for the preceding layers by propagating the errors backwards using

$$\delta_i^L = f'(u_i^L)(t_i - y_i) \quad (11.2\text{a})$$

$$\delta_j^{l-1} = f'(u_j^{l-1})\sum_{i=1}^{N_l} w_{ij}\delta_i^l \quad 1 \le i \le N_l \quad , \quad 1 \le l \le L \quad (11.2\text{b})$$

where δ are the error terms, t the targets and f' the derivative function of f

3 Weight update step. Update the weights using

$$^m w_{ij}^l = {}^{m-1} w_{ij}^l + {}^m \Delta w_{ij}^l \quad (11.3\text{a})$$

$$^m \Delta w_{ij}^l = \eta^m \delta_i^l y_j^{l-1} \quad 1 \le i \le N_l \quad , \quad 1 \le l \le L \quad (11.3\text{b})$$

where η is the learning factor and Δw the variation of weight

All the elements in (3) are given at the same time as the necessary elements for the error calculation step; therefore, it is possible to perform these last two steps simultaneously (during the same clock cycle) in this on-line version and to reduce the number of steps to two: forward step (1) and backward step (2) and (3). However, in the batch version, the weight update is performed at the end of an epoch (set of training patterns) and this approximation would be impossible.

11.2.2 Pipeline versus non-pipeline

11.2.2.1 Non-pipeline. The algorithm takes one training pattern m. Only when the forward step is finished in the output layer can the backward step for this pattern occur. When this step reaches the input layer, the forward step for the following training pattern can start (Fig. 2).

In each step s only the neurons of each layer can perform simultaneously, and so this is the only degree of parallelism for one pattern. However, this disadvantage means we can share the hardware resources for both phases, because these resources are practically the same (matrix-vector multiplication).

11.2.2.2 Pipeline. The algorithm takes one training pattern m and starts the forward phase in layer i. The following figure shows what happens at this moment (in this step) in all the layers of the multilayer perceptron. Fig.3 shows that in each step, every neuron in each layer is busy working simultaneously, using two degrees of parallelism: synapse-oriented parallelism and forward-backward parallelism. Of course, in this type of implementation, the hardware resources of the forward and backward phases cannot be shared in one cycle. Evidently, the pipeline carries an important modification of the original backpropagation algorithm [16, 17]. This is clear because the alteration of weights at a given step interferes with computations of the states a_i and errors δ_i for patterns taken from different steps in the network. For example, we are going to observe what happens with a pattern m on its way to the network during the forward phase (from input until output). In particular, we will take into account the last pattern that has modified the weights of each layer. We can see: For the layer I the last pattern to modify the weights of this layer is the pattern m-5.

When our pattern m passes the layer J, the last pattern to modify the weights of this layer will be the pattern $m-3$.

Finally, when the pattern reaches the layer L the last pattern to modify the weights of this layer will be the pattern m-1.

Of course, the other patterns also contribute. The patterns which have modified the weights before patterns m-5, m-3 and $m-1$, are patterns m-6, m-4 and $m-2$ for the layers I, J and L respectively. In the pipeline version, the pattern

FORWARD PHASE

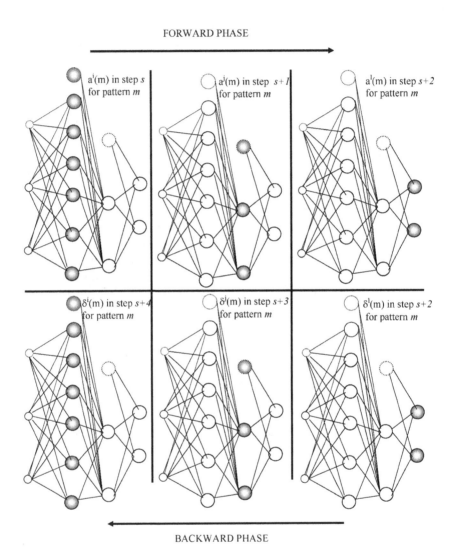

BACKWARD PHASE

Figure 11.2.   Non-pipeline version

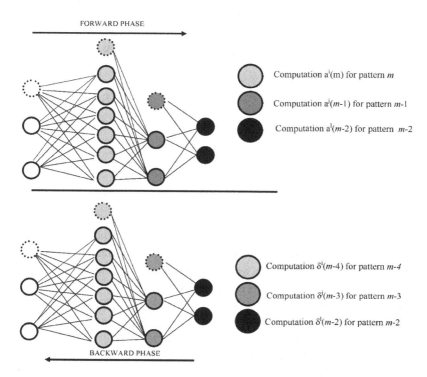

FORWARD PHASE

Computation a^i(m) for pattern *m*

Computation a^j(*m*-1) for pattern *m*-1

Computation a^l(*m*-2) for pattern *m*-2

Computation δ^i(*m*-4) for pattern *m-4*

Computation δ^i(*m*-3) for pattern *m*-3

Computation δ^l(*m*-2) for pattern *m*-2

BACKWARD PHASE

Figure 11.3. Pipeline Version

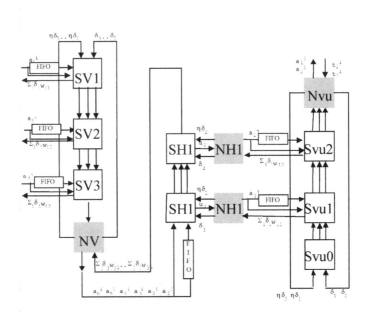

Figure 11.4.   Arquitecture of MLP

$m-1$ is always the last pattern to modify the weights of the all layers. It is curious to note that when we use the momentum variation of the backpropagation algorithm with the pipeline version, the six patterns immediately before the current pattern contribute to the weight updates, while with the non-pipeline version, only the last two patterns contribute before the current pattern.

Therefore, we have a variation of the original on-line backpropagation algorithm that consists basically in a modification of the contribution of the different patterns of a training set in the weight updates, and in the same line as the momentum variation. The main advantage of this modification is that real-time learning is enabled because of a reduction in the throughput of the neural network.

11.3   Synthesis and FPGAs

11.3.1   Digital architecture of the ANN

We assume that we have a MLP (multilayer perceptron) with three layers (Fig.1) and the following characteristics:

NE = number of inputs. N1O = number of neurons in the first hidden layer. N2O = number of neurons in the second hidden layer. NS = number of outputs.

Fig. 4 shows the 'alternating orthogonal systolic array' of an MLP with two hidden layers [18]. This architecture can implement the following structure (2-N1O-2-NS) and is useful for the XOR problem.

We can observe that NE+1 (3) determines the number of vertical synapses (SV) in the first layer, and N2O (2) determines the dimensions of the horizontal layer and the last vertical layer; that is to say, the number of horizontal synapses (SH) and horizontal neurons (NH) and the number of last vertical synapses (SVU). The size of N1O will determine the size of the weight memories of the vertical and horizontal synapses, and the size of NS will determine the size of the weight memories of the synapses in the last vertical layer.

The design entry of the pipelined on-line BP is accomplished in VHDL. It is very important to make these system descriptions independent of the physical hardware (technology) because our future objective is to test our descriptions on other FPGA's and even on ASIC's. The design flow is shown in Fig. 5.

The VHDL description of the 'alternating orthogonal systolic array' (always the unit under test) is totally configurable by means of generics and generates statements whose values are obtained from three ASCII files:

- Database file: number of inputs, number of outputs, training, and validation patterns.

- Learning file: number of neurons in the first hidden layer, number of neurons in the second hidden layer, type of learning (on-line, batch-line or BLMS), value of learning rate η, value of momentum rate, and type of sigmoid.

- Resolution file: resolution of weights (integer and decimal part), resolution of activations, accumulator resolution (integer and decimal part), etc.

11.3.2   Design of Synapses

The main problem in the incorporation of the forward-backward parallelism (pipeline version) is the design of the synapses (white blocks of Fig. 4). We can specify this problem in two aspects: a larger arithmetic and a more complex memory of weights. We can see the necessary hardware resources for a synapse in Fig. 6 when working in one cycle.

The structure of the Memory of Weights represented in Fig. 6 can be showed in detail in Fig. 7. Evidently the control of the addresses are not a problem (design of simple counters); however, it doesn't occur the same with the dual-port RAM. We observe in Fig. 7 the necessary simultaneous operations in the RAM

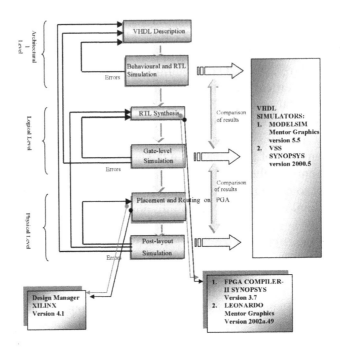

Figure 11.5.   Flow of Design and Tools

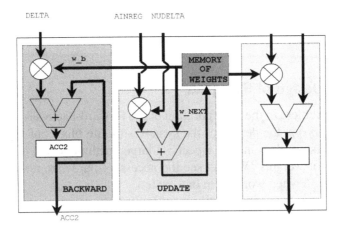

Figure 11.6.   Structure of Synapse Units

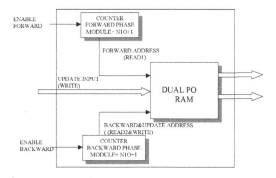

Figure 11.7. Structure of memory

memory of synaptic weights: two read operations (one for forward hardware and one for backward hardware), and one write operation from update hardware. The working of the pipeline algorithm means that the addresses for read operations of the backward phase and write operations of the update phase are the same. However the read operation of the forward phase can work with a different address – as is usual. This behaviour has to be taken into account when we give solutions for the weights.

11.3.2.1   Memory of weights.

When we want to give a real solution for the weight storage in multilayer perceptrons with the on-line pipeline backpropagation algorithm we have different alternatives in the world of FPGA. For our implementations we have studied the technologic alternatives of Table 1 in order to implement this memory, and we have got different results and interesting conclusions:

CLB or Slices (XILINX).   If we employ the DP_RAM Macro from Logi-BLOX we can implement our dual-port RAM for the weight storage without functional limitations. The worse case, when the forward and backward addresses are the same, is resolved without conflicts in data storage.

The only problem with this solution is the inherent problem of using distributed RAM blocks. These small RAM blocks must be connected together to make weight storage of manageable size. For example, if we want to implement a memory module of a vertical synapse of our "alternating orthogonal systolic array", we need N1O CLBs for weight storage (being N1O the number of neurons in the first hidden layer). These CLBs are connected together using

Table 11.1.   Alternatives of implementation of memory of weights

FAMILY	Resource	Organization
XILINX	CLB o Slices Block SelecRAM+	16x1 (4000 &Virtex) 256x16 (Virtex) 1024x16 (Virtex II)
ALTERA	EAB EAB ESB M4K	256x8 (FLEX10K) 256x16(FLEX10KE) 256x8 (APEX II) 256x16 (STRATIX)

multiplexers implemented with more logic blocks. These extra multiplexers cause extra delay and extra routing problems, which slow down the weight storage block. In contrast, non-distributed resources from the following can be used to implement large, dedicated blocks of RAM that eliminate these timing and routing concerns.

Block SelecRAM+(XILINX-VIRTEX-VIRTEX II).   If we employ Block SelecRAM+ we cannot implement our dual-port RAM for the weight storage in one cycle. This dual-port RAM can perform two operations simultaneously (read-read, read-write, write-write), but in the last two cases, the address for these two operations must not be the same. Evidently, the case write-write is not relevant for our problem, but the case of read-write is fundamental for a one-cycle solution. If the address of the forward phase and backward phase is the same the data of the forward phase is incorrect. Of course the probability of this coincidence is low and we can employ this situation for introducing a random noise in the working of the neural network.

Our contributed solution works in two cycles. In the first, we perform the forward and backward read operations. In the second, we perform the update write operation. For example, a Xilinx XC2V250 device has 24 blocks that can implement 1024 weights with 16-bits of resolution. The architecture shown in Fig. 5 only needs 8 of these embedded RAMs; one for each synapse. This supposes that the size of N1O and NS is between 1 and 1024, and therefore we can implement a (2-1024-2-1024) network with the same hardware resources.

Embedded Array Block (ALTERA).   If we employ EAB we must distinguish between the FLEX10K family, and the FLEX10E family. The embedded array blocks of FLEX10K cannot implement our dual-port RAM for the weight storage in one cycle. This RAM can perform only one operation (read or write) in each cycle, and therefore we need three cycles to perform the necessary operations.

The FLEX10KE EAB can act in dual-port or single-port mode. When in dual-port mode, separate clocks can be used for EAB read and write sections, which allows the EAB to be written and read, as well as different rates. Additionally, the read and write sections can work with the same address, which is the main issue for our application.

To perform the second read operation we propose one of two solutions: either using another cycle and multiplexing in time; or storing the weights in duplicate. Of course, this second solution requires two EABs for synapse: one for the forward phase and another for the backward phase.

M4K (ALTERA). The Stratix M4K offer a true dual-port mode to support any combination of two port operations. Between them, this mode permit perfectly the specifications of our weight memories. The biggest configuration of these M4K is 256x16. This mode also existed in ESB (Embedded System Blocks) of APEX II but their biggest organization is 256x8 . It is a good perspective for our next implementations.

11.3.2.2   Arithmetic resources.   The arithmetic resources of the synapses are practically three multiplier-accumulators (MAC); one for each phase of the training algorithm. The size of the synapses is increased by 40% when we want to manage the forward and backward phases simultaneously. When we want to work in one cycle, the arithmetic blocks of both phases cannot be shared. Some 24 multipliers (16x8) for a whole neural network as in Fig.5 are necessary if we work with a precision of 16-bits for the weights and 8-bits for the activations and deltas – and this is a very large resource for each FPGA.

Before the Virtex II and Stratix families, the only solution for these multipliers would be a lot of implementing slices or logic cells. Fortunately, we now find new FPGA families with embedded multipliers that enable considerably larger neural networks; but we need to show that these new families open the perspectives of implementation of neural networks in the world of FPGA .

For example , for our neural network of Fig. 4 we can observe easily in Table 2 the importance of the contributed resources for the new families of XILINX FPGA. The family VIRTEX II incorporates embedded multipliers (18x18) that considerably reduce the number of slices necessary – and it is much faster.

11.3.3   Design of Neurons

Fig. 8 shows the structure of the neuron units of the neural network illustrated in Fig. 4.

Table 11.2.   Design summary for MLP (2-6-2-2)

RESOURCES	XCV300E	XCV300E	X2V250
32x1 RAMs:	640	72	72
Slice Flip Flops:	653	896	846
4 input LUTs:	4712	3713	1931
Number of slices	2731	2519	1299
Number of block RAMs:	**0**	**8**	**8**
Number of Mult18x18s:	**0**	**0**	**24**
Maximum frequency MHz:	39,77	40.22	41.55

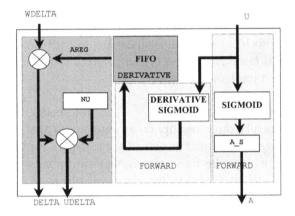

Figure 11.8.   Structure of Neuron Units

The main digital design problem is the implementation of non-linear functions (See equation 1). Common examples of activation functions include hard-limiter, pseudo-linear functions; hyperbolic tangent functions; and sigmoid functions. Of these, we will consider the bipolar sigmoid function (4) that is commonly used in the implementation of a multilayer perceptron model. We will also consider the first derivative of this function (5) necessary for the training equation (2).

$$f(x) = \frac{2}{1 - e^x} - 1 \qquad (11.4)$$

$$f'(x) = \frac{1}{2}((1 + f(x))(1 - f(x)) = \frac{1}{2}(1 - f^2(x)) \qquad (11.5)$$

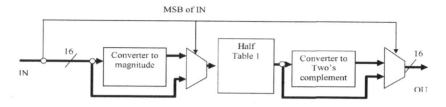

Figure 11.9. Implementation of the Sigmoid Function

11.3.3.1   Sigmoid Function.   In digital implementations, there are a number of approaches for implementing the sigmoid function in FPGA technology.

(look-up table). The sigmoid function can be stored in RAM or ROM and the appropriate values can be read either directly from the stored values, or from interpolation of the stored values [19]. This method requires a large silicon area for the implementation of the memory required. For example, if we work with 16-bits of resolution for input data, we would need 128Kbytes of memory. The number of bytes can be obtained by using the expression $(n/8)2^n$; n being the number of bits of resolution. For FPGAs, we would need 256 EAB for FLEX10KE of ALTERA, or 64 Block SelecRAM+ of VIRTEX II of XILINX. Evidently, these numbers are currently impossible. However, these quantities change if we work with 8-bits of resolution. With this resolution we would need only 256 bytes, and of course, we would use only one embedded RAM memory of these families of FPGA for each non-linear function. In this paper, we have not considered this solution because it is inadequate to work (section 4.1) with a precision of less than 14 or 16 bits for the input of the sigmoid function and 8 bits for the output of this function. Therefore for a NN as shown in Figure 4 (with four neuron units) we would need 128-512 EAB for the implementation of all sigmoid functions if we worked with 14 or 16 bits

Bipartite tables is a method that has recently been developed for approximating functions [20]. With this technique, two tables are accessed in parallel. The outputs of the two tables provide an approximation to the function in carry-save form. These two tables must be combined with a carry propagate adder two's complement approximation to the function. Compared with the above solution, bipartite tables require less memory. For 16 bit input and 8 bits output, the sigmoid function can be implemented with a 2 EAB and 1 adder of 8 bits. We increase the logic cells used and the implementation delay to perform the addition; but it may be a good alternative for a conventional look-up table.

Piecewise linear (PWL) approximation is achieved by approximating the sigmoid function using a piecewise linear approximation PWL [21, 22]. Different alternatives have been evaluated in these papers and it seems that the best solution (for performance and hardware approaches) is the PWL of 15-segment

Table 11.3.   PWL approximation for the 15 segment bipolar sigmoid function

Input Range	Input	Output	Output Range
-8 → -7	1000.abcdefghijklmno	1.0000000abcdefg	-1 → -0.9921875
-7 → -6	1001.abcdefghijklmno	1.0000001abcdefg	-0.9921875→ -0.984375
-6 → -5	1010.abcdefghijklmno	1.000001abcdefgh	-0.984375 → -0.96875
-5 → -4	1011.abcdefghijklmno	1.00001abcdefghi	-0.96875 → -0.9375
-4 → -3	1100.abcdefghijklmno	1.0001abcdefghij	-0.9375 → -0.875
-3 → -2	1101.abcdefghijklmno	1.001abcdefghijk	-0.875 → -0.75
-2 → -1	1110.abcdefghijklmno	1.01abcdefghijkl	-0.75 → -0.5
-1 → 0	1111.abcdefghijklmno	1.1abcdefghijklm	-0.5 → 0
0 → 1	0000.abcdefghijklmno	0.0abcdefghijklm	0 → 0.5
1 → 2	0001.abcdefghijklmno	0.10abcdefghijkl	0.5 → 0.75
2 → 3	0010.abcdefghijklmno	0.110abcdefghijk	0.75 → 0.875
3 → 4	0011.abcdefghijklmno	0.1110abcdefghij	0.875 → 0.9375
4 → 5	0100.abcdefghijklmno	0.11110abcdefghi	0.9375 → 0.96875
5 → 6	0101.abcdefghijklmno	0.111110abcdefgh	0.96875 → 0.984375
6 → 7	0110.abcdefghijklmno	0.1111110abcdefg	0.984375 → 0.9921875
7 → 8	0111.abcdefghijklmno	0.1111111abcdefg	0.9921875→ 1

[23]. If we work with the input range of –8 to 8; and outside of these limits we approximate the outputs to the values 1 or –1; we can see the results in Table 3. We call this solution PWL_sigmoid (15) mode. The solution is symmetric if we use a sign-magnitude representation, therefore we can implement only half of Table 3 by means of the structure in Figure 9.

Seven step approximation. This implementation is made to avoid the multipliers of the forward phase. This solution is found in [24] for the binary sigmoid function. The proposed function for our bipolar sigmoid is the following equation (6):

$$
f(x) = \begin{cases}
1, & \text{if } x \leq y \\
1/2, & \text{if } 1 \leq x < 2 \\
1/4, & \text{if } 0 < x < 1 \\
0, & \text{if } x = 0 \\
-1/4, & \text{if } -1 < x < 0 \\
-1/2, & \text{if } -2 < x \leq -1 \\
-1, & \text{if } x \leq -2
\end{cases} \tag{11.6}
$$

As f(x) takes only seven values. The multiplications of (1) are made with a combination of shift and logical circuitry, which is much simpler than a conventional multiplier. We call this solution the STEP_sigmoid (7) mode.

11.3.3.2   Derivative Function.   In the digital implementations, there are a number of approaches for implementing the derivative of sigmoid function in FPGA technology.

1   Implementation of the expression (5) . We call this the Equation alternative. This is obtained by computing the f(.) from the PWL (15) approximation of the sigmoid, and the derivative is obtained by basically one multiplication (See equation 5). The error generated has an average value of 0.00736 and a maximum value of 0.0325, occurring in the intervals [3,4] and [-4,-3]. It is very important to have a good approximation of the derivative around x=0. This approximation for the new FPGA architectures with embedded multipliers could be very important in the future.

2   Powers-of-two piecewise linear approximation (similar to f(x). This is obtained by approximating the derivative by using piecewise approximations.. We call this the PWL (15) alternative.

3   Power-of-two step function approximation.   This is obtained by a straightforward differentiation of the PWL of the sigmoid function. The gradient of each section is represented by a step function. We call this the STEP (15) alternative. In this case, the approximation around x=0 is very poor and this behaviour will have consequences in the training and, therefore, the generalization.

11.4   Implementation on FPGA

We have examined the possibilities of FPGAs in the logic domain for neural networks. From the point of view of physical domain, the main problem is discovering if the FPGA has sufficient hardware resources to implement a whole neural network. For showing these possibilities, we are going to plan the implementation of a useful application of this type of neural network with the pipeline version of the backpropagation algorithm.

11.4.1   Word ranges, precision and performance

If a digital approach is considered for implementing our pipelined BP algorithm, one of the most important parameters to be fixed is the precision necessary for making the individual building blocks of the array processing units. We have to find a good balance between the system performance indicators (processing speed and area occupied by the implementation) and the quality of the solution that this system is capable of offering for real world tasks (convergence and generalisation properties). This is why internal precision is one of the first parameters to be fixed for a digital implementation of artificial neural networks.

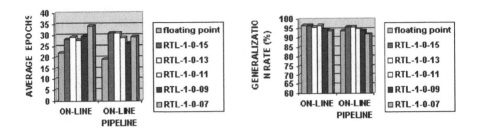

Figure 11.10. Precision analysis of inputs

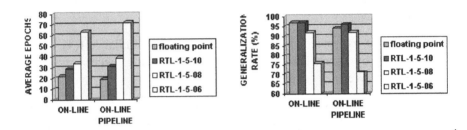

Figure 11.11. Precision analysis of weights

We shall use in our analysis the sign-magnitude format for all the variables involved. This format requires the individual variable to be represented by a bit string divided into three fields. The first field indicates the sign associated with the represented number. The second field is composed of a variable number of bits that represent the number integer part. Finally, the third field represents the decimal part of the considered variable.

Bearing in mind equations (1) (2) and (3), it is easy to deduce the parameters whose precision will influence the performance of the algorithm: the synaptic weight vector; the input and delta vectors; and finally, the precision used for storing the partial products. However, from the point of view of the architecture, we centre our analysis in the synaptic weight and the input vectors. We show the vectors for speed of convergence and generalisation in the results obtained for the vector weights and input (activation and delta) in Figures 10 and 11.

In these figures, we compare the behaviour between the pipeline version (our approach) and the non-pipeline version. We can observe differences between both versions when we work with 32 bits (floating point). However, we cannot observe differences between both versions when we work with limited precision. Therefore, the hardware implementation of the pipeline version

Table 11.4. Summary of performance results for MLP (2-6-2-2)

Family	VIRTEXE 0.18 um, six-layer metal silicon process		VIRTEX2 0.15 um, eight-layer metal silicon process	
Device	XCV300E Max frequency: 40 MHz		X2V500 Max frequency: 60 MHz	
Version	Pipeline	Non-pipeline	Pipeline	Non-pipeline
1/Throughput recall phase	0.18 us	0.18 us	0.12 us	0.12 us
1/Throughput training phase	0.18 us	0.78 us	0.12 us	0.51 us
Performance	217 MCPS 217 MCUPS	217 MCPS 49 MCUPS	326 MCPS 326 MCUPS	326 MCPS 73 MCUPS

works as well as the implementations of classic versions of BP. The other conclusion of our analysis is that the size of 8-bits for the activation data and 16-bits for the vector weight are adequate for the performance of the ANN.

With these word ranges and with the neural network of Fig. 4 we have reached the speed performance showed in the Table 4 for different families of XILINX. The ANN has been analysed in pipeline mode and in non-pipeline mode. The performance of the training phase is measured with the number of connections updated per second (CUPS). The processing in the recall phase (after training) is measured with the number of processed connections per second (CPS). Of course, the pipeline mode will only affect the training phase.

The throughput values for this implementation are satisfactory for most real-time applications and the performance obtainable with a single FPGA easily competes with other neurocomputers with standard Integrated Circuits or with full or semi-custom ASICs.

11.4.2   Implementation of Non-Linear Predictor

After deciding the word ranges and precision we need to know the topology of the neural network.

The application we have chosen is important for our research in image compression based on wavelet transform. In this application, it is common to introduce non-linear predictors based on neural networks. These neural network predictors may be multiplayer perceptrons with one output and with a number of inputs that equal the context size. We chose one of the biggest solutions in

Table 11.5.   Design summary for MLP(42-128-50-1) with PWL_sigmoid(15)

	X2V8000BF957-5	
Number of 4 input LUTS:	82,895 out of 93184	88%
For 32x1 RAMs:	18,560	
As route-thru:	7,417	
As LUTs:	52,313	
Number of Slices:	46,590 out of 46,592	99%
Number of Flip-Flops:	17,454 out of 93,184	18%
Number of block RAMs:	144 out of 168	55%
Number of Mult18x18s:	168 out of 168	100%

this area of application because we want to know if it can implemented by an FPGA without problems of hardware resources (area considerations).

Christopher, Simard and Malvar [25] make an exhaustive experimenting task using a non-linear predictor as an element of an image compression system based on the wavelet transform. They experiment with rectangular and pyramidal context, with completely connected and convolutional networks, with static and dynamical training sets.

The first size of the multiplayer perceptron we want to implement for a non-linear predictor in image compression is obtained of these experiments : NE=42 Inputs, N1O=128, N2O=50, NS=1. We chose the largest available FPGA devise from an XILINX vendor: the VIRTEX II X2V8000. The results of this implementation are shown in Table 5.

We can observe that this neural network consumes a large percentage of the VIRTEX II device. There is scarcely enough area for other applications and therefore it is very difficult to justify this type of implementation. Evidently, we tested the worst case when we used the PWL_sigmoid (15) option for the sigmoid function. If we use the STEP_sigmoid (7) option, we can considerably reduce the area for each synapse because the multiplication of the forward phase is unnecessary. Another, more radical, solution to reduce the hardware resources of our neural network is to not do the training on the fly. We can first train the neural network predictor without other applications on the chip, and then, with the fixed weights of the first step obtained, reconfigure the FPGA with the neural network and the other applications – without the learning structures. This neural network without training needs fewer resources. The implementation results of both solutions are shown in the Tables 6 and 7.

Table 11.6.   Design summary for MLP(42-128-50-1) with STEP_sigmoid(7)

	X2V8000BF957-5	
Number of 4 input LUTS:	78,290 out of 93184	84%
For 32x1 RAMs:	18,560	
As route-thru:	7,417	
As LUTs:	52,313	
Number of Slices:	46,590 out of 46,592	99%
Number of Flip-Flops:	17,454 out of 93,184	18%
Number of block RAMs:	144 out of 168	55%
Number of Mult18x18s:	168 out of 168	100%

Table 11.7.   Design summary for MLP(42-128-50-1) with PWL_sigmoid(15) without training

	X2V8000BF957-5	
Number of 4 input LUTS:	21,645 out of 93184	23%
For 32x1 RAMs:	0	
As route-thru:	1,958	
As LUTs:	19,696	
Number of Slices:	13,735 out of 46,592	29%
Number of Flip-Flops:	7,283 out of 93,184	7%
Number of block RAMs:	144 out of 168	55%
Number of Mult18x18s:	71 out of 42	100%

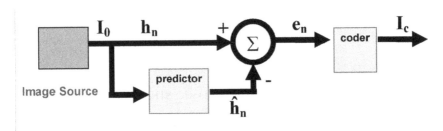

Figure 11.12.   Structure of the predictor system

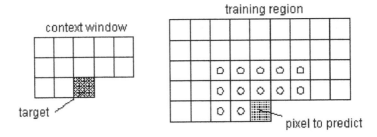

Figure 11.13.   Area for adaptive training

We can conclude that to implement a normal and complete neural network with the current FPGA technology and the proposed degrees of parallelism is almost impracticable; however, it is achievable without training. This conclusion can seem pessimistic, but we must think that we have assumed the implementation of the largest predictor found in the image compression references.

One alternative is experimenting with the adaptive non linear prediction approach proposed by Marusic and Deng [26] in order to obtain a complete lossless compression system (Fig. 12).

The prediction stage is an adaptive predictor that uses a dynamic training region for each pixel to be estimated. The training region has twelve context windows, each one with twelve pixels (Fig. 13).

The image is scanned, pixel by pixel, from left to right and from top to bottom. On each pixel, the training region is represented by twelve training vector; these are used to train a 12-10-1 two layer perceptron. Once the network has been trained, the pixel is predicted and the current weight matrix is used as the initial one for the next pixel (Fig.13).

Table 11.8. Results of prediction with differents topologies with Lena256x256 Image

Hidden Neurons	10	20	10	10	20
Inputs	12	12	4	12	12
PSNR	27,85	27,95	27,49	28,56	28,55
Entropy e_n	4,98	4,97	5,04	4,85	4,86
Entropy h_n	7,5888	7,5888	7,5888	7,5888	7,5888
Version Learning	Batch line	Batch line	Batch line	On line	On line

Table 11.9. Design summary for MLP(12-10-1) in XILINX

	XC2V6000-BF957-6 (VIRTEX2)	
Number of Slices:	3,601 out of 33,792	10%
Number of Flip-Flops:	2,185 out of 67,584	3%
Number of Mult18x18s:	75 out of 144	52%
Number of RAMB16s:	48 out of 144	33 %

We can evaluate a predictor method by its ability to reduce the correlation between pixels (entropy) as well as the PSNR for the predicted image (h_n). We can observe a summary of these parameters for different topologies of the MLP of the Adaptive Non-linear Predictor in Table 8. We obtain, with the on line version of the Backpropagation algorithm, better results than with batch line version and we show that the number of hidden layers (20 or 10) is not relevant for the prediction performance. These two ideas are very important for our hardware implementation, both in throughput and in area.

The results of the implementation of the MLP(12-10-1) are shown in Table 9 and 10. This implementation works, both in recall and learning modes, with a throughput of 50 MHz (only possible with our pipeline version), reaching the necessary speed for real-time training in video applications and enabling more typical applications (wavelet transform and run-length coder) to be added to the image compression.

11.5   Conclusions

We believe this paper contributes new data for two *classic contentions*.

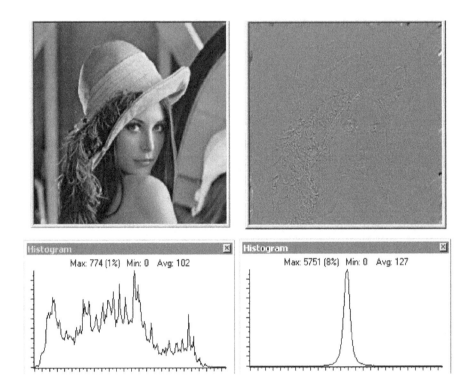

Figure 11.14.   Original (h_n) and residue (e_n) images, and its histograms

Table 11.10.   Design summary for MLP(12-10-1) in ALTERA

	EP1S60B856C7 (STRATIX)	
Number of Logic Cells:	7,362 out of 57,120	12%
Number of Flip-Flops:	2,381 out of 59,193	4%
Number of DSP blocks:	133 out of 144	92%
Number of m512s:	47 out of 574	8%

Firstly; for researchers who work with a specific hardware implementation for artificial neural networks and those working with software approaches and general purpose processors. Until now, software solutions offered the preferred method to obtain quick, flexible designs for different topologies, algorithms, connectivity, activation and base functions, etc. Now, we can see that to exploit all the degrees of parallelism and fault tolerance, we can use hardware designs with several fine-grained processors without degrading flexibility, quick design, and reusability – thanks to the combination of the reconfiguration properties of FPGA and a design flow based on VHDL .

Secondly; until now, mask ASICs offered the preferred method for obtaining large, fast, and complete neural networks for designers who implement neural networks with full and semi-custom ASICs and those implementing with FPGA. Now, we can exploit all the embedded resources of the new programmable ASICs (FPGA) and the enormous quantities of logic cells to obtain useful neural networks, with real-time training on the fly, and with topologies that were impossible to achieve just two years ago.

References

[1] S. Hauck, "The Roles of FPGAs in Reprogrammable Systems" *Proceedings of the IEEE,* 86(4), April 1998, pp. 615-638.

[2] C.E. Cox, W.E. Blanz, "GANGLION- A fast field-programmable gate array implementation of a connectionist classifier" *Journal of Solid State Circuits,* Vol.27, no. 3, March 1992, pp. 288-299.

[3] V. Jean, B. Patrice, R. Didier, S. Mark, T. Herve, B. Philippe. "Programmable active memories: reconfigurable systems come of age", *IEEE Transactions on VLSI Systems*, Vol 4, No 1, March 1996, pp. 56-69.

[4] P. Lysaght, J. Stockwood, J. Law and D. Girma, "Artificial Neural Network Implementation on a Fine Grained FPGA". *Proc of FPL 94*, pp.421-432

[5] V. Salapura, M. Gschwind, and O. Maischberger, "A fast FPGA implementation of a general purpose neuron", *Proc. of the Fourth International Workshop on Field Programmable Logic and Applications,* September 1994.

[6] S.L. Bade and B.L. Hutchings, "FPGA-based stochastic neural network implementation", *IEEE Workshop on FPGAs for Custom Computing Machines,* April 1994, pp. 189-198.

[7] K. Kollmann, K. Riemschneider, and H.C. Zeider, "On-chip backpropagation training using parallel stochastic bit streams" *Proceedings of the IEEE International Conference on Microelectronics for Neural Networks and Fuzzy Systems MicroNeuro'96,* pp. 149-156.

[8] J.G. Elredge and B.L. Hutchings, "RRANN: A hardware implementation of the backpropagation algorithm using reconfigurable FPGAs", *IEEE World Conference on Computational Intelligence,* June 1994, pp. 77-80.

[9] J-L. Beuchat, J-O. Haenni and E. Sanchez, "Hardware reconfigurable neural networks. Parallel and Distributed Processing, Lecture Notes in Computer Science, Springer-Verlag, Vol. 1388, 1998, pp. 91-98.

[10] N. Izeboudjen, A. Farah, S. Titri, H. Boumeridja, "Digital implementation of artificial neural networks: from VHDL description to FPGA implementation", *Lecture Notes in Computer Science ,* Vol.1607, June 1999, pp. 139-148.

[11] Titri, S.; Boumeridja, H.; Lazib, D.; Izeboudjen, N. "A reuse oriented design methodology for artificial neural network implementation", *ASIC/SOC Conference, 1999. Proceedings.* 1999, pp. 409 –413.

[12] R. Gadea, A. Mochol", "Forward-backward parallelism in on-line backpropagation", *Lecture Notes in Computer Science ,* Vol. 1607, June 1999, pp. 157-165.

[13] C.R. Rosemberg, and G. Belloch, "An implementation of network learning on the connection machine", *Connectionist Models and their Implications*, D. Waltz and J Feldman, eds., Ablex, Norwood, NJ. 1988

[14] A. Petrowski, G. Dreyfus, and C. Girault, "Performance analysis of a pipelined backpropagation parallel algorithm", *IEEE Transaction on Neural Networks,* Vol.4 , no. 6, November 1993, pp. 970-981.

[15] B. Burton, R.G. Harley, G. Diana, and J.R. Rodgerson, "Reducing the computational demands of continually online-trained artificial neural networks for system identification and control of fast processes" *IEEE Transaction on Industry Applications,* Vol 34. no.3, May/June 1998, pp. 589-596.

[16] D.E. Rumelhart, G.E. Hinton, and R.J. Williams, "Learning internal representations by error backpropagation, *Parallel Distributed processing*, Vol. 1, MIT Press. Cambridge, MA, 1986, pp. 318-362.

[17] S.E. Falhman, "Faster learning variations on backpropagation: An empirical study", *Proc. 1988 Connectionist Models Summer School,* 1988 , pp. 38-50.

[18] P. Murtagh, A.C. Tsoi, and N. Bergmann, " Bit-serial array implementation af a multilayer perceptron ", *IEEE Proceedings-E,* Vol. 140, no. 5, 1993, pp. 277-288.

[19] J. Qualy, and G. Saucier, "Fast generation of neuro-ASICs", *Proc. Int. Neural Networks Conf,* vol. 2, 1990, pp. 563-567.

[20] H.Hassler and N. Takagi, "Function Evaluation by Table Look-up and Addition", *Procedings of the 12th Symposium on Computer Arithmetic*, 1995, pp. 10-16.

[21] D.J. Myers, and R.A. Hutchinson, " Efficient implementation of piece-wise linear activation function for digital VLSI neural networks", *Elec. Lett.*, vol.25, (24), 1989, pp. 1662-1663.

[22] C. Alippi, and G. Storti-Gajani, "Simple approximation of sigmoidal functions: realistic design of digital neural networks capable of Learning", *Proc. IEEE Int. Symp. Circuits and Syst.*, 1991, pp. 1505-1508.

[23] P. Murtagh, and A.C.Tsoi, "Implementation issues of sigmoid function and its derivative for VLSI digital neural networks", *IEE Proceedings*, V.139, 1992, pp.201-214.

[24] H. Hikawa, "Improvement of the learning performance of multiplierless multiplayer neural network*", IEEE Intenatr.l Synposium on Circuits and Systems"*,1997, pp. 641-644.

[25] Christopher, J.C., Simard, P., Malvar, H.S.: Improving Wavelet Image Compression with Neural Networks. 2000

[26] Marusic, S., Deng, G. "A Neural Network based Adaptive Non-Linear Lossless Predictive Coding Technique". *Signal Processing and Its Applications, 1999. ISSPA '99.*

Chapter 12

THE REMAP RECONFIGURABLE ARCHITECTURE: A RETROSPECTIVE

Lars Bengtsson,[1] Arne Linde,[1] Tomas Nordstrøm,[2] Bertil Svensson,[3] and Mikael Taveniku[4]

[1] *Chalmers University of Technology, Sweden;*
[2] *Telecommunications Research Center Vienna (FTW), Austria;*
[3] *Halmstad University, Sweden;*
[4] *XCube Communication, Inc., Westford MA, USA, and Gøteborg, Sweden.*

Abstract   The goal of the REMAP project was to gain new knowledge about the design and use of massively parallel computer architectures in embedded real-time systems. In order to support adaptive and learning behavior in such systems, the efficient execution of Artificial Neural Network (ANN) algorithms on regular processor arrays was in focus. The REMAP-β parallel computer built in the project was designed with ANN computations as the main target application area. This chapter gives an overview of the computational requirements found in ANN algorithms in general and motivates the use of regular processor arrays for the efficient execution of such algorithms. REMAP-β was implemented using the FPGA circuits that were available around 1990. The architecture, following the SIMD principle (Single Instruction stream, Multiple Data streams), is described, as well as the mapping of some important and representative ANN algorithms. Implemented in FPGA, the system served as an architecture laboratory. Variations of the architecture are discussed, as well as scalability of fully synchronous SIMD architectures. The design principles of a VLSI-implemented successor of REMAP-β are described, and the paper is concluded with a discussion of how the more powerful FPGA circuits of today could be used in a similar architecture.

Keywords: Artificial neural networks, parallel architecture, SIMD, field-programmable gate arrays (FPGA).

A. R. Omondi and J. C. Rajapakse (eds.), FPGA Implementations of Neural Networks, 325–360.
© 2006 *Springer. Printed in the Netherlands.*

12.1   Introduction

Ten years is a long time in the computer industry. Still, looking back and with the power of hindsight, many valuable conclusions can be drawn from a project that old. We will therefore in this chapter look back on our REMAP research project that ran between 1989 and 1996. A major part of this project was to implement artificial neural networks (ANN) onto a parallel and reconfigurable architecture, which we built out of field programmable gate arrays (FPGA).

The chapter is organized as follows. First, we make a short retrospect of the state of art in research and industry when we started this project. Then follows some computational considerations regarding the simulation of ANN in hardware. This then forms the background to the architecture decisions made in the REMAP project. We then present the REMAP-β reconfigurable architecture, which is based on FPGA technology. Next we look at the implementation and mapping of some important and popular ANN algorithms onto REMAP-β. We present further development in the VLSI-implemented REMAP-γ architecture, designed for scalability in both array size and clock speed. Finally, conclusions of the project are given, including reflections on possible designs using today's improved technology.

12.1.1   The REMAP Research Project

The REMAP project started in 1989 as "Reconfigurable, Embedded, Massively Parallel Processor Project" but later evolved into "Real-time, Embedded, Modular, Adaptive, Parallel processor project." This change reflects the change of focus in the project. The *reconfigurability* became less emphasized, even if it still was there. As discussed in [13, 29, 42], we instead found *modularity*, *adaptability*, and *real-time operations* important for this type of massively parallel computer. It also became apparent that building several highly parallel modules better fit the intended application area than building a monolithic massively parallel computer.

The REMAP project was carried out as a joint effort between three research departments, the Centre for Computer Systems Architecture at Halmstad University, the Department of Computer Science and Engineering at Lulea University of Technology, and the Department of Computer Engineering at Chalmers University of Technology, all in Sweden.

The REMAP project (supported by NUTEK, the Swedish National Board for Industrial and Technical Development), had the overall goal of gaining new knowledge about the design and use of massively parallel computer architectures in embedded real-time systems. Since adaptive (learning) behavior was important in this context, Artificial Neural Network (ANN) algorithm execution in hardware was in focus.

Two research architectures (the REMAP-β and the REMAP-γ^1) resulted from the project. REMAP-β was implemented using field programmable logic (FPGAs), and REMAP-γ in semi-custom VLSI.

12.1.2 Historical Perspective

In the beginning of the REMAP project, that is, in the early 1990's, the first larger field-programmable gate arrays (FPGAs) started to appear. With the arrival of Xilinx XC4005 one had 5000 gates available as soft hardware. Even if this is minuscule compared to the devices with 8 million gates now on the market, it enabled us to start exploring reconfigurable processor architectures. However, the limited amount of gates also restricted us in what architectures we could implement.

In the late 1980's we also saw a number of massively parallel computer architectures of SIMD type (Single Instruction stream, Multiple Data streams [16]) appearing on the market; following the tracks of exploratory research that showed the power of data-parallel, bit-serial processing [6, 15]. Perhaps the most famous product was the Connection Machine [18], the most powerful supercomputer of the late 80's. It was a massively parallel SIMD computer with up to 64k (bit-serial) processors clocked at 20 MHz. At the time, it sparked a lot of application research that explored the applicability of the SIMD computer concept.

We also should keep in mind that the typical PC in 1990 was using an Intel 80386SX processor at 16 MHz (even if the first 33 MHz 486 was announced). We note that we didn't have any clock speed difference between on-chip and off-chip communication as we have in most modern processors/computers. We also note that a fully synchronous machine with tens of thousands of processors (like the Connection Machine) could be clocked with the same speed as a single microprocessor. This is not easily achieved anymore, and this has been a limiting factor in further development of SIMD architectures, as noted by Bengtsson and Svensson in [11].

12.2 Target Application Area

The target application area for our REMAP project was so called action-oriented systems [4, 5]. These systems interact in real-time with their environments by means of sophisticated sensors and actuators, often with a high degree of parallelism, and are able to learn and adapt to different circumstances and environments. These systems should be trainable in contrast to the programming of today's computers.

---

[1] Earlier studies and implementations, known as REMAP and REMAP-α, were performed as Master thesis projects.

Action-oriented systems were studied mainly by focusing on separate ANN modules (algorithms) and on separate hardware modules, but all these software and hardware modules would then be parts in the concept of a modular and heterogeneous system.

Even ten years ago the artificial neural networks area contained many different algorithms. In this section we will review some of the most common ANN algorithms, in order to be able to discuss their implementation on our architecture. A more thorough discussion of these algorithms will follow in Section 12.4.

- *Multilayer perceptron* (MLP) is maybe the most known ANN algorithm. It is a multilayer feedforward network with supervised learning typically using a so called *error back-propagation* (BP) [39]. In each layer every node (neuron) computes a weighted sum of the output of the previous layer and then applies a non-linear function to this sum before it forwards this activation value to the next layer. The weight update can be seen as a generalization of Widrow-Hoff error correction rule. In [39] Rumelhart, Hinton, and Williams give a recipe for how weights in a hidden layer can be updated by propagating the errors backwards in a clever way.

  MLPs can thereby be used as efficient non-linear pattern classifiers or feature detectors, meaning that they can recognize and separate complex features or patterns presented to their inputs.

- *Feedback networks* (also referred to as *recurrent networks*) use a single layer topology where all the nodes are completely interconnected. Different variations in node topology and node characteristics have been proposed. For example, symmetric connectivity and stochastic nodes: *Boltzmann machines* [20]; symmetric connectivity and deterministic nodes: *Hopfield nets* [21–23]; and nonsymmetric connectivity and deterministic nodes: *Recurrent back-propagation* (RBP) [3, 36].

  The feedback models can be used as hetero- or autoassociative memories, but also for solving optimization problems. Using an ANN as an autoassociative memory means that whenever a portion or a distorted version of a pattern is presented, the remainder of the pattern is filled in or the pattern is corrected.

- *Self-organizing maps* (SOM), also called *self-organizing feature maps* (SOFM) or topological feature maps, are models developed by Kohonen [26, 27] which learn by using competitive learning. That is, when an input is presented all nodes compete and the closest (using some measure) to the current input is declared winner. This winner and a neighborhood around it are then updated to even better resemble the input. Thus, they learn without specific supervision (or teaching). With these models the responses from the adapted nodes (i.e., their weight vectors) tend to become localized. After appropriate training, the nodes specify clusters or codebook vectors that in fact approximate the probability density functions of the input vectors.

The SOM model, with its variations *learning vector quantization* (LVQ1-3), is used for vector quantization, clustering, feature extraction, or principal component analysis [27].

■ *Sparse distributed memory* (SDM), developed by Kanerva [25] may be regarded as a special form of a two-layer feedforward network, but is more often – and more conveniently – described as an associative memory. It is capable of storing and retrieving data at an address referred to as a "reference address". A major difference compared to conventional Random Access Memories (RAMs) is that, instead of having, e.g., 32 bit addresses, SDMs may have 1000 bit addresses. Since it is impossible to have a memory with 2^{1000} bits, SDMs have to be sparse. Also, data are stored in counters instead of one-bit cells as in RAMs. Note that the address as well as the data can be of lengths of hundreds of bits, still there are only a small number (like thousands) of actual memory locations. The SDM algorithm has a comparison phase, in which the sparsely distributed locations that are closest to the reference address are identified, and an update (write) or retrieval (read) phase, in which a counter value in each of these locations is used.

The SDM model has been used, e.g., in pattern matching and temporal sequence encoding [24]. Rogers [38] has applied SDM to statistical predictions, and also identified SDM as an ideal ANN for massively parallel computer implementation [37].

In Section 12.4 we will generalize and visualize the commonalities between all the models in the latter two groups into a localized learning system (LLS) concept introduced by Nordstrøm [32].

Given that we want to design a digital parallel computer suitable for ANN there are still many trade-offs. The main questions are:

■ What form and level of execution autonomy should the processing elements (PE) have?

■ What is a suitable size and complexity of the PEs? How many PEs should be available and what amount of memory should each PE have?

■ How should the PEs be interconnected?

These questions correspond to the building blocks of a parallel computer, cf. Figure 12.1: the *control unit*, the *processing elements* and their memory, and the *interconnection network* (ICN), which will be discussed in the following sub-sections.

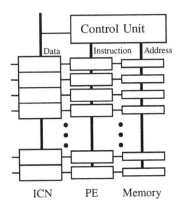

Figure 12.1.   Structure of a computer using SIMD control

How can this (still large) design space be further reduced? The most important aspect is of course the intended application, which in our case was action-oriented systems. That is, the architecture should be able to run a number of different ANN models well. We first start by analyzing the inherent parallelism in these ANN computations.

12.2.1   Parallelism in ANN Computations

For implementation on a parallel computing structure, parts of the algorithm that can be run in parallel must be identified. Unfolding the computations into the smallest computational primitives reveals several dimensions of parallelism.

12.2.1.1   Unfolding the Computations.
A typical ANN algorithm has the following structure:

    For each training session
        For each training example in the session
            For each layer (going Forward and Backward)
                For all neurons (nodes) in the layer
                    For all synapses (weights) of the node
                    For all bits of the weight value

This shows that there are (at least) six different ways of achieving parallelism:

Training session parallelism
    Training example parallelism
        Layer and Forward-Backward parallelism
           Node (neuron) parallelism
                Weight (synapse) parallelism
                Bit parallelism

12.2.1.2   Dimensions of Parallelism.

Training session parallelism means starting different training sessions on different PEs. Different sessions may have different starting values for the weights, and also different learning rates.

Training example parallelism. When training a network, the number of training examples used is usually very large, typically much larger than the number of nodes in the network. The parallelism of the training set can be utilized by mapping sets of different training examples to different PEs and have each PE calculate the outputs for its training example(s). The weight changes are then summed.

Training example parallelism is easy to utilize without communication overhead. This gives an almost linear speedup with the number of PEs. However, a corresponding reduction in training time should not be taken for granted.

Layer parallelism. In a multilayer network the computations may be pipelined, which introduces a small amount of parallelism.

Node (neuron) parallelism. The parallel processing performed by many nodes in each layer is perhaps the most obvious form of parallelism in an ANN. Each node computes a weighted sum of all its inputs. This form of parallelism corresponds to viewing the calculations as matrix operations and letting each row of the matrix map onto one processor.

Weight (synapse) parallelism. At each input to a neuron the arriving activation value is multiplied by the weight of the specific input. This can be done simultaneously at all inputs to the neuron. The subsequent summation of all the products may also be parallelized using a suitable communication structure.

Bit parallelism. Utilizing the full degree of bit parallelism (i.e., treating all bits in a data item simultaneously) is often taken for granted. However, giving up this form of parallelism, and treating data bit-serially, increases the opportunities for using the other forms.

12.2.1.3   Degrees of Parallelism.

The typical degree of parallelism varies widely between the six different kinds, as Table 12.1 shows.

Table 12.1.   Types and ranges of parallelism in ANN

Parallelism	Typical range
Training session	10 – 1 000
Training example	10 – 1 000 000
Layer and Forward-Backward	1 – 6
Node (neuron)	100 – 1 000 000
Weight (synapse)	10 – 100 000
Bit	1 – 64

The table gives an indication of what dimensions should be utilized in a massively parallel implementation. Such an implementation is capable of performing at least thousands of elementary computations simultaneously. Hence an ANN implementation that is to utilize the computing resources efficiently must utilize at least one of the following dimensions:

> Training session parallelism
> Training example parallelism
> Node parallelism
> Weight parallelism

The use of the two first-mentioned types is of interest only in batch processing situations in order to train a network. In real-time applications where the ANN is interacting with the outside world, these two forms are not available. In those cases, node and/or weight parallelism must be chosen, maybe in combination with, e.g., bit and layer parallelism.

12.2.2   Computational Considerations

The computations involved in neural network simulations show great similarities from one model to another. In this section we discuss topics that are of general interest and not specific to one single model. The reader is referred to the survey by Nordström and Svensson [31] for more details.

12.2.2.1   Basic Computations.

For feedforward and feedback network algorithms the basic computation is a matrix-by-vector multiplication, where the matrices contain the connection weights and the vectors contain activation values or error values. Therefore, an architecture for ANN computation should have processing elements with good support for multiply, or even multiply-and-add, operations and a communication structure and memory system suitable for the access and alignment patterns of matrix-by-vector operations.

Assuming N units per layer, the matrix-by-vector multiplication contains N^2 scalar multiplications and N computations of sums of N numbers. The fastest possible way to do this is to perform all N^2 multiplications in parallel, which requires N^2 processing elements (PEs) and unit time, and then form the sums by using trees of adders. The addition phase requires $N(N-1)$ adders and $O(\log N)$ time.

The above procedure means exploitation of both node and weight parallelism. For large ANNs, this is unrealistic, depending on both the number of PEs required and the communication problems caused. Instead, a natural approach is to basically have as many PEs as the number of neurons in a layer (node parallelism) and storing the connection weights in matrices, one for each layer.

Many algorithms (such as the self-organizing map) require finding the maximum or minimum value of a set of values. The efficiency of this operation is strongly dependent on the communication topology, but may also depend on the characteristics of the PEs. Later we will demonstrate how bit-serial processors offer specific advantages. After a maximum (or minimum) node is found, its neighbors are selected and updated. The selection time will depend on the communication topology, and the update time on the length of the training vectors.

In algorithms with binary valued matrices and vectors (like in the sparse distributed memory) the multiplications are replaced with exclusive-or and the summation by a count of ones.

Thus, to be efficient for ANN computations, computers need to have support for matrix-by-vector multiplications, maximum and minimum finding, spreading of activity values, count of ones, and comparisons.

12.2.2.2   Numerical Precision.

In order to optimize utilization of the computing resources, the numerical precision and dynamic range in, e.g., the multiply-and-add operations should be studied with care. Unfortunately, one of the most used algorithms, back-propagation, is very sensitive to the precision and the range used. Using ordinary back-propagation with low precision without modifications will lead to instability and poor learning ability. There are modifications to back-propagation which improve the situation, and, for most problems, $8 - 16$ bits per weight seems to be efficient. (See [31] for a more thorough discussion).

Finally, it should be noted that there is a trade-off between using few weights (nodes) with high precision and using many weights (nodes) with low precision.

12.2.2.3   Bit-Serial Calculations.

Many massively parallel computers have been using bit-serial PEs. For the majority of operations, processing times on these computers grow linearly with the data length used. This may be regarded as a serious disadvantage (e.g., when using 32- or 64-bit floating-point numbers), or as an attractive feature (use of low-precision data speeds up the computations accordingly). In any case, bit-serial data paths simplify communication in massively parallel computers.

In simple bit-serial processors the multiplication time grows quadratically with the data length. However, bit-serial multiplication can be performed in linear time the time required to read the operands (bit by bit, of course) and store the result. The method, based on the carry-save adder technique, requires as many full adders as the length of the operands. Further details will be given when we present the REMAP-β implementation below.

Floating-point calculations raise special problems on bit-serial computers based on the SIMD principle. Additions and subtractions require the exponents to be equal before operations are performed on the mantissas. This alignment process requires different PEs to take different actions, and this does not conform with the SIMD principle. The same problem appears in the normalization procedure. However, these issues may be solved with a reasonable amount of extra hardware, as we show later in this chapter.

Some operations benefit from the bit-serial working mode and can be implemented very efficiently. Search for maximum or minimum is such an operation. Assuming that one number is stored in the memory of each PE, the search for maximum starts by examining the most significant bit of each value. If anyone has a one, all PEs with a zero are discarded. The search goes to the next position, and so on, until all bit positions have been treated or there is only one candidate left. The time for this search is independent of the number of values compared; it depends only on the data length (provided that the number of PEs is large enough).

12.2.3   Choice of Architectural Paradigm

The survey of parallel architectures used or designed for ANN computations by Nordstrøm and Svensson [31] drew the major conclusion that the regularity of ANN computations suits SIMD architectures perfectly; in none of the implementations studied was a division in multiple instruction streams required. It also showed that the demands on the inter-PE communication, for the requirements of ANN algorithms, could be met with surprisingly simple means

The choice of SIMD architecture in the REMAP project was based on this fact, as well as on several detailed studies of algorithm-to-array mappings (to be reviewed later in this chapter). Arrays of bit-serial PEs can be very efficiently used for ANN computations. Since multiplication is the single most important operation in these computations, there is much to gain in the bit-serial architecture if support for fast multiplication is added. This will be demonstrated in the PE design for REMAP shown below.

The REMAP-β was designed and implemented as a flexible research platform (completed in 1991), in which various PE architecture/ algorithm trade-offs could be studied [9]. The architecture follows the strict SIMD model where a Control Unit (CU) broadcasts instructions to the PE array at the lowest (clock cycle) level.

The PEs are bit-serial and each has a substantial amount of (off-chip) memory for local storage. A typical configuration has 128 PEs running at 10 MHz, each having 256 kbit of local memory. Each PE is connected to its neighbors in a ring, but also to a common one-bit bus to enable easy broadcast of data to all other PEs.

FPGAs are used to house the PEs (and a part of the I/O section), enabling changes in the PE architecture by downloading different configuration files.

12.3 REMAP-β – design and implementation

In this section, the architecture and implementation of REMAP-β is described, both its overall structure and its constituent parts. Following this comes a discussion of its use as an architecture laboratory, which stems from the fact that it is implemented using FPGA circuits. As an architecture laboratory, the prototype can be used to implement and evaluate, e.g., various PE designs. A couple of examples of PE architectures, including one with floating-point support, are given, and programming on low level is shortly discussed. Finally, we briefly review our work on mapping of neural network algorithms on processor arrays of this kind and discuss possible tuning of the architecture to meet specific processing demands.

12.3.1 Overall Structure

The logical structure of the parallel computer is shown in Figure 12.2. The machine is a SIMD organized architecture meaning that a large number of *PEs* are working in parallel with different data, but all doing the same operations on this data. In each clock cycle it is the task of the *Control Unit* to broadcast the microinstruction (PE instruction and address) to all PEs in the system.

The PEs form an array of processors which are linearly connected (with wraparound connections to form a ring). Each PE may communicate with its two nearest neighbors (NORTH, SOUTH). Also, *broadcast* is possible where one PE may send information to all PEs in the system. This feature is very useful in neural network execution. Each PE has its own memory (256 k * 1 bit) and the mode of operation is *bit-serial*, meaning that each operation (microinstruction) operates on one or two bits of data only. An array-wide I/O interface is present which can input or output one array-wide bit-slice per clock cycle. A second I/O interface is handled with a *corner-turner* section, which transforms the byte parallel data from the outside world (i.e., the master processor) to the bit-slice parallel world of the array.

Since the PEs are implemented in *FPGAs*, large parts of the machine are reconfigurable. Configuration data is downloaded whenever the PEs need to be reconfigured. Different configurations may thus be used for different applications and algorithms. One configuration may suit a specific algorithm in image processing, another may suit some special neural network execution and a third may be used for a signal processing application. Detailed research on the perfect combination of architecture/algorithm is thus possible on this machine.

At the top, a *Master Processor* controls it all. It handles not only instruction instantiation but also FPGA configuration downloading, and data I/O. In our implementation, a conventional microprocessor (Motorola 68000) was used for this purpose. A link to a Host computer (Workstation or PC) enables the use of a convenient environment for software development for the parallel array.

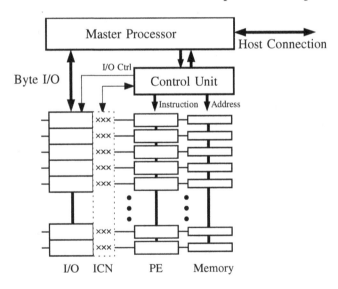

Figure 12.2. Overall structure of REMAP-β: Each horizontal slice represents one element of the processor array with associated memory and I/O. An interconnection network (ICN) connects neighboring PEs with each other. The I/O unit to the left is used for array-parallel-bit-serial or sequential, byte-parallel I/O

12.3.2   The Control Unit

The principal task of the Control Unit (CU) [8], outlined in Figure 12.3, is to generate instructions and addresses to the processor array. The address field (maximum 32 bits) is generated by the *Address Processor* and the instruction field (maximum 32 bits) by the *Controller*.

The CU interfaces to the master processor through control and status registers in the controller's sequencer and through writing into the address processor's register file. A standard VME bus was used for this interface.

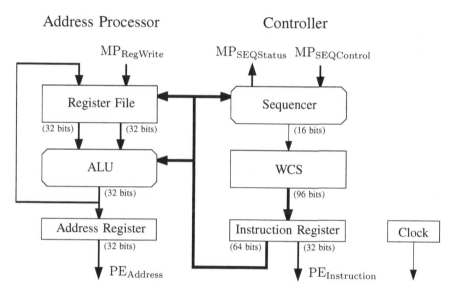

Figure 12.3.   The control unit (CU) with the address processor to the left and the controller to the right

The controller was implemented using a conventional microprogrammed architecture. The microprogram execution was controlled by means of an AM29C331 sequencer chip. This chip can be thought of as an advanced program counter, including features like a stack for subroutine calls, a counter register for looping, and a breakpoint register. The principal output of this chip is a 16-bit address, addressing the microprogram memory.

The microprogram memory (the *Writable Control Store - WCS*) was an 8k deep and 96 bits wide, static RAM. The WCS output was stored in a *pipeline-register*, holding the current microinstruction while the next one was being calculated. Out of these 96 bits in the pipeline-register, 64 bits were controlling the control unit itself and 32 bits were used as the PE array instruction. Microprograms could be downloaded from the master processor under complete software control.

The address processor was composed of three major parts – the *Register File*, the *ALU*, and the *Address Register.* The register file contained 64 registers with 32 bits each, constructed from two AM29C334 chips. There were two read-ports and two write-ports available for register access. The two read register values were sent to the ALU while the ALU output value could be written back through one of the write ports. The second write port was actually shared with the master processor (MP) interface in such a way that the MP could write into this register file. This facilitated parameter passing between the MP and the CU.

The ALU was implemented using the AM29C332 chip, capable of full 32-bit arithmetic and logical operations. The address register was holding the current address while the address processor was calculating the next.

Using this control unit design it was possible to generate a new microinstruction every 100 nanoseconds. An important feature of this address processor was its ability to do a *read-modify-write* operation in one clock cycle. This was especially important for the over-all performance, given that the PEs work bit-serially.

12.3.3   The Processor Array Board

The Processor Array executes the microinstructions generated by the Control Unit. One microinstruction is executed every clock cycle. Seen from the PE, the microinstruction consists of a number of control signals used to set the data-paths in the PEs, memory and I/O-modules. These data-paths determine the function to be performed. The functions available are simple one-bit operations, thus making the PEs small and fast. As mentioned earlier, several studies have shown that this type of PE is well suited for high-speed signal processing and neural network algorithms. Support for fast bit-serial multiplication is well motivated, and it is described later how this is implemented.

Figure 12.4 shows one board (of four) of the REMAP-β prototype. This prototype was mainly built with FPGA chips (Xilinx XC4005 and XC3020) and memory. Depending on the configuration downloaded to the FPGAs the number of PEs varies. However, implementing 32 PEs (i.e., 8 in each XC4005 chip) was the intended major use of the resources.

To each of the FPGAs hosting the PEs (i.e., the XC4005s) we connected 8 pieces of 256 kbit static RAM. Each memory was connected with one read and one write bus. These two (one-bit) buses were also connected to the PE neighbors as an ICN and to the data transposer for I/O. In total there were 32 pieces of 256 kbit static RAM on each board.

The data transposer (Corner Turner, CT) was used to convert byte-wide data from the Master Processor, or other devices requiring byte-wide data, to the

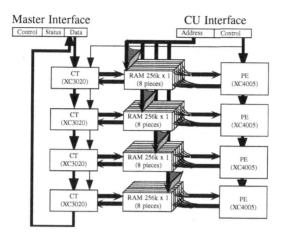

Figure 12.4.   Chip level view of a single REMAP-β processor board. 32 PEs assumed

array-wide parallel format. Being reconfigurable, it could, of course, also be programmed to perform other tasks.

The 32-bit instruction sent from the CU was divided into 26 PE-control signals and 6 I/O-control signals. The master interface was primarily used for data transfers between the master computer and the PEs, but also for loading the configuration to the FPGAs.

The module has two built-in inter-PE communication paths, nearest neighbor and broadcast communication (broadcast not shown in the figure). The nearest neighbor communication network was implemented so that each PE can read its neighbors' memories, i.e., $PE(n)$ can read from $PE(n + 1)$ and $PE(n - 1)$ (the first and the last PEs are considered neighbors). At any time, one of the PEs can broadcast a value to all other PEs or to the CU. The CU can also broadcast to or read from the PEs. It also has the ability to check if any of the PEs are active. (To disable a processing element is a common technique used in SIMD processing in order to enable data dependent operations). If several PEs are active at the same time and the CU wants one PE only to broadcast, the CU simply does a *select first* operation, which selects the first active PE and disables the rest. These communication and arbitration operations can be used to efficiently perform matrix, search and test operations, which are the core functions of many application areas, especially Artificial Neural Networks.

To be useful in real-time applications, where interaction with a constantly changing environment is crucial, the processor array needs a powerful I/O-system. To meet these demands, the processor array was equipped with two I/O-channels, one for 8-bit wide communication and the other for array-wide data. The byte-wide interface can run at speeds up to 80 MHz, with a maximum transfer rate of 80 Mbyte/s. The array-wide interface follows the clock rate of the PE array, i.e. 10 MHz, implying a maximal transfer rate of 156 Mbyte/s.

12.3.4   An Architecture Laboratory

The PEs – reconfigurable through the use of FPGA technology – have a fixed hardware surrounding, which consists of the Control Unit interface, the Master Processor interface, the byte-parallel and array-parallel I/O interfaces, and the external memory interface, which also can connect to an external interconnection network. The primary use of the Corner Turner (CT) was to communicate with devices requiring 8-bit wide data, such as the Master Processor. The CU interface provides 32 control signals, of which 26 are used for PE control. As the computer modules are primarily designed for bit-serial PEs, arranged as a one-dimensional array, the boards can be split into 32 equal parts (one of which is shown in Figure 12.5), consisting of a data-transposer, memory and a processing element.

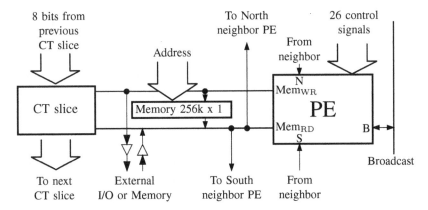

Figure 12.5.   One processing element with its I/O, memory, and a slice of the corner turner (CT)

12.3.4.1   Designing with Xilinx XC4005 FPGA Circuits.

The PEs were housed in Xilinx XC4005 FPGA circuits, which are based on SRAM technology. This enables fast reprogramming as well as high density and speed. The XC4005 chip, the internal structure of which is shown in Figure 12.6, consists of a number of combinatorial logic blocks (CLB), input-output blocks (IOB) and an interconnection network (ICN), all user programmable. The configuration was loaded from an off-chip memory or from a microprocessor. In the REMAP computer, programming/reprogramming was done from the Master Processor. The programming sequence takes about 400 ms, thus enabling the Master Processor to dynamically change the architecture of the PEs during the execution of programs ("soft hardware").

The XC4005 circuit has a 14 by 14 matrix of configurable logic blocks (CLBs), giving an approximate usable gate count of 5000 logic gates. One

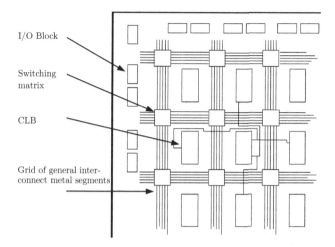

Figure 12.6. Xilinx FPGA overview. The I/O Blocks connect the I/O-pads to the internal interconnection network (ICN). These blocks can be configured as input, output or bidirectional blocks. The CLBs are configurable logic blocks consisting of two 16-bit (and one 8-bit) "lookup table" for logic functions and two flip-flops for state storage. These blocks are only connected to the ICN. The ICN connects the different blocks in the chip, it consist of four kinds of connections: short-range connections between neighboring blocks, medium-range connections connecting blocks on slightly larger distances, long-lines connecting whole rows and columns together, and global nets for clock and reset signals that are broadcast throughout the whole chip

PE of the kind shown in Figure 12.7 occupies 21 CLBs, which makes it possible to implement 8 PEs in each Xilinx chip.

We note that, with current FPGA technology (year 2003), it would be possible to implement on the order of a thousand bit-serial PEs in one chip. It is not immediately clear if the best use of the added resources would be to implement more PEs of the same kind or to make the PEs more powerful. In both cases, while the internal PE-to-PE communication between neighbors should be quite easily solved inside the FPGA circuits, the chip I/O would present a major challenge. This will be further discussed in the end of this chapter.

12.3.4.2 Using FPGAs to Build Computers.

One of the findings in our implementation work was that, in order to use the FPGA circuits efficiently and get high performance, the signal flow is crucial [28]. Unfortunately, the Xilinx EDA software did not support this at the time of implementation (1991-1992), and the signal flow design had to be made by hand. The need to permanently assign the I/O pins to memory and CU further restricted the reconfigurability. Thus, even if the processing elements were simple and regular, which made it fairly easy to implement them with the XACT editor, the possibility to reconfigure was not used to the extent that was anticipated when the project started. On the other hand, the implemented PE design fulfills the requirements for most ANN models, and thus the need to change the PEs was actually limited. The design method also gave the PEs high performance, with clock rates up to 40-50 MHz. These issues were discussed by Taveniku and Linde in [43].

The positive side of using FPGAs is that they allow you to think of the computer as "modeling clay" and you feel free to really change the architecture towards the application and not the other way around. With today's better and more evolved tools, this kind of technology also has the potential to allow architectural variations to be tested and evaluated on real applications.

Our detailed studies of artificial neural network computations resulted in a proposal for a PE that is well suited for this application area. The design is depicted in Figure 12.7. Important features are the bit-serial multiplier and the broadcast connection. Notably, no other inter-PE connections than broadcast and nearest neighbor are needed. The PE is quite general purpose (and the hypothesis is that this is also a useful PE in several other application areas). In this version, the PE consists of four flip-flops (R, C, T, and X), eight multiplexers, some logic and a multiplication unit. The units get their control signals directly from the microinstruction word sent from the control unit. The multiplication unit may be considered a separate device that can be substituted for something else (e.g. a counter) when it is advantageous.

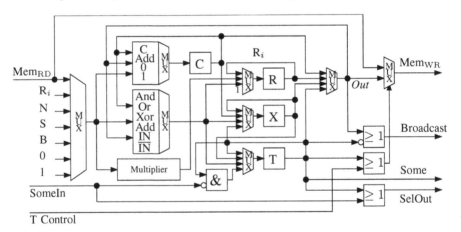

Figure 12.7. The sample PE: The ALU has inputs from memory input (Mem_{RD}), the previous result (R_i), communication network (North, South, Broadcast) as well as the constants 0 and 1. The SomeIn, Some and SelOut creates a chain used in global comparison operations

In a simple bit serial PE without support for multiplication, the multiplication time grows with the square of the operand size. This is a major drawback in multiplication-intensive computations, such as many ANN algorithms. In our design, we included hardware that makes the time of bit-serial multiplication grow linearly with the data length. This means that multiplication is as fast as addition. It simply takes the time it takes to read the operands and store the result (bit-serially). The design, developed in the LUCAS project [15, 35] and based on carry-save addition techniques, is shown in Figure 12.8. In the Xilinx implementation the multiplier (8-bit) takes up about 50% of the PE.

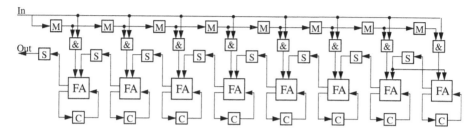

Figure 12.8. Design of a two's-complement bit-serial multiplier. It is operated by first shifting in the multiplicand, most significant bit first, into the array of M flip-flops. The bits of the multiplier are then successively applied to the input, least significant bit first. The product bits appear at the output with least significant bit first

As shown by Nordstrøm [29], the incorporation of a counter instead of a multiplier in the PE design may pay off well when implementing the Sparse Distributed Memory (SDM) neural network model. In a variation of the PE design we have therefore exchanged the multiplier with a counter, which takes up about the same space in the FPGA.

12.3.4.3   A PE Design With Floating-Point Support.
Some low-level operations may require different actions to take place in different PEs, a fact that conflicts with the SIMD principle. Floating-point calculations belong to this category of operations. The processes of alignment and normalization are data dependent, because the different mantissas may need to be shifted a different number of steps at different places.

Adding shift registers to each PE provides the PE with some autonomy in the addressing [1, 2]. Thus we overcome the rigidity in the addressing capability of the SIMD model. The shift registers are still controlled by microinstructions broadcasted by the central control unit, but a *local condition* in each PE determines whether these instructions should be performed, or not.

A floating-point support unit, based on the addition of shift registers, is shown in Figure 12.9. It supports addition, subtraction and multiplication operations, as well as (with additional connections not shown in the figure) multiply-and-add. In addition to the units described above for addition, subtraction and multiplication, it contains three bi-directional shift registers. Two of these hold the mantissas, which are to be shifted for alignment, and the third register is used for holding the mantissa in the normalization stage. An exponent comparator generates local conditions for the shift registers in order to control the alignment operations. (The local condition for the normalization operation is generated directly by one of the shift registers). The exponent comparator also calculates the resulting exponent.

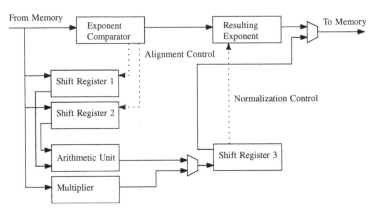

Figure 12.9. Floating-point support for PE

The combination of a bit-serial multiplier and a set of shift registers yields a very powerful bit-serial arithmetic unit for floating-point operations. The multiplier is capable of multiplying two numbers as fast as they can be delivered from and stored to memory. The shift register based units are able to perform alignment and normalization "on the fly" in pipelined operations

12.3.4.4   Programming.
With the "floating" environment resulting from our very flexible micropro-gramming concept used for the CU, and the reconfigurability for the PE ar-chitecture, it became difficult to handle programming. It was decided that two programming levels should be defined, one at the microinstruction level, the other at a "C - Assembly" level.

For the microinstruction level we specified and implemented a micropro-gram assembler (MASS) [7], designed to be adaptable to changes in the under-lying architecture. An instruction in MASS is one microinstruction word; this word can be (from MASS point of view) of arbitrary length, divided into any number of fields, each field defining one or more bits of the microinstruction word. As the PE hardware is modifiable, it is essential that as much of the software as possible can be translated from one PE implementation to another. Therefore, MASS allows the user to define fields and the bits assigned to them in an arbitrary fashion.

In MASS we then could write small microprograms implementing array instructions such as ADD_FIELDS, MULTIPLY_FIELD_WITH_CONSTANT, and COMPARE_FIELDS. Also, specially tuned algorithms were written directly in microcode like MATRIX_MULT and CALC_DIST (from the SDM algorithm).

Then such array instructions/routines were called from the "C - Assembly" level. This level, executing on the master processor, was implemented as an ordinary C program, where function calls to array instructions were made available. A function call here means that the master processor instructs the CU to start executing at a point in the WCS where the microprogram subroutine resides. MASS, or more correct, the MASS linker, automatically generated the needed .c and .h files to make the connection between the MASS and the "C - Assembly" level. Before running a "C - Assembly" program the master processor needed to download the assembled MASS instructions into the WCS.

An example MASS program, which adds two fields and places the result in the first, is shown below.

```
OpParAddS2_1L: /* Micro procedure OpParAdd2_1L definition */
DECRA (R0);
LOOP (R2), PE_C = ZERO,      LOADB_AR (R1); // Loop R2 times
            PE_LOAD (R,MEM), INCRA (R0);    // Mem(R1) -> Rreg
            PE_ADDC (MEM),   LOADB_AR (R0); // Rreg+Mem(R0)->Rreg
                                            // Carry -> Creg
            PE_STORE (R),    INCRA (R1),    // Rreg -> Mem(R0)
END_LOOP ();
```

An example "C - Assembly" program is shown below. It adds all the columns of the matrix arg2 and places the result (which is a vector) in arg1. The arguments arg1 and arg2 are pointers to parallel variables in the array.

```
/* Example add a number of vectors and place them in arg1 */
/* *arg1,*arg2 are pointers to parallel variables */
/* 16 is the length of integer (it could be of any length)*/
void SumVect (PARINT *arg1,PARINT *arg2,unsigned len)
{ int count;
  OP_PAR_CLEAR1_1L(arg1,16); // set arg1 to all zeroes
  for(count=0; count < len; count++){
    OP_PAR_ADD2_1L(arg1,arg2[count],16);
  }
}
```

12.4   Neural networks mapped on REMAP-β

In the survey by Nordstrøm and Svensson [31] we identified the most commonly used ANN algorithms and the parallel computer architectures that had been used to implement them. The computational and communication needs were analyzed for the basic ANN models. The different dimensions of parallelism in ANN computing were identified (as described earlier), and the possibilities for mapping onto the structures of different parallel architectures were analyzed.

We were mainly focusing our research on how various ANN modules (algorithms) could be implemented on separate hardware modules, but all these software and hardware modules would then be parts in our vision of an action-oriented, modular, and heterogeneous system.

The concept of localized learning systems (LLSs), introduced by Nordstrøm [33, 34, 32], makes it possible to combine many commonly used ANN models into a single "superclass." The LLS model is a feedforward network using an expanded representation with more nodes in the hidden layer than in the input or output layers. The main characteristics of the model are local activity and localized learning in active nodes. Some of the well known ANN models that are contained in LLS are generalized radial basis functions (GRBF), self-organizing maps (SOM) and learning vector quantization (LVQ), restricted Coulomb energy (RCE), probabilistic neural network, sparse distributed memory (SDM), and cerebellar model arithmetic computer (CMAC). The connection between these models as variations of the LLS model was demonstrated. Two of the LLS models were studied in greater detail (SDM and SOM) [34, 32]. Furthermore, in [17] and [41] the mapping of two well known ANNs not contained in the LLS class, the multilayer perceptron (MLP) with error back-propagation and the Hopfield network, were studied. Thus, our studies cover the mapping of both feedforward and feedback neural nets onto parallel architectures. The characteristics of these models and system implementation aspects are summarized below.

These results were then used to restrict the hardware design space (finding what extensions to the basic concept that was needed).

12.4.1   Localized Learning Systems (LLS)

In this section we briefly summarize the main aspects of the LLS model. A more complete description can be found in [32]. As mentioned in the previous section the LLS is a *feedforward* ANN forming an *expanded representation* (the largest number of nodes are in the hidden layer). The feed-forward phase can be visualized as in Figure 12.10a. In Figure 12.10b the data flow and data organization of the feedforward phase are shown.

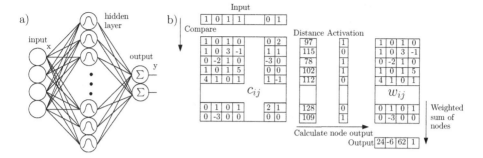

Figure 12.10. The LLS model is a feedforward neural network model with localized activity at the hidden nodes. b) The data flow and data organization of the LLS model (feedforward phase)

The main characteristics of the model are the *local activity* (only a subset of nodes are active at the same time in the hidden layer), and the *localized learning* (only active nodes are updated). We are mainly interested in variations that allow training to take place after each new training data, that is, the LLS is used in an *on-line* fashion.

The feedforward phase for LLS with M nodes and multiple outputs can be written as:

$$F_j(\mathbf{x}_p, \Theta) = \sum_{i \in A} w_{ij} \varphi(r_i), \qquad (12.1)$$

where $\varphi(r_i)$ is the ith node output, $A = A(\mathbf{x}_p) = \{\, i \mid \varphi(r_i(\mathbf{x}_p)) > \alpha \,\}$ is the *set of active nodes*, α is a preset threshold, $\mathbf{x}_p$ is the input, and w_{ij} is the weight connecting node i with output j. The node output ($\varphi(r_i)$) will depend on the distance measurement used to calculate the distance r_i to some centers (templates) c_i, the size and form of the receptive field S_i, and type of kernel function φ. One general form of distance measure r_i can be defined as $r_i^2 = (\mathbf{x}_p - \mathbf{c}_p)S_i(\mathbf{x}_p - \mathbf{c}_p) = \|\mathbf{x}_p - \mathbf{c}_p\|_S$, where S_i is a $d \times d$ positive definite matrix. This measure is also called the Mahalanobis distance. However, the more specialized cases with $S_i = diag[s_i, \ldots, s_d]_i$, $S_i = s_i I$, or $S_i = I$ are the commonly used receptive fields LLSs. A more complete discussion on various receptive fields can be found in [33].

Training (if used) of the free parameters $\Theta = \{\mathbf{w}_i, \mathbf{c}_i, S_i\}_{i=1}^M$, can be done in many ways; one common way is to use a *gradient descent* method as described in [33]; another common way to update the kernel centers is to use *competitive learning* (CL) which we will describe in the self-organizing map (SOM) section below.

The characteristics of an LLS model can then be completely described by nine features as shown below:

Feature:	Variants
Input type:	Real, Integer, Boolean
Distance measure:	L_1 (cityblock), L_2 (Euclidean), L_∞, Dot product, Hamming distance
Type of receptive field:	1, sI, s_iI, $\text{diag}[s_j]$, $\text{diag}[s_j]_i$, S_{ij}, Hierarchical, Sample/Hash
Kernel function:	Radial, Threshold logic unit, Min/Max, exp
Initiation of c_i:	Random, Uniform, Subset of data, All data
Update method of c:	Fixed, Gradient, Competitive learning (CL), CL+Topology, Incremental addition, Genetic Algorithm
Update method of w:	*Pseudo-inverse*, Gradient, Occurrence (Hebb)
Update method of S:	Fixed, Gradient, RCE
Output type:	Real, Integer, Boolean

Two of the LLS variations were studied in more detail through implementation on the REMAP-β. They were the Sparse Distributed Memory and the Kohonen's Self-Organizing (Feature) Map, described in more detail below.

12.4.1.1 Sparse Distributed Memory.

Sparse Distributed Memory (SDM) [25] is a neural network model which is usually described as a memory. Instead of having (e.g.) 32-bit addresses as an ordinary RAM, an SDM may have as large addresses as 1000 bits. Since it is impossible to have 2^{1000} memory locations, an SDM must be sparsely populated. The key property of this memory is that data is stored not in one position but in many.

Using our LLS characterization we can identify SDM as the following LLS variation:

LLS feature:	SDM and some SDM variations (in italic)
Input type:	Boolean
Distance measure:	Hamming distance, L_∞
Type of receptive field:	sI, s_iI, $\text{diag}[s_j]$, *Sample/Hash*
Kernel function:	Threshold logic unit
Initiation of c_i:	Random, *Subset of data, All data*
Update method of c:	Fixed, *Competitive learning*, Genetic Algorithm
Update method of w:	Occurrence (Hebb)
Update method of S:	Fixed
Output type:	Boolean

The algorithm for training the network (i.e., writing to the memory) is as follows (cf. Figure 12.11):

1 The location addresses are compared to the address register and the distances are calculated.

2 The distances are compared to a threshold and those below are selected.

3 In all the selected rows, if the data register is "1" the counter is incremented, and if the data register is "0" the counter is decremented.

The corresponding algorithm for reading from the memory is:

1 The location addresses are compared to the address register and the distances are calculated.

2 The distances are compared to a threshold and those below are selected.

3 The values of the up-down counters from the selected rows are added together column-wise. If the sum is below "0" a zero is returned, otherwise a "1".

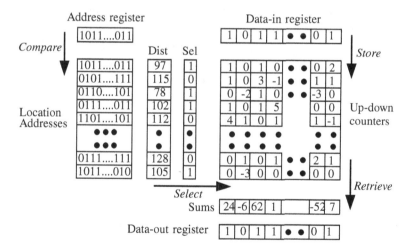

Figure 12.11. Sparse Distributed Memory. Note that even when the address is hundreds of bit, there are only a small number of memory locations, some hundreds of thousands or so

Thus, the SDM model requires a number of distance calculations, comparisons, and summations of vectors. Nordstrøm [29] shows extremely efficient mappings of these computations on the REMAP-β architecture. He uses a "mixed mapping", meaning that, during the comparison phase, each PE computes the distance from "its" location address to the reference address and compares to the threshold value, but, during the update (or readout) phase, the computation is "turned 90 degrees" so that all counters corresponding to a certain row are updated simultaneously, one in each PE.

Due to this efficient mapping a 128 PE REMAP-β with counters in the PEs is found to run SDM at speeds 5–15 times that of an 8k PE Connection Machine CM-2 [18, 19] (same clock frequency assumed). Already without counters (then the PEs become extremely simple) a 128 PE REMAP outperforms a 32 times larger CM-2 by a factor of between 2 and 5. Even if this speed-up for REMAP can be partly explained by the more advanced control unit, the possibility to tune the PEs for this application is equally important.

12.4.1.2 Self-Organizing Maps.

Self organizing maps (SOM), also called self organizing feature maps (SOFM) or topological feature maps, are competitive learning models developed by Kohonen [26, 27]. For these models a competition finds the node (kernel centers c in the LLS terminology) that most resembles the input. The training then updates the winning node and a set of nodes that are (topologically) close to the winner.

In a refined form (rival penalized competitive learning (RPCL) [44], only the node closest to the input (node k) is moved towards the input, while the second best node (the runner up) r is moved away. To involve all nodes, the distances are weighted with the number of inputs assigned to a certain node. We can note that the active set A, in this case, only contains two nodes (k and r) and is determined in a slightly modified way compared to the original SOM.

Using our LLS characterization we can identify SOM as the following LLS variation:

LLS feature:	SOM (and CL) variation
Input type:	Real
Distance measure:	Dot product
Type of receptive field:	$s_i I$
Kernel function:	Threshold logic unit
Initiation of c_i:	Subset of data
Update method of c:	Competitive learning + Topology
Update method of w:	Gradient
Update method of S:	Fixed
Output type:	Real

In [32] Nordström describes different ways to implement SOM on parallel computers. The SOM algorithm requires an input vector to be distributed to all nodes and compared to the weight vectors stored there. This is efficiently implemented by broadcast and simple PE designs. The subsequent search for minimum is extremely efficient on bit-serial processor arrays. Determining the neighborhood for the final update part can again be done by broadcast and distance calculations. Thus, for SOM and CL, it was found that broadcast is sufficient as the means of communication. Node parallelism is, again, simple to utilize. Efficiency measures of more than 80% are obtained (defined as the number of operations per second divided by the maximum number of operations per second available on the computer).

12.4.2   Multilayer Perceptron

Multilayer perceptron (MLP) is the most commonly used ANN algorithm that does not fall into the LLS class. This is actually a feedforward algorithm using error backpropagation for updating the weights [39, 40]. (Therefore this ANN model is commonly referred to as a *back propagation* network.)

The nodes of the network are arranged in several layers, as shown in Figure 12.12. In the first phase of the algorithm the input to the network is provided ($O^0 = I$) and values propagate forward through the network to compute the output vector O. The neurons compute weighted sums $net_j = \sum_j w_{ji}^l o_i^{l-1}$, which are passed through a non-linear function $o_j^l = f(net_j + b_j)$ before leaving each neuron.

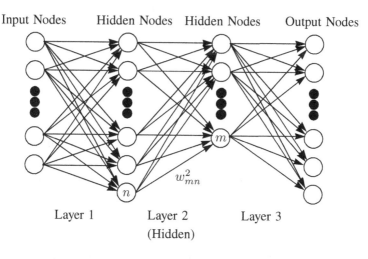

Input Nodes   Hidden Nodes   Hidden Nodes   Output Nodes

Layer 1   Layer 2   Layer 3
                    (Hidden)

Figure 12.12.   A three-layer feedforward network

The output vector of the network is then compared with a target vector, T, which is provided by a teacher, resulting in an error vector, $E = T - O$. This part of the computation is easily mapped on the array using node parallelism and either broadcast or ring communication.

In the training phase the values of the error vector are propagated back through the network. The error signals for hidden units are thereby determined recursively: Error values for layer l are determined from a weighted sum of the errors of the next layer, $l + 1$, again using the connection weights

– now "backwards". The weighted sum is multiplied by the derivative of the activation function to give the error value,

$$\delta_j^l = o_j^l \left(1 - o_j^l\right) \sum_i \delta_i^{l+1} w_{ij}^{l+1}.$$

Here we have used the fact that we can use a sigmoid function $f(x) = 1/(1 + \exp(-x))$ as the non-linear function which has the convenient derivative $f' = f(1 - f)$.

This back-propagation phase is more complicated to implement on a parallel computer architecture than it might appear at first sight. The reason is that, when the error signal is propagated backwards in the net, an "all-PE sum" must be calculated. Two solutions are possible on the REMAP-β architecture: one is based on an adder tree implemented in the corner turner (CT) FPGAs (used in combination with broadcast), while the other one uses nearest-neighbor communication in a ring and lets the partial sum shift amongst all PEs. Both methods give about the same performance [41].

Now, finally, appropriate changes of weights and thresholds can be made. The weight change in the connection to unit i in layer l from unit j in layer $l - 1$ is proportional to the product of the output value, o_j, in layer l, and the error value, δ_i, in layer $l - 1$. The bias (or threshold) value may be seen as the weight from a unit that is always on and can be learned in the same way. That is:

$$\begin{aligned} \Delta w_{ij}^l &= \eta \delta_i^l o_j^{l-1}, \\ \Delta b_i^l &= \eta \delta_i^l. \end{aligned}$$

The REMAP architecture with an array of 128 PEs can run training at 14 MCUPS (Million Connection Updates Per Second) or do recall (forward phase) at 32 MCPS (Million Connections Per Second), using 8-bit data and a clock frequency of 10 MHz.

12.4.3   Feedback Networks

In addition to the feedforward ANN algorithms there are also algorithms using feedback networks. (Hopfield nets, Boltzmann machines, recurrent nets, etc). As reported in [17] and [41] we found that a simple PE array with broadcast or ring communication may be used efficiently also for feedback networks.

A feedback network consists of a single set of N nodes that are completely interconnected, see Figure 12.13. All nodes serve as both input and output nodes. Each node computes a weighted sum of all its inputs: $net_j = \sum_i w_{ji} o_i$. Then it applies a nonlinear activation function to the sum, resulting in an activation value – or output – of the node. This value is treated as input to the network in the next time step. When the net has converged, i.e., when the output no longer changes, the pattern on the output of the nodes is the network response.

This network may reverberate without settling down to a stable output. Sometimes this oscillation is desired, but otherwise the oscillation must be suppressed.

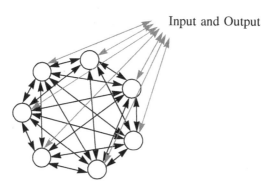

Figure 12.13. A seven-node feedback network

Training or learning can be done in supervised mode with the delta rule [40] or back-propagation [3], or it can be done unsupervised by a Hebbian rule [40]. It is also used "without" learning, where the weights are fixed at start to a value dependent on the application.

The MCPS performance is, of course, the same as a for one-layer feed-forward phase of the back-propagation algorithm above. Thus an array of 128 PEs runs recall (forward phase) at 32 MCPS (Million Connections Per Second) using 8-bit data at 10 MHz.

12.5   REMAP- γ architecture

During the design of the REMAP-β machine, a number of important observations were made regarding the SIMD architecture. One of these is the speed bottleneck encountered in the broadcasting of data values on the common data broadcast bus. Several 10 MHz clock cycles must be used when transmitting data on this bus. Other observations include the latency in address generation and distribution, the latency in the control signal broadcast network, and the importance of clock skew – all these are factors that contribute to the limitation in clock frequency. These observations led us to examine fundamental clock speed bottlenecks in a SIMD architecture (see Bengtsson's analysis in [12]). It was found that the SIMD concept suffered from two major bottlenecks: the signal delay in the common data broadcast bus, and the global synchronism required. Both these get worse as the array size increases (in other words, it shows bad scalability). In addition, Bengtsson found that shrinking the chip geometries in fact emphasizes these speed bottlenecks.

To overcome the discovered limitations a hierarchical organization of the control path was proposed. Two levels of control was suggested: one global CU for the whole array and one local CU per PE. The REMAP-γ design (aimed for VLSI implementation) was started in order to thoroughly analyze the feasibility and performance of such a solution. REMAP-γ was a 2D array (in which each PE was connected to its four neighbors) designed using a semi-custom design style with VHDL synthesis at the front end and VLSI place&route of standard cells at the back-end.

The hierarchical organization of the control path offered the possibility of using only nearest-neighbor PE-to-PE communication, even for broadcasts. A broadcast was implemented as a pipelined flow of bits, transmitted using the nearest-neighbor links. With only nearest-neighbor connections, and no array-wide broadcasts, array size scalability regarding clock speed was maintained.

Also, the possibility of abandoning the rigid synchronous SIMD style, with a single global clock, was investigated. Instead, local PE clocks, synchronized to their immediate neighbors, were used, making it possible to solve the clock skew problems, independently of the array size [11].

In addition, a new type of SIMD instruction was introduced, namely the pipelined array instructions, examples of which are the row and column multiply-and-accumulate (i.e., RMAC and CMAC) instructions. These work similar to the pipelined broadcast instruction, but during the flow of bits between the PEs, local products are added to the bit flow as it passes through each PE, creating a sum-of-products across each row (RMAC) or column (CMAC). Other instructions belonging to this category were RMIN/RMAX and RMIN/CMIN, which searched and found the maximum and minimum values across the PE rows and columns, respectively. This instruction type was found to be very useful when executing ANN algorithms. (Bengtsson's thesis [10] gives a more thorough description of this).

12.6 Discussion

Even if the REMAP-β implementation reached impressive performance for some algorithms, also when compared to some of the fastest computers of its time, the main goal of the REMAP project was not to build a machine that achieved as high performance as possible for some specific applications. Rather, we wanted to explore the design space for massively parallel architectures in order to find solutions that could offer modularity, scalability and adaptability to serve the area of action-oriented, real-time systems. The architecture was designed to take benefit from the common principles of several ANN algorithms, without limiting the necessary flexibility. In addition to this algorithm – generality tradeoff, there is always a technology tradeoff, which, as technology develops, influences the position of the optimal point in the design space. Therefore, after our retrospective on the REMAP project, a natural question is: How would we have done it if we started today?

Most of our observations on how to efficiently map ANNs onto highly parallel computers are still valid. From the point of view of mapping the algorithms efficiently, there is no reason to abandon the SIMD paradigm. However, as noted in the previous section, the inherent total synchronism of the SIMD paradigm creates problems when increasing the clock frequency. Keeping the array size limited and instead increasing the performance of each individual PE seems to be one way to handle this. The techniques described by Bengtsson [10] (such as hierarchical control and pipelined execution) would also, to some extent, alleviate these issues and allow implementation of large, high-speed, maybe somewhat more specialized, arrays.

The design challenge is a matter of finding the right balance between bit and node parallelism in order to reach the best overall performance and general applicability to the chosen domain, given the implementation constraints. Of course, when implementing the array in an FPGA, the tradeoff can be dynamically changed – although the necessary restrictions in terms of a perhaps fixed hardware surrounding must be kept in mind.

One effect of the industry following Moore's law during the last decade is that we today can use FPGAs with up to 8 million gates, hundreds of embedded multipliers, and one or more processor cores. We have also seen the speed difference between logic and memory growing larger and so has also the mismatch between on-chip and off-chip communication speeds. However, for FPGA designs, DRAM and FPGA clock-speeds are reasonably in parity with each other. An FPGA design can be clocked at around 200 MHz, while memory access time is in the 10 ns range with data rates in the 400/800 MHz range.

A 1000 times increase in FPGA size compared to the Xilinx XC4005 used in the REMAP-β, enables a slightly different approach to PE design. Instead of implementing, let's say, 4000 bit-serial processing elements, a more powerful processing element can be chosen. In this way, the size of the array implemented on one chip will be kept at a reasonable level (e.g., 128). Similarly, the processor clock speed could be kept in the range of external memory speed, which today would be somewhere around 200–400 MHz. In the same way, the latency effects of long communication paths and pipelining can be kept in a reasonable range.

In addition, when implemented today, the control unit can very well be implemented in the same FPGA circuit as the PEs. The block-RAM in a modern FPGA can hold the micro-code. Furthermore, to keep up with the speed of the PEs, the address generation unit could be designed using a fast adder structure.

An hierarchical control structure with one global unit and local PE control units (as in the REMAP-γ project), can be used to cope with control signal distribution latencies and the delay in the data broadcast bus. However, this scheme imposes extra area overhead (about 20% extra in PE size was experienced in the REMAP-γ design), so here is a tradeoff between speed and area that must be considered in the actual design. An alternative solution, with less area overhead, is to use a tree of pipeline registers to distribute control signals and use no local PE control. However, the issue with a slow common data broadcast bus would remain. Selecting the most suitable control structure is dependent on both technology (speed of internal FPGA devices, single FPGA or multiple connected FPGAs etc) and array size.

There is a tradeoff between ease of use and efficiency when mapping algorithms onto the array, and this will influence the optimal processor array size. For most of the ANN algorithms studied in the REMAP project, an array size in the 100's of nodes is acceptable, much larger array sizes makes mapping algorithms harder, and edge effects when ANN sizes grow over array size boundaries become increasingly costly. Once more, this implies that it seems to be advantageous to increase the complexity of the PE to keep clock frequency moderate (to cope with control signal generation, memory speed, and synchronization) and network sizes in the low hundreds (to deal with communication latency issues).

We see a similar tradeoff between general applicability and SIMD array size in modern general purpose and DSP processors, for example the PowerPC with Altivec from Motorola/IBM and the Pentium 4 processors from Intel. In these, the SIMD units are chosen to be 128 bits wide, with the option to work on 8, 16, 32, or 64 bit data. The size is chosen so that it maximizes the general usability of the unit, but still gives a significant performance increase. For the next generation processors the trend seems to be to increase the number of SIMD (as well as other) units instead of making them wider. This has to do with (among other reasons) the difficulty with data alignment of operands in memory.

Finally, it should be noted that, with hundreds of times more processing power in one chip, we also need hundreds of times more input/output capacity. While the REMAP-β implementation in no way pushed the limits of I/O capacity in the FPGA chips, an implementation of a similar architecture today definitely would. Here the high-speed links present in modern FPGAs probably would be an important part of the solution to inter-chip as well as external I/O communication. The available ratio between I/O and processing capacity in the FPGA circuits will, of course, also influence the choice of interconnection structure in the SIMD array.

12.7 Conclusions

In this chapter we have summarized an early effort to efficiently perform ANN computations on highly parallel computing structures, implemented in FPGA. The computational model and basic architecture were chosen based on a thorough analysis of the computational characteristics of ANN algorithms. The computer built in the project used a regular array of bit-serial processors and was implemented using the FPGA circuits that were available around 1990. In our continued research, also briefly described in this chapter, we have developed ways to increase the scalability of the approach - in terms of clock speed as well as size. This issue is, of course, very important, considering that several VLSI generations have passed during these years. The techniques described can be applied also to FPGA implementations using today's technology. In the discussion towards the end of this chapter we discuss the implications of the last decade's technology development. We also outline some general guidelines that we would have followed if the design had been made today (as well as the new problems we then would encounter).

Acknowledgments

The work summarized in this chapter was partially financed by NUTEK, the Swedish National Board for Industrial and Technical Development. We also acknowledge the support from the departments hosting the research, as well as our present employers who have made resources available to complete this retrospective work. Among the master students and research engineers that also were involved in the project, we would like to particularly mention Anders Ahlander for his study and design of bit-serial floating-point arithmetic units as well as his contributions to the implementation and programming of the machine.

References

[1] Ahlander, A. "Floating point calculations on bit-serial SIMD computers: problems, evaluations and suggestions." (Masters Thesis), University of Lund, Sweden, 1991. (in Swedish)

[2] Ahlander, A. and B. Svensson, "Floating point calculations in bit-serial SIMD computers," Research Report, Centre for Computer Architecture, Halmstad University, 1992.

[3] Almeida, L. D., "Backpropagation in perceptrons with feedback" In NATO ASI Series: Neural Computers, Neuss, Federal Republic of Germany, 1987.

[4] Arbib, M. A., *Metaphorical Brain 2: An Introduction to Schema Theory and Neural Networks*, Wiley-Interscience, 1989.

[5] Arbib, M. A., "Schemas and neural network for sixth generation computing," *Journal of Parallel and Distributed Computing*, vol. 6, no. 2, pp. 185-216, 1989.

[6] Batcher, K.E., "Bit-serial parallel processing systems", *IEEE Trans. Computers*, Vol. C-31, pp. 377-384, 1982.

[7] Bengtsson, L., "MASS - A low-level Microprogram ASSembler, specification", Report CCA9103, Centre for Computer Systems Architecture - Halmstad, Oct. 1991.

[8] Bengtsson, L., "A control unit for bit-serial SIMD processor arrays", Report CCA9102, Centre for Computer Systems Architecture - Halmstad, Oct. 1991.

[9] Bengtsson, L., A. Linde, B. Svensson, M. Taveniku and A. Ahlander, "The REMAP massively parallel computer platform for neural computations," *Proceedings of the Third International Conference on Microelectronics for Neural Networks (MicroNeuro '93)*, Edinburgh, Scotland, UK, pp. 47-62, 1993.

[10] Bengtsson L., "A Scalable SIMD VLSI-Architecture with Hierarchical Control", PhD dissertation, Dept. of Computer Engineering, Chalmers Univ. of Technology, Goteborg, Sweden, 1997.

[11] Bengtsson L., and B. Svensson, "A globally asynchronous, locally synchronous SIMD processor", *Proceedings of MPCS'98: Third International Conference on Massively Parallel Computing Systems*, Colorado Springs, Colorado, USA, April 2-5, 1998.

[12] Bengtsson L., "Clock speed limitations and timing in a radar signal processing architecture", *Proceedings of SIP'99: IASTED International Conference on Signal and Image Processing*, Nassau, Bahamas, Oct 1999.

[13] Davis, E. W., T. Nordstrom and B. Svensson, "Issues and applications driving research in non-conforming massively parallel processors," in *Proceedings of the New Frontiers, a Workshop of Future Direction of Massively Parallel Processing*, Scherson Ed., McLean, Virginia, pp. 68-78, 1992.

[14] Fahlman, S. E. "An Empirical Study of Learning Speed in Back-Propagation Networks." (Report No. CMU-CS-88-162), Carnegie Mellon, 1988.

[15] Fernstrom, C., I. Kruzela and B. Svensson. *LUCAS Associative Array Processor - Design, Programming and Application Studies*. Vol. 216 of *Lecture Notes in Computer Science*. Springer Verlag. Berlin. 1986.

[16] Flynn, M. J., "Some computer organizations and their effectiveness," *IEEE Transactions on Computers*, vol. C-21, pp. 948-60, 1972.

[17] Gustafsson, E., A mapping of a feedback neural network onto a SIMD architecture, Research Report CDv-8901, Centre for Computer Science, Halmstad University, May 1989.

[18] Hillis, W. D., *The Connection Machine*, MIT Press, 1985.

[19] Hillis, W. D. and G. L. J. Steel, "Data parallel algorithms," *Communications of the ACM*, vol. 29, no. 12, pp. 1170-1183, 1986.

[20] Hinton, G. E. and T. J. Sejnowski. "Learning and relearning in Boltzmann machines." *Parallel Distributed Processing; Explorations in the Microstructure of Cognition Vol. 2: Psychological and Biological Models*. Rumelhart and McClelland ed. MIT Press, 1986.

[21] Hopfield, J. J., "Neural networks and physical systems with emergent collective computational abilities". *Proceedings of the National Academy of Science USA. 79*: pp. 2554-2558, 1982.

[22] Hopfield, J. J. "Neurons with graded response have collective computational properties like those of two-state neurons". *Proceedings of the National Academy of Science USA. 81*: pp. 3088-3092, 1984.

[23] Hopfield, J. J. and D. Tank. "Computing with neural circuits: A model." *Science*. Vol. 233: pp. 624- 633, 1986.

[24] Kanerva, P. "Adjusting to variations in tempo in sequence recognition." In *Neural Networks Supplement: INNS Abstracts*, Vol. 1, pp. 106, 1988.

[25] Kanerva P., *Sparse Distributed Memory*, MIT press, 1988.

[26] Kohonen, T. *Self-Organization and Associative Memory*. (2nd ed.) Springer-Verlag. Berlin. 1988.

[27] Kohonen, T., The self-organizing map, *Proceedings of the IEEE*. Vol. 78, No. 9. pp. 1464-1480, 1990.

[28] Linde, A., T. Nordstrøm and M. Taveniku, "Using FPGAs to implement a reconfigurable highly parallel computer," Field-Programmable Gate Array: Architectures and Tools for Rapid Prototyping; Selected papers from: *Second International Workshop on Field-Programmable Logic and Applications (FPL'92)*, Vienna, Austria, Grønbacher and Hartenstein Eds. New York: Springer-Verlag, pp. 199-210, 1992.

[29] Nilsson, K., B. Svensson and P.-A. Wiberg, "A modular, massively parallel computer architecture for trainable real-time control systems," *Control Engineering Practice*, vol. 1, no. 4, pp. 655-661, 1993.

[30] Nordstrøm, T., "Sparse distributed memory simulation on REMAP3," Res. Rep. TULEA 1991:16, Lulea University of Technology, Sweden, 1991.

[31] Nordstrøm, T. and B. Svensson, "Using and designing massively parallel computers for artificial neural networks," *Journal of Parallel and Distributed Computing*, vol. 14, no. 3, pp. 260-285, 1992.

[32] Nordstrøm, T., "Highly Parallel Computers for Artificial Neural Networks," Ph.D. Thesis. 1995:162 D, Lulea University of Technology, Sweden, 1995.

[33] Nordstrøm, T., "On-line localized learning systems, part I - model description," Res. Rep. TULEA 1995:01, Lulea University of Technology, Sweden, 1995.

[34] Nordstrøm, T., "On-line localized learning systems, part II - parallel computer implementation," Res. Rep. TULEA 1995:02, Lulea University of Technology, Sweden, 1995.

[35] Ohlsson, L., "An improved LUCAS architecture for signal processing," Tech. Rep., Dept. of Computer Engineering, University of Lund, 1984.

[36] Pineda, F. J. "Generalization of back-propagation to recurrent neural networks." *Physical Review Letters*. Vol. 59(19): pp. 2229-2232, 1987.

[37] Rogers, D. "Kanerva's sparse distributed memory: an associative memory algorithm well-suited to the Connection Machine." (Technical Report No. 88.32), RIACS, NASA Ames Research Center, 1988.

[38] Rogers, D. "Statistical prediction with Kanerva's sparse distributed memory." In *Neural Information Processing Systems* 1, pp. 586-593, Denver, CO, 1988.

[39] Rumelhart, D. E. and J. L. McClelland. *Parallel Distributed Processing; Explorations in the Microstructure of Cognition.* Vol. I and II, MIT Press, 1986.

[40] Rumelhart, D. E. and J. L. McClelland., *Explorations in Parallel Distributed Processing*, MIT Press, 1988.

[41] Svensson, B. and T. Nordstrøm, "Execution of neural network algorithms on an array of bit-serial processors," *Proceedings of 10th International Conference on Pattern Recognition, Computer Architectures for Vision and Pattern Recognition*, Atlantic City, New Jersey, USA, vol. II, pp. 501-505, 1990.

[42] Svensson, B., T. Nordstrøm, K. Nilsson and P.-A. Wiberg, "Towards modular, massively parallel neural computers," *Connectionism in a Broad Perspective: Selected Papers from the Swedish Conference on Connectionism - 1992*, L. F. Niklasson and M. B. Boden, Eds. Ellis Horwood, pp. 213-226, 1994.

[43] Taveniku, M. and A. Linde, "A Reconfigurable SIMD Computer for Artificial Neural Networks," Licentiate Thesis 189L, Department of Computer Engineering, Chalmers University of Technology, Sweden, 1995.

[44] Xu, L., A. Krzyzak and E. Oja, "Rival penalized competitive learning for clustering analysis, RBF net, and curve detection," *IEEE Transactions on Neural Networks*, vol. 4, no. 4, pp. 636-649, 1993.